One Billion Hungry

One Billion Hungry

Can We Feed the World?

GORDON CONWAY

with Katy Wilson

Foreword by Rajiv Shah

Published with the support of
Agriculture for Impact, Imperial College, London

Comstock Publishing Associates, a division of

Cornell University Press | Ithaca and London

This publication is based on research funded in part
by the Bill & Melinda Gates Foundation.
The findings and conclusions contained within
are those of the authors and do not necessarily reflect
positions or policies of the Bill & Melinda Gates Foundation.

First published 2012 by Cornell University Press
First printing, Cornell Paperbacks, 2012

Printed in the United States of America and the United Kingdom

Library of Congress Cataloging-in-Publication Data
Conway, Gordon.
One billion hungry : can we feed the world? / Gordon Conway with
Katy Wilson ; foreword by Rajiv Shah.
p. cm.
"Published with the support of Agriculture for Impact, Imperial College, London."
Includes bibliographical references and index.
ISBN 978-0-8014-5133-1 (cloth : alk. paper)
ISBN 978-0-8014-7802-4 (pbk. : alk. paper)
1. Agricultural innovations—Developing countries.   2. Green Revolution—
Developing countries.   3. Food supply—Developing countries.
4. Food security—Developing countries.   I. Wilson, Katy.
II. Conway, Gordon. Doubly green revolution.   III. Title.

S494.5.I5C663 2012
338.1'6091724—dc23

2012014403

Cornell University Press strives to use environmentally responsible suppliers
and materials to the fullest extent possible in the publishing of its books.
Such materials include vegetable-based, low-VOC inks and acid-free papers
that are recycled, totally chlorine-free, or partly composed of nonwood fibers.
For further information, visit our website at www.cornellpress.cornell.edu.

Cloth printing   10   9   8   7   6   5   4   3   2   1
Paperback printing   10   9   8   7   6   5   4   3   2   1

To the memory of
Cyril and Thelma Conway

Contents

Foreword

by Rajiv Shah

Administrator of the United States Agency for International Development

By the late 1990s, global food security had mostly fallen off the world's agenda. The success of the Green Revolution had helped hundreds of millions of people in Latin America and Asia avoid a life of extreme hunger and poverty. Governments—developed and developing alike—assumed this success would spread and cut their investments in agriculture, allowing them to turn their attention elsewhere.

But while many had lost sight of the importance of agricultural development, Gordon Conway stayed focused. In his book, *A Doubly Green Revolution*, published in 1997, Gordon issued a pressing call for the development community to recommit to the goals of fighting hunger and malnutrition around the world.

Gordon argued that the lasting elimination of hunger required us to do more than transform the production of food—the hallmark achievement of the Green Revolution. We also had to help smallholder farmers build resilience to natural disasters and climate change, use advances in science and technology to boost yields, and partner with the private sector to get those crops to market. Thanks to Gordon's leadership, the world is once again delivering a global commitment to strengthening food security.

In 2009, President Obama established a global food security initiative called Feed the Future. Spearheaded by the U.S. Agency for International Development, Feed the Future helps countries develop productive agriculture sectors so they can feed themselves and transform their economies over the long term. In partnership with local and international private companies, the global initiative supports country-led plans to focus investment in regions most likely to flourish, and in crop and livestock systems with the greatest chance of fighting poverty and malnutrition.

These efforts are more important now than ever before. In the Horn of Africa, the worst drought in 60 years put more than 13.3 million people in the path of hunger and disease during 2011. In Somalia, where decades of civil war and disorder have contributed to the complete breakdown of governance, drought led to a famine that threatened the security and economic growth of the entire region.

In coordination with humanitarian assistance, long-term approaches to strengthen resilience—for example, safety nets, frontline health programs, and vaccination campaigns—helped mitigate the worst of the drought impacts across the Horn. But the reality is that we must do more as a global community to prevent these crises from occurring in the first place.

We have to harness the power of science and technology to transform agriculture around the world. Our efforts must begin by strengthening investments in breakthrough technologies, such as climate-resilient seeds, and breakthrough services, such as climate-based agricultural insurance.

In today's difficult growing conditions, smallholder farmers from Kenya to India are improving their yields with new varieties of drought- and disease-resistant seeds and other technologies. To continue the development of crop and livestock advances that boost farm income and improve nutrition, we must support the full range of research techniques, including both conventional and genetically engineered approaches, and look beyond the issues centered on production alone.

An invaluable voice in the fight against hunger, Gordon's new book, *A Billion Hungry: Can We Feed the World?*, calls on the global community to do more—guiding our way forward while building on the messages of *The Doubly Green Revolution*. As global population grows toward 10 billion people, the challenges ahead remain daunting. Nonetheless, strong community leadership, renewed international commitment, and new opportunities, from mobile technology to drought-resistant seeds, are dramatically expanding the possibilities for success. By delivering meaningful results more effectively and efficiently than ever before, we can share Gordon's optimism that we will finally end hunger.

Acknowledgments

In some respects this is a second edition of my book *The Doubly Green Revolution: Food for All in the 21st Century,* first published in 1997 by Penguin Books in the United Kingdom and then a year later by Cornell University Press (and in a Portuguese translation, *Produção de Alimentos no Século XXI: Biotecnologia e Meio Ambiente,* Estação Liberdade, São Paulo, Brazil in 2003).

As in the previous edition, the book draws on a wide variety of sources and is indebted to the work of those with whom I have been associated over the past fifty years. Many are named in the text. They include postgraduate students and faculty at the Imperial College of Science and Technology, the University of Sussex, the Universities of Chiang Mai and Khon Kaen in Thailand, Padjajaran University, Indonesia, and the University of the Philippines, Los Baños, Philippines; colleagues at institutes where I have worked or been a director—the International Institute for Environment and Development (IIED), the Institute of Development Studies (IDS), the International Food Policy Research Institute (IFPRI), the Ford and Rockefeller Foundations, the UK Department for International Development (DFID); and workers in numerous nongovernmental organizations including the Aga Khan Rural Support Programme (AKRSP) in Pakistan and India, Action Aid and Mysore Relief and Development Agency (MYRADA) in India, Winrock International in Nepal, Concern Worldwide, the Ethiopian Red Cross, the Mo Ibrahim Foundation, the Alliance for a Green Revolution in Africa (AGRA), and the African Agricultural Technology Foundation (AATF); and the many scientists at the International Agricultural Research Centers (IARCs) of the Consultative Group on International Agricultural Research (CGIAR).

I have drawn freely on their writings and am grateful for their advice, experience, and friendship. Perhaps the greatest influence on my thinking has been the group of scientists at the University of Chiang Mai, Thailand, notably Benjavan Rerkasem, Kanok Rerkasem, Phrek Gypmontasiri, Rapeepan Jaisaard, Methi Ekasingh, Manu Seetisarn, Nakorn Na Lampang, and Ian Craig.

I am also particularly indebted to Tim Wheeler at DFID, Prabhu Pingali at the Bill & Melinda Gates Foundation, Rattan Lal at Ohio State University, Lawrence Haddad at the Institute for Development Studies, Peter Hazell at SOAS and visiting professor at Imperial College London, Margaret Catley-Carlson at CGIAR, Gary Toenniessen at the Rockefeller Foundation, Christopher Delgado at the World

Bank, and K. L. Heong at IRRI, who have read and commented on parts of the book.

The conception of the *Doubly Green Revolution* was an outcome of the deliberations of a small panel, which I chaired, commissioned to develop a vision statement for the CGIAR—the organization that supports the international research centers, which, over the past fifty years, have provided the research base that spearheaded not only the original Green Revolution (detailed in Chapter 3) but attempts to replicate its successes.[1] The vision was presented and adopted at a meeting of ministers of overseas development from the developed countries and of agriculture and natural resources from the developing countries, held in Lucerne in February 1995. My colleagues on the panel were Uma Lele, of the University of Florida, Martin Piñeiro (formerly director of the Inter-American Institute for Cooperation on Agriculture [IICA]), Jim Peacock, Chief of the Division of Plant Industry of Australia's Commonwealth Scientific and Industrial Research Organization (CSIRO), Selçuk Özgediz of the World Bank, Johan Holmberg of the Swedish Agency for Research Cooperation with Developing Countries (SAREC), Henri Carsalade of the French Centre for International Cooperation in Development-Oriented Agricultural Research (CIRAD), Michel Griffon of CIRAD, and Peter Hazell of IFPRI. I am grateful to all of them, to Paul Egger of the Swiss Development Cooperation and Robert Herdt of the Rockefeller Foundation, both on the CGIAR's Oversight Committee, and to Ismail Serageldin, then chair of CGIAR, for their collegial support and encouragement. Many of the ideas in this book are theirs.

In 1998, not long after *Doubly Green Revolution* was published in the United States, I became the twelfth president of the Rockefeller Foundation, centered in New York. The Foundation has a distinguished history in funding agricultural research and development, going back to the origins of the Green Revolution in the early 1940s (which I describe in Chapter 3). During my tenure as president we continued to fund pioneering programs in agriculture, under the direction of first-class international staff led by Bob Herdt, Gary Toenniessen and Peter Matlon. The team included Joe DeVries, Akin Adesina, Ruben Puentes, John O'Toole, Bharati Patel, and John Lynam.

I returned to the United Kingdom at the end of 2004 and became the chief scientific adviser to the UK Department for International Development (DFID). Again this was, and still is, an organization supplying major funding to agricultural research. I benefited from the experience of many of the DFID staff, in particular Jonathan Wadsworth, Alan Tollervey, and Steve Hillier, and the chief economist Tony Venables.

I have also greatly benefited from Sir John Beddington, the UK government chief scientist; a former holder of the post, Lord Robert May; and another departmental chief scientist, Sir Roy Anderson. I have known all three of them for many years and been stimulated by their ideas.

While at DFID I helped to set up the UK Collaborative on Development Sciences (UKCDS) under the very able direction of Andrée Carter. She was also heavily involved in the production of a book entitled *Science and Innovation for Development*, authored by Jeff Waage, of the London International Development Centre, Sara Delaney, and myself, published by the UKCDS and funded by DFID (free to download from the UKCDS website). I have drawn heavily from some of its contents. Once it was complete, Sara Delaney spent several months working with me on the early drafts of this book.

I stepped down from the chief scientist post in 2010 and returned to Imperial College as Professor of International Development running an advocacy grant from the Bill & Melinda Gates Foundation. Under the grant, we have set up a Montpellier Panel consisting of distinguished European and African agricultural experts: Tom Arnold, Joachim von Braun, Henri Carsalade, Louise Fresco, Peter Hazell, Namanga Ngongi, David Radcliffe, Lindiwe Majele Sibanda and Ramadjita Tabo. I have gained much from their wisdom, and also from the leaders of a sister Gates Foundation grant, the Global Agricultural Development Initiative of the Chicago Council for Global Affairs, Marshall Bouton, Catherine Bertini, and Dan Glickman; and from the staff of the Gates Foundation, Brantley Browning, Mark Suzman, Joe Cerell, and Laurie Lee from the advocacy program and Rajiv Shah, Sam Dryden, Prabhu Pingali, Rob Horsch, Roy Steiner, and Lutz Goedde from the agriculture program.

I benefited in 2011 from a stay at the Rockefeller Foundation's villa at Bellagio on Lake Como in Italy, accompanying my wife, Susan, who was working on her book on the Shan. It provided much-needed time to dedicate to this book at a critical juncture. I also owe much to Susan for her encouragement and sound advice.

My team at Imperial College has provided invaluable counsel and guidance: Yvonne Pinto, Calum Handforth, and Sara Delaney, who were with us for various periods, and the core team, Jo Seed, who has provided much support, Katy Wilson, who did a great deal of the research for the book and wrote many sections, and my deputy, Liz Wilson, who is a continuing source of wisdom and practicality.

I am grateful to Heidi S. Lovette, science acquisitions editor at Cornell University Press, for her support and enthusiasm and also to the team at Cornell University Press for their ongoing dedication and assistance. I would also like to thank Ros Morley and Sarah Starkey for their work in the production of the book.

Editorial Note

For supplementary material, please visit www.canwefeedtheworld.org.

Regions

The *developed regions* (as per UN regional country groupings shown in Plate 1) comprise Europe (including the transition countries but not the Commonwealth of Independent States [CIS]), Australia, Canada, Japan, New Zealand, and the United States (US), shown in maroon.

The CIS (Commonwealth of Independent States), which include the territories of the former Soviet Union in Europe and Asia, is sometimes included in the developing regions.

However, in this book the *developing regions* are the rest of the world and can be subdivided into:

Middle-income countries—such as Brazil, Vietnam, and South Korea, and including the *emerging economies* or BRICs (a term derived from the first letters of Brazil, Russia, India, and China but now used generally to describe countries with rapidly emerging industry and improving per capita wealth);

Low-income countries—such as Kenya, Ghana, and Honduras, and including the 48 *least developed countries* such as Cambodia, Mali, and Haiti;

Fragile states include those who are conflict ridden or recently emerging from conflict such as Afghanistan, Sierra Leone, and Nepal.

Groups

G7   Finance ministers from France, Germany, United Kingdom, Italy, Canada, United States, and Japan (European Union [EU] also present)—established in 1975

G8   Heads of government for the G7, plus Russia established in 1997

G20 Finance ministers and central bank governors—Argentina, Australia, Brazil, Canada, China, European Union, France, Germany, India, Indonesia, Italy, Japan, Mexico, Russia, Saudi Arabia, South Africa, Republic of Korea, Turkey, United Kingdom, United States—established in 1999

Measures

Dollars ($) are U.S. dollars
1 billion = 1,000 million
1 trillion = 1,000 billion
1 kg = 1 kilogram = 2.2 pounds (lbs)
1 ton = 1 tonne = 1 metric ton
    = 1,000 kg = 1 million grams (Mg)

1 ha = 1 hectare = 10,000 square meters
    = 2.47 acres
1 Mha = 1 million hectares (Mha)
1 km = 1 kilometer = 0.62 miles
calories = kilocalories (kcal)
ppb = parts per billion

Grain means cereals, e.g., wheat, barley, rice, maize, oats, sorghums, millets, and other coarse grains. (It does not include grain legumes or pulses.)

minimum dietary energy requirement (MDER or r_L)

Institutional Acronyms

CCAFS   CGIAR Research Program on Climate Change, Agriculture and Food Security

CGIAR   Consultative Group on International Agricultural Research (Washington, DC)

CIAT   Centro Internacional de Agricultura Tropical (Cali, Columbia)

CIFOR   Center for International Forestry Research (Bogor, Indonesia)

CIMMYT   Centro Internacional de Mejoramiento de Maiz y Trigo (Mexico City, Mexico)

CIP   Centro Internacional de la Papa (Lima, Peru)

DFID   Department for International Development (London)

FAO   Food and Agriculture Organization of the United Nations (Rome)

IARCs   International Agricultural Research Centers (funded by the CGIAR)

ICARDA   International Center for Agricultural Research in the Dry Areas (Aleppo, Syria)

ICLARM   World Fish Center (formerly International Center for Living Aquatic Resource Management [Penang, Malaysia])

ICRAF   World Agroforestry Centre (formerly International Center for Research in Agro-forestry [Nairobi, Kenya])

ICRISAT   International Crops Research Institute for the Semi-Arid Tropics (Patancheru, Andhra Pradesh, India)

IDS   Institute of Development Studies (Brighton, UK)

IFAD   International Fund for Agricultural Development (Rome)

IFPRI   International Food Policy Research Institute (Washington, DC)

IIED   International Institute for Environment and Development (London)

IITA   International Institute of Tropical Agriculture (Ibadan, Nigeria)

ILRI   International Livestock Research Institute (Nairobi, Kenya)

IMF   International Monetary Fund (Washington, DC)

IPC   International Planning Committee for Food Sovereignty

IPCC   Intergovernmental Panel on Climate Change

IRRI   International Rice Research Institute (Manila, Philippines)

ISAR   Institut des Sciences Agronomiques du Rwanda

IWMI   International Water Management Institute (Colombo, Sri Lanka)

KARI   Kenya Agricultural Research Institute

NGOs   Nongovernmental organizations

MYRADA   Mysore Relief and Development Agency (Bangalore, India)

OECD   Organisation for Economic Co-operation and Development (Paris)

UN   United Nations (New York)

UNEP   United Nations Environment Programme (Nairobi, Kenya)

UNHCR   United Nations High Commissioner for Refugees (Geneva)

UNICEF   United Nations Children's Fund (New York)

UNRISD   United Nations Research Institute for Social Development (Geneva)

USAID   U.S. Agency for International Development (Washington, DC)
USDA   U.S. Department of Agriculture (Washington, DC)
WFP   World Food Programme (Rome)
WHO   World Health Organization (Geneva)
WTO   World Trade Organization (Geneva)

One Billion Hungry

Part I

1

Acute and Chronic Crises

On Wednesday, police cleared away torched cars and other debris left by two days of looting and rioting. But helicopters circled the air amid black smoke rising from intersections as protesters continued to set tires ablaze, and gunfire was heard throughout Pétionville . . .

Associated Press, Haiti, April 9, 2008[1]

Hunger (from the Old English *hungor*) is an evocative, old Germanic word meaning "unease or pain caused by lack of food, craving appetite, debility from lack of food."[2] In the developed countries it is a feeling of slight discomfort when a meal is late or missed. By contrast, in the developing countries hunger is a chronic problem. Television images convey the realities of hunger—emaciated and starving children—in war-torn countries or in the aftermath of droughts, floods, or other calamities. Yet for a billion people—men, women, and children—hunger in the developing countries is a day-to-day occurrence, both persistent and widespread.

Food security, which implies the absence of hunger, is one of those apparently straightforward concepts that appears amenable to common-sense definition. But, somewhat surprisingly, it has been the subject of much debate. A review by staff of the Institute of Development Studies (IDS) at Sussex University identified some two hundred different definitions.[3] It is a highly politicized concept (I discuss this further in Chapter 4).[4] Nevertheless, in 1986 after further debate, the World Bank adopted the following definition which, with various elaborations, is now largely accepted: "Food security is access by all people at all times to enough food for an active, healthy life."[5]

The controversy over the definition arises, in part, because food security operates at many different levels and over different time scales: It can apply to the globe as a whole, to a region such as Sub-Saharan Africa, or to an individual country, community, or household. These different levels relate only very loosely to each other. A country can be food secure, but a household may not. Sometimes lack of sufficient food can be temporary, although devastating; at others, it is persistent and seemingly intractable.

Mrs. Namarunda

How chronic food insecurity comes about and what it means in practice is well illustrated by the conditions under which an African woman farmer, such as Mrs. Namarunda (who represents a composite of situations existing in Africa), struggles to feed herself and her family (Box 1.1).

Although Mrs. Namarunda lives at the end of a poorly maintained track, far from a town or markets, she is not immune to the events in the larger world. The high price of food affects her; the small sums her son irregularly sends her are insufficient to keep her family adequately fed.

Box 1.1 A One-Hectare Farm in Kenya, Africa

Several years ago, Mrs. Namarunda's husband died from HIV/AIDS. Her eldest son inherited the family farm, a single hectare running up one side of a hill, in the Siaya district near Lake Victoria. The soils are moderately deep and well drained, but they are acidic, highly weathered, and leached. Mrs. Namurunda's first son married and moved to Nairobi, where he is a casual laborer with children of his own.

Mrs. Namarunda was left on the farm—still officially owned by her absent son—with four younger children and the responsibility to produce food, fetch water, gather fuel, educate the children, and care for the family. But shortages of almost everything—land, money, labor, plant nutrients in soil exhausted from many years of continual crop production—mean that she is often unable to provide her family with adequate food. The two youngest children suffer from undernourishment and persistent illnesses.

Like many others in Africa, Mrs. Namarunda's smallholding provides an "insecure" livelihood (Figure 1.1). Fertilizer is expensive and she can't get credit, so she starts each growing season with a maximum potential harvest of only about 2 tons from mixed cropping on her 1 hectare of land. To survive, her family requires a harvest of about 1 ton, so if everything goes right and the maximum harvest is achieved, it would be sufficient to meet their needs and to generate a modest income. But, during the course of every growing season, she faces innumerable threats to her crops that reduce her yields.

Weeds are her most persistent and pervasive problem. It takes 40 to 50 days of weeding each crop, by her and the children, to keep the weeds under control. Her staple crop, maize, is attacked by:

- *streak virus*, where leaves develop long, white, chlorophyll-depleted lesions
- *Striga*, the parasitic weed that sucks nutrients from the roots and poisons its host
- *boring insects*, which weaken the stem
- *fungi*, which rot the ears that do develop, before and after harvest

(continued)

4

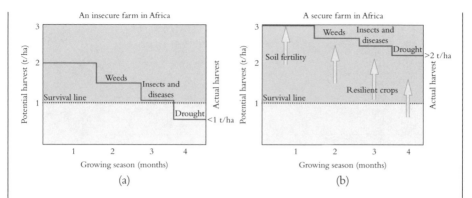

Figure 1.1 Harvest from an insecure and a secure farm.[6]

Mrs. Namarunda has tried growing cassava as an "insurance crop." But it, too, was attacked, first by mealy bugs and green mites, and then it was devastated by a new, supervirulent strain of African cassava mosaic virus.

The banana suckers she obtained from neighbors were already infected with weevils, nematodes, and the fungal disease Black Sigatoka when she bought them. Her beans, which are intended as a source of protein for the family and nitrogen for the soil, suffer from fungal diseases that rot the roots, deform leaves, shrivel pods, and lower nitrogen fixation. She also faces periodic drought that reduces yields. At the end of each season, what she actually harvests is usually less than 1 ton. She and her children are often hungry, and there is no money for schooling or health care.

Food Price Spikes

When food prices rose in 2007 there were street demonstrations in Mexico City, protesting at the high price of tortillas. Subsequent protests, many associated with violence, erupted throughout Latin America, Africa, and Asia. The prime minister of Haiti was deposed following the riots in 2008. Even in Italy there were popular demonstrations triggered by the rise in the price of pasta.

The Food and Agriculture Organization of the United Nations (FAO; based in Rome) keeps track of the prices of basic food staples. They began to rise in late 2006, apparently following a rise in oil prices.[7] Initially it looked like a new, maybe long-term, trend but eventually in mid-2008 the food prices peaked and fell down again (Figure 1.2). To some extent it was a classic, short-lived commodity price spike:

first, a commodity becomes, or is perceived to be, scarce;
second, prices rise;

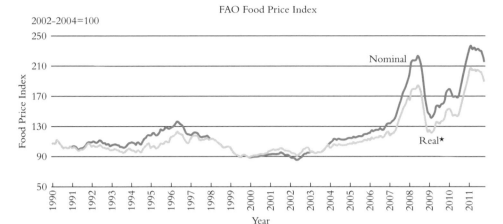

*The real price index is the nominal price index deflated by the World Bank Manufactures Unit Value Index (MUV)

Figure 1.2　The FAO Food Price Index as from 1990 to 2011.[8]

third, producers respond by producing more of the commodity and finally, prices fall.

But while food prices fell at the end of 2008, they were still 20 percent higher than 2006 prices and were growing again in 2009, tracking the 2007 rise. They leveled off in 2010 and then rose again, producing another spike. People in the developed countries felt the consequences, with higher food prices compounding the effects of the economic recession, but for the developing countries the outcomes were devastating. Grain prices remained high, even after the peak.[9] The FAO and the Organisation for Economic Co-operation and Development (OECD; based in Paris) estimated that, globally, average crop prices would be 10 to 20 percent higher in real terms relative to 1997–2006 for the next ten years, while for vegetable oils they would be more than 30 percent higher.[10]

I discuss how such food price volatility can be dampened in Chapter 4. Yet the spikes were not simple transitory events. They grew out of an underlying chronic crisis of global food insecurity and made this insecurity deeper and probably more persistent. According to the FAO, by the end of 2009 the number of chronically hungry people had risen to 1.02 billion, greater than at any time in human history.[18]

Box 1.2   The Causes of the Food Price Spikes

There has been much analysis of the proximal causes of the 2007–2008 food price spike.[11] A major factor was the fall in grain stocks. Demand had been steadily rising, and, while supply was also growing, adverse weather produced a run of poor harvests, enough to take stocks to dangerously low levels. Global consumption of grains and oilseeds exceeded production in seven of the eight years since 2000. The Australian drought and other weather events led to the very poor harvest of 2006. By 2007, stocks were only 14 percent of use (Figure 1.3). Importing countries became anxious about satisfying their future food needs and the prices began to spike.

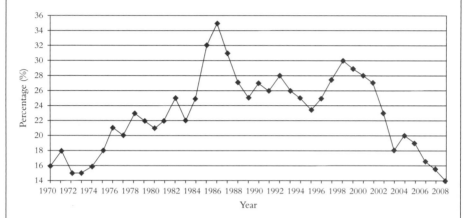

Figure 1.3   The stock to use percentage for all grains and oil seeds.[12]

The situation was compounded by speculation and by export bans imposed by Argentina, India, Kazakhstan, Pakistan, Ukraine, Russia, and Vietnam.[13] For rice, where the market is very thin (only 7 percent or 30 million tons of total production is traded), there was resort to panic buying. The Philippines purchased large amounts for domestic consumption at very high prices. Another factor was the depreciation of the U.S. dollar; some countries had accumulated considerable dollar reserves and were able to purchase large stocks of grain.[14] The energy price rise also played a role, as did the shift of land to biofuel production, especially in the United States.[15]

The subsequent spike in 2010 was triggered by a severe heat wave in Russia, which forecast a 30 percent reduction in the wheat harvest compared with the previous year, leading to an export ban that extended into 2011. Writers in such journals as the *Economist* and the *Financial Times* said there was plenty of wheat in stock and little to worry about[16]—"prices are unlikely to surge to the all time highs of 2007-08."[17] But then Egypt and other North African states purchased large quantities of wheat and prices began to rise. Various newspaper stories, ranging from possible locust outbreaks

(continued)

Box 1.2 (*continued*)

in Australia to heavy rains in Canada, heightened perceptions of impending shortage and future price rises. The actual and perceived shortages were enough to trigger a spike that eventually peaked above the 2007–2008 spike. Unlike that spike, the rises were not confined to food staples but extended initially to cotton and then to palm oil, cocoa, sugar, and rubber.

The Chronic Crisis

This book is concerned with three challenges: (1) the probability of repeated food price spikes and a continuing upward trend in food prices; (2) the persistence of a billion or more people suffering from chronic hunger; and (3) how we will feed a growing global population in the face of a wide range of adverse factors, including climate change. How will these challenges be manifest, both globally and in the developing countries, in the lives of the poor? What are their drivers? How do they operate at global, regional, and local levels? What can be done to mitigate or reverse their effects? Where do the barriers to action and the opportunities lie? How can we utilize the social and natural sciences to find solutions? How can we harness both government and the private sector to ensure solutions are applied on a large enough scale to achieve significant impact?

In the rest of this chapter I briefly look, in turn, at the fundamental drivers of

- the *increasing demand for food:* population increases, rising per capita incomes, and the competing demand for biofuel crops; and
- the *deficiencies in supply* caused by rising input prices, land and water scarcity and deterioration, slowing productivity gains and climate change.

Rising Populations

The first key driver of demand is the growth in the world's population.[19] According to the latest United Nations (UN) estimates, the global population is set to rise from close to 7 billion in 2010 to about 9.15 billion by 2050.[20] Then it is expected to stabilize. Virtually all of the additional 2.15 billion will be in the developing countries. The good news is that over forty percent of the world's population is living in countries where the fertility rate is at or under 2.1. The *fertility rate* represents the number of children an average woman is likely to have during her childbearing years, which in conventional international statistical usage is ages 15 to 49. A rate of 2.1 represents the *replacement level*; that is, the level of fertility at which a couple has only enough children to replace themselves, taking account of daughters who die before childbearing years.[21] For countries with this fertility rate, population growth will slow down and eventually stabilize.

The UN expects fertility in the less-developed regions as a whole to drop from 2.73 children per woman in 2009 to 2.05 in about 2050, and the reduction projected for the group of forty-eight least developed countries is even steeper: from 4.39 children per woman to 2.41 children per woman.[22] This optimism is based on the experience of recent decades. Today's fertility decline in developing countries is similar to the European experience during nineteenth and early twentieth century industrialization, but even faster.[23] In Bangladesh, for example, the rate halved from 6.0 in 1980 to 3.0 by 2000—a mere 20 years.

There is a close link between standard of living and fertility rate. Fertility starts to drop at an annual income per person of $1,000 to $2,000 and falls until it hits the replacement level of 2.1 at an income per head of $4,000 to $10,000 a year. Thereafter fertility continues at or below replacement level. The link between living standards and fertility also exists within countries. India's poorest state, Bihar, has a fertility rate of about 4.0, while the richer states of Tamil Nadu and Kerala have rates below 2.0. In effect there are two interlinked transitions: one demographic, from high growth to population stability, and the other developmental, from an agrarian to an industrial society.[24]

Nevertheless, the fertility rate decline has not been universal. In many Sub-Saharan countries fertility rate declines have stalled at rates over 5.0 after gradually decreasing for several years.[25] The reasons are complex, but a common feature appears to be the decreased funding for family planning programs. According to data from thirty-one countries, on average 30 percent of women in Sub-Saharan Africa have an unmet need for modern family planning methods, a proportion that has not declined in the last decade.[26] In nineteen of these countries, it is as high as nearly 50 percent. If fertility were to remain constant at current levels, the population of less-developed regions would increase to 9.8 billion in 2050 instead of the projected 7.9 billion.[27]

A popular misconception is that providing the developing countries with more food will serve to increase populations; in other words, it is a self-defeating policy.[28] The more food women have, the more children they will have and the greater will be their children's survival, leading to population growth, so goes the argument. However, the experience of the demographic transition described above suggests the opposite. As people become more prosperous, which includes being better fed and having lower child mortality, the fewer children women want. Providing they then have access to family planning methods, the fertility rates will drop and the population will cease to grow.[29]

Rising Per Capita Incomes

A population rise to 9.15 billion—that is, an increase of about 30 percent—will require increased food production of a similar amount.[30] However, this assumes no change in consumption patterns or in the proportion of the population who remain malnourished and outside the market. It also ignores the many threats to food security we are facing.

Consumption patterns are changing fast. In recent decades, per capita incomes have increased in most countries of the world, leading to a greater demand for improved quality and, to some extent, larger quantity of food.

The growth in per capita incomes in the high-income OECD countries has been over threefold in constant U.S. dollars since the 1960s. Income growth has been even faster in East Asia and the Pacific (over sixfold), although from a much lower base. It has doubled in the Middle East and North Africa. Only in Sub-Saharan Africa has there been little growth (Figure 1.4).[31]

With rising incomes, growing urbanization, and exposure to so-called diet globalization, a significant proportion of people in both the developed and developing countries are buying more processed and higher value foods, while reducing their purchases of raw agricultural commodities. In Asia, the consumption of traditionally "Western" foods—wheat and wheat-based products, temperate-zone vegetables, and dairy products—is growing, while per capita consumption of rice is declining.[33] Per capita rice consumption in India, Indonesia, and Bangladesh peaked in the early

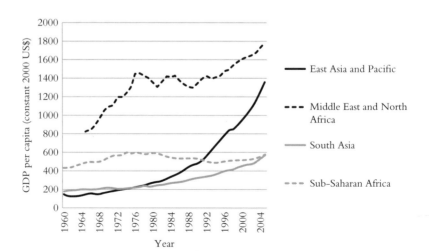

Figure 1.4   Growth in per capita incomes in low- and middle-income countries.[32]

1990s and has since fallen. Global rice consumption is also projected to fall from 441 million tons in 2010 to 360 million tons in 2050.[34]

One of the most significant outcomes has been the so-called Livestock Revolution, which many argue is having as important an effect on developing countries as the Green Revolution (see Chapter 10).[35] In recent decades, developing countries' meat consumption has been growing at over 5 percent per annum, and that of milk and dairy products at 3.5 percent to 4 percent.[36] The most dramatic increase has been in China, where meat consumption has risen to about 60 kg per person per year in 2005 from 20 kg per person per year in 1985.[37]

The correlation between income and consumption of meat and dairy products is strong, although religious and other cultural practices produce significant diversity. In Figure 1.5, China is above the line, showing the relationship between income and meat consumption, primarily because of the traditional consumption of pig meat (half of all the pig meat consumed in the world), and India is beneath, reflecting the continuing high levels of religious-based vegetarianism.[38]

Livestock, especially those reared intensively in confined forms of housing, are very demanding of cereals, oilseeds, and other feedstuffs. As a rule of thumb, it takes about eight kg of grain to produce one kg of meat product, although there is a considerable range from production on grazed grassland where little, if any, of the feed is brought in, to raising the livestock in stalls where the feed is entirely cereals and oilseeds.

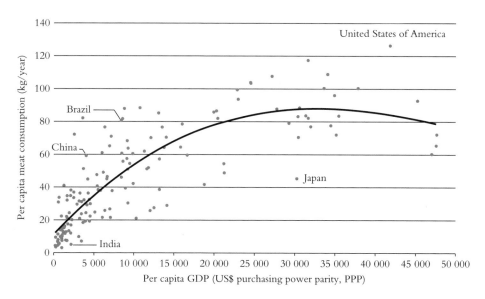

Figure 1.5   The relationship between meat consumption and per capita income.[39]

It is reasonable to expect that the Livestock Revolution will generate a greatly increased demand for feed, although this has not been borne out by recent history.[40] In the future that may change. China purchased 35 million tons of soybeans for its livestock and poultry industry in 2007, increasing oilseed prices. Although a one-off purchase, it may be a harbinger of the longer run impact of future demand.

The FAO expects there to be a slowdown in growth of meat consumption after 2030 but believes production will need to double from about 230 million tons in 1999–2001 to over 460 million tons by 2050, the great bulk of which will be produced in the developing countries.[41] As this happens, developing countries will shift from free-range to more intensive stall-fed systems of production, thus greatly increasing the demand for feed.[42] An estimate for 1999–2001 put the total feed use of cereals (note this excludes oilseeds) at 666 million tons, or 35 percent of world total cereal use. By 2050 this is likely to have doubled.[43]

Demand for Biofuels

Biofuel crops, or crops grown for products that can be converted to ethanol, biodiesel, or other fuels, can, in the right circumstances, reduce carbon and other greenhouse gas emissions (See Chapter 16). However, for many countries, especially the United States and China, the prime objective may be to reduce dependency on imported oil and petroleum. These objectives have driven a rapid rise in biofuel crop production in the developed countries and in some developing countries, notably China and Brazil. Global ethanol production has increased from under 20 billion liters in 2000 to around 85 billion liters in 2010. Global biodiesel production increased tenfold to almost 11 billion liters between 2000 and 2008, and in 2009 was almost 17 billion liters.[44]

The impact of this surge on the 2007–2008 food price spike is the subject of debate.[45] Donald Mitchell, the senior commodity economist at the World Bank, put the contribution as high as 75 percent (i.e., the greatest single cause of the spike) because growth in the demand for maize both for consumption and biofuels was not matched by growth in production (resulting in a shortfall of 27 million tons) and, as a consequence, a doubling of maize prices. However, Mark Rosegrant of the International Food Policy Research Institute (IFPRI) concluded the effect was much less, at under 40 percent.[46]

The discrepancies arise, according to Peter Timmer of Stanford University, because the estimates do not fully capture the linkages between maize, which is the primary grain used to make ethanol, and other commodities such as soybeans, wheat, and other feed grains. Thus, the 7 percent increased planting of maize in 2007 led to reductions in soybean acreage of 4 percent, as maize and soybeans tend to compete for land area, increasing the price of soy oil.[47] China is not a significant importer of maize, but it is a huge importer of vegetable oils, and the higher price of soy oil in 2007 caused a switch to palm oil, which, in turn, increased in price.[48] The increased growing of biofuel crops is likely to make land substitution between maize, soybean, and wheat more frequent and intense. Prices of these crops will move together and also with dairy and meat products. The same phenomenon will apply to biodiesel crops such as oilseed rape.

The range of estimated impacts of biofuels on global food production and prices has also been linked to different model assumptions regarding, for example, land availability, the substitutability of biofuels and food, and associations between food and energy markets.[49] Biofuels and the poor, a research project of IFPRI funded by the Bill & Melinda Gates Foundation, estimated that biofuel development in the United States, European Union, and Brazil will cause an increase in the world average export price of maize, soybean, other oilseeds and sugar of 17.7 percent, 13.6 percent, 27.6 percent, and 11.3 percent, respectively, by 2020.[50] The bottom line is that growing crops for biofuels will tend to increase the prices of basic food crops; biofuel production needs to be decoupled from food production either through policies or through cellulosic technologies.

I will discuss biofuel crops more fully in Chapter 16. For the rest of this chapter, I want to turn to the key long-term factors affecting supply.

Rising Oil and Fertilizer Prices

Rising oil prices were one of the key elements of the 2007–2008 food price spike. Indeed, their rise was a precursor to the food price increase. Their effect was, and continues to be, felt through the demand for biofuels, through the increased costs of transportation that affects both agricultural input and output prices, and in the manufacturing costs of fertilizers and other synthetic chemical inputs. In the developed countries the combined energy, chemical, and fertilizer costs typically account for a high proportion of crop production costs (16 percent for soybean, 27 percent for wheat, and 34 percent for maize in the United States).[51] Since the United States accounts for 40 percent of global grain exports and 25 percent of the global oilseeds exports, these costs are quickly reflected in global food prices.[52]

Energy-based inputs are also high in the irrigated lands of developing countries and growing rapidly where water is being depleted and water tables declining.[53] There, and on rain-fed lands, use of fertilizers and pesticides is often excessive and inefficient. Nevertheless, fertilizer and other inputs are crucial for small farmers trying to achieve higher yields of food crops. At the time of the food price spike, one

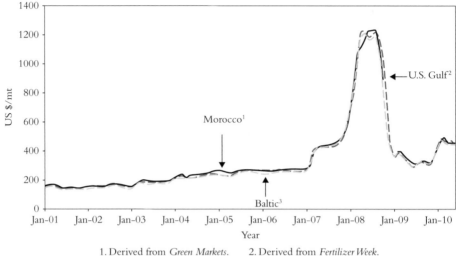

1. Derived from *Green Markets*.　2. Derived from *Fertilizer Week*.
3. Derived from *FMB Weekly Fertilizer Report*.

Figure 1.6　Increasing fertilizer prices (Diammonium phosphate, freight on board, monthly prices are all represented by a single line).[54]

of the biggest fertilizer price increases was in diammonium phosphate (DAP), a commonly used source of nutrients in developing countries (Figure 1.6). Its price rose nearly sixfold in early 2008, partly due to the energy prices involved in the production of the ammonium, but also because of shortages in both sulfur and phosphate, key elements in its manufacture. Although the subsequent fall was significant, the price of DAP fertilizer began to grow again from the middle of 2009.

The Scarcity of Land

Before 1960 greater food supplies were obtained by taking more land into cultivation; since then raising crop yields has been far more important.[55] However, there has been a continuing controversy over how much potential cultivable land remains for use. In the 1990s three apparently thorough analyses were conducted using the data provided by the FAO's *Soil Map of the World* and their annual *Production Yearbooks* and subsequent *Statistical Yearbooks*.[56] These indicated considerable areas of land available for cultivation: Asia–Pacific with 778 million hectares (Mha) available (27.1 percent of total land area), Sub-Saharan Africa with 1.2 billion ha available (45.8 percent of total), and South and Central America with one billion ha (50.1 percent).[57]

Yet these seem to be overestimates. The results have been trenchantly criticized by Anthony Young, who has a long and extensive experience of soil and land surveys.[58] He believes the estimates suffer from the following flaws:

- Overestimation of cultivable land (not accounting for features such as hills and rock outcrops when the maps are reduced in scale)
- Underestimation of presently cultivated land (illegal land occupation; e.g., forest incursions, not recorded)
- Failure to take sufficient account of land required for purposes other than cultivation (underestimates of human settlements and industrial use)

A more recent FAO analysis in 2000 accepts these criticisms as possibly valid and acknowledges that much of the cultivable but uncultivated land is under rainforest or needed for purposes such as grazing land and ecosystem services.[59] Probably the most telling data is the area harvested over time. Total global cropland has increased by only 10 percent over the past fifty years, while population has grown by 110 percent.[60] Given the pressures to increase food production, we would expect to see much greater land expansion if it were readily available. The only exceptions are for oil crops (Figure 1.7). Soybeans and oil palms have each increased by over 300 percent in area and by over 700 percent and over 1,400 percent, respectively, in production over the past fifty years. Presumably this is a result of clearing the Cerrado in Brazil and rainforests in the Amazon, Africa, and Southeast Asia. Permanent meadows and pastures (the land used to grow herbaceous forage crops) have also increased somewhat, by nearly 9 percent in area from 1961 to 2008.[61]

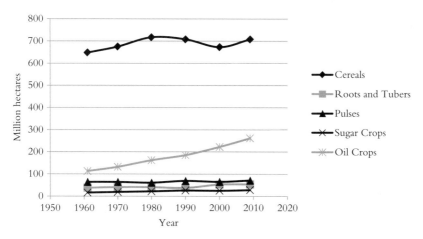

Figure 1.7   Trends in harvested area for selected food crops, million hectares.[62]

15

More land could be brought into cultivation by clearing the tropical rainforests, but this would be at the expense of biodiversity and would add considerably to greenhouse gas emissions (see Chapter 16). A further factor in the equation is the large amount of land being degraded as a result of erosion, loss of fertility, and desertification. According to an essentially qualitative *Global Assessment of Human Induced Soil Degradation (GLASOD),* about 300 million hectares (Mha), or 5 percent of the formerly usable land in developing countries, has been lost by severe soil degradation as of 1991; that is, more than has been brought into production.[63] The current rate of loss is claimed to be not less than five Mha per year.[64]

"Land Grabs"

Despite this apparent lack of available arable land, or maybe because of it, foreign or national investors have been purchasing large tracts of land in developing countries, often for the production of crops destined for export. In 2009, hedge funds and other speculators bought or leased almost 60 Mha of land in Africa.[65] Ethiopia, a country known for its famines and reliance on foreign aid, began to sell millions of hectares of land to investors in 2008.[66] The impacts of such investments on host countries are unclear given the diversity of land acquisition agreements and investor commitments. Increased investment may boost gross domestic product (GDP) growth and benefit economic development through employment and infrastructure. Indeed, the Ethiopian government claims land deals will bring in much-needed foreign currency and facilitate technology transfer from large agribusinesses to smallholder farmers.[67]

But while communities may benefit from these developments, access to land crucial for food security may be taken from local people, particularly because land users in developing countries are often marginalized in the land ownership system.[68] Although large areas of land "look empty," they are not. Often local communities have a range of traditional rights over the land, and many countries do not have legal or procedural mechanisms to protect these rights. Furthermore, a lack of transparency and regulation can foster corruption. Millions of local people have already been displaced from areas purchased by investors, the majority without any form of compensation.

Large-scale land grabs could also create environmental conflicts if incentives such as priority rights over water are offered to encourage investment or if inappropriate resource-intensive farming practices are employed over large areas. In a 2011 review of twelve land deals in Africa, such incentives and long-term rights to land were, in some cases, offered for very little in return (e.g., jobs or investment in infrastructure) and with little consideration of social and environmental issues.[69] Nevertheless, large-scale land acquisition in Africa will likely increase in the future, in part spurred by concerns over food security, competing land uses, rising food prices, and human expansion in investor countries.[70]

Scarcity of Water

Water is, of course, crucial to agricultural production and, like land, is similarly in short supply and for similar reasons—overuse, inefficient use, and degradation through pollution.[71] Today, much of Asia's grain harvest comes from the irrigated, annual double- and triple-crop, continuous rice systems in the tropical and subtropical lowlands of Asia and from irrigated, annual rice-wheat, double-crop systems in northern India, Pakistan, Nepal, and southern China. Large areas of this land and of the Middle East and North Africa are now maintaining irrigated food production through unsustainable extractions of water from rivers or the ground.[72] In China the groundwater overdraft rate exceeds 25 percent, and the rate is over 56 percent in parts of northwest India.[73] Groundwater levels in India have declined by nearly 18 cubic km per year over the past decade.[74] Much of this excessive water use results from water pumps using subsidized or free electricity.[75]

By contrast, Sub-Saharan Africa has large untapped water resources for agriculture. Only 4 percent of cultivated land (5.3 Mha) is irrigated, of which 70 percent is in Madagascar, South Africa, and Sudan. The potential exists to bring an additional 20 Mha of land under irrigation but, so far, technical, financial, and socioeconomic constraints have slowed this expansion. At the same time, almost a quarter of the African population lives in water-stressed countries, and the share is rising.

Many river basins in the world do not have enough water to meet all the demands; about a fifth of the world's people, more than 1.2 billion, live in areas of physical water scarcity. Rivers are drying up, groundwater levels are declining rapidly, freshwater fisheries are being damaged, and salinization and water pollution are increasing.

I discuss soil and water conservation in more detail in Chapters 13 and 14. In combination, declining amounts and quality of cultivable land and of utilizable water will make agriculture less productive and food more expensive.

Slowing Productivity Increases

As I stated earlier, global consumption of grains and oilseeds exceeded production in seven of the years between 2000 and 2008. This was partly due to declines in production resulting from extreme climatic events. Deteriorating land and water supplies may have had an effect. But there is also some evidence of declines in the rate of growth in yields of the major cereal and oilseeds. Figure 1.8, which measures the

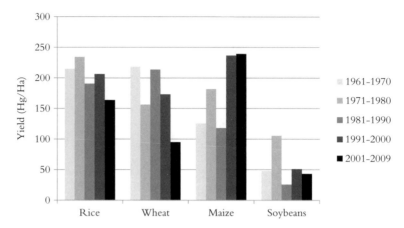

Figure 1.8   Average decadal increase in yield for rice, wheat, maize, and soybeans between 1961 and 2009.[76]

average decadal increase in yields, shows the high rates of growth in the Green Revolution years of the 1960s and 1970s for all four crops—rice, wheat, maize, and soybeans. Subsequently there have been declines in rates of growth for rice and wheat, although yields of maize and soybeans have both grown strongly in the past two decades.

The reasons for the decline are not clear, but one factor may have been the reduced investment in agricultural research and development by government.[77] Low stable food prices since the late 1970s have led to complacency about global food supply and to a reduction in research and development (R&D) funding. There has also been a partly ideologically based belief that agriculture is best served by the private sector. Indeed, private funding of agricultural research has grown but has tended to focus on the crops of developed countries where the returns have been greater. This may be the reason why yield growth in maize and soybeans has been strong in recent years.

The Impact of Climate Change

In the longer term climate change is likely to have a bigger effect on food supply than any other factor. Moreover, agriculture will probably be affected more than any other economic sector in the developing countries (this may also be true for the developed countries).

As yet, the effects are localized and mostly lie within the range of natural climatic variation. The 2007–2008 price spike was not a result of climate change, although there is some evidence that the severity of the catastrophic 2006 drought in Australia, which had a major effect on world grain supplies, was a consequence of global warming.[78] Australia naturally has a highly variable climate, and the 400,000 square mile Murray-Darling basin, the country's prime food-growing region, is subject to periodic severe droughts, largely influenced by the El Niño phenomenon in the Pacific (see Chapter 15). The year 2006 had the lowest rainfall for one hundred years, but it also experienced the highest temperature. The subsequent higher evapotranspiration coupled with drought resulted in even lower yields than might otherwise have occurred. This trend is likely to continue to have an increasingly adverse effect on Australia's crop production.

A similar phenomenon may have occurred in Russia in the summer of 2010. There, temperatures reached 5°C above the norm for July, leading to a doubling in Moscow's death rate and the burning of 30 percent of the nation's grain crops. Notably this was combined with the worst floods in eighty years in Pakistan, which killed over 1,600 people and submerged a fifth of the country, including about 14 percent of Pakistan's cultivated land. Global warming is expected to increase the frequency and duration of extreme weather events but climate models are not yet sophisticated enough to determine whether the abnormal airflow in the region in 2010 was the result of increasing greenhouse gas emissions.[79]

In Chapter 15 I will show how future climate change will have an increasingly adverse effect, worldwide, on food security and discuss what can be done to adapt.

Outline of the Book

In some respects this is a second edition of my book *The Doubly Green Revolution: Food for All in the 21st Century,* first published in 1997.

As I began to write I realized how much had changed in the intervening years. Most of the concepts and arguments remain as relevant today as then, in some cases more so. I have retained some of the elements of the original structure—after all agriculture retains some constant truths. Some of the original examples are present, where they remain pertinent. But the book reflects what has happened globally in the past few years—the growing number of people who are chronically hungry, the devastating impact of the food price spikes and the challenges we face if we are to feed a world subject to climate change. Alongside these challenges there have been enormous strides in technical achievements—revolutionary improvements in crop and livestock breeding, successes in application of ecological principles to agricultural production and real examples of exciting farmer innovation. Out of the concept of a Doubly Green Revolution has grown an emphasis on Sustainable

Intensification, partly based on the recognition of the increasing shortage of land and water and of the threats posed by climate change. As a consequence there are new chapters and much new material. In essence it is a new book.

The first part of the book sets out the context more fully. In Chapter 2, I describe what is meant by *hunger*, how we measure it, and how it relates to poverty. Chapter 3 discusses the lessons from the Green Revolution of the 1960s and 1970s, one of the great technological success stories of the twentieth century. In Chapter 4 I discuss the political economy of food security and agriculture and how it is changing over time.

In the second part, I introduce the concept of a Doubly Green Revolution (Chapter 5) and then discuss its various components: the challenges of sustainable intensification (Chapter 6), the need for appropriate research and development, including technologies (Chapter 7), and the creation of markets (Chapter 8).

Part 3 examines specific areas in detail. Chapter 9 discusses the role of breeding to produce better crop varieties and livestock breeds, drawing on the full range of traditional, conventional, and biotechnology approaches. Chapter 10 is devoted to the topic of the Livestock Revolution, examining the role of domesticated animals as important components of agricultural livelihoods. Chapter 11 explores the contribution of farmers as innovators.

In Part 4 I discuss some of the environmental challenges facing the Doubly Green Revolution. Chapter 12 focuses on pest control, arguing for a combination of integrated pest management techniques with the use of biotechnology. Chapters 13 and 14 discuss soil and water conservation, emphasizing the importance of restoring good organic matter to the soil and efficiently using available water resources. In Chapters 15 and 16 I analyze the relationship between climate change and agriculture, where agriculture is presented as both victim and culprit.

Finally in chapter 17 I summarize the previous chapters and return to the question posed in the title of the book: Can we feed the world?

2

What Is Hunger?

To die of hunger is the bitterest of fates.

Homer, *The Odyssey*[1]

Homer, in his epic poem *The Odyssey,* recounts how Odysseus and his companions have resisted the lure of the Sirens, sailed safely between Scylla and Charybdis, and have come to the island of Thrinacia where the "Sun-god's cattle and plump sheep graze." Odysseus has been warned the animals are not to be harmed, but his companions succumb to the temptation. "To die of hunger," declares Eurylochus "is the bitterest of fates." They kill the cattle and feast. No sooner have they set sail again than Zeus sends a hurricane as a punishment. All perish except for Odysseus.

Today, there are a billion people who, like Odysseus's companions, live in a world where food is relatively plentiful yet it is not available to them. If we were to add up all of the world's production of food and then divide it equally among the world's population, each man, woman, and child would receive a daily average of over 2,800 calories[2]—enough for a healthy lifestyle. But of course, food is not divided in this way (nor is income) and it is unrealistic to expect it will happen in the near, or even distant, future.

Unlike Odysseus's companions, many of the hungry are women and children. In addition to their own needs, women require an adequate diet if they are to give birth to, and raise, healthy children. More than 130 million children under five years of age are underweight; that is, they are well below the standard weight for their age. This represents nearly a quarter of the under fives in the developing countries.[3] Children are especially vulnerable from their time in the womb until the age of five. During this period of rapid physical and cognitive development, even short-term dietary deprivation can have lasting effects such as stunted growth, poor brain development, low educational attainment, and decreased life expectancy. Girls grow into women who give birth to sickly children, so that ill health and poverty are passed from one generation to the next. In this way, hunger and poverty are inextricably linked, one resulting in the other, so creating a trap from which escape is very difficult.

In this chapter I address these questions:

- What is hunger and how is it measured?
- What is the balance between food supply and access to food?
- How does hunger relate to poverty?

The Nature of Hunger

Hunger comes in many forms: [4]

- *Malnutrition* is a general term for a condition caused by improper diet or nutrition. It means "badly nourished," but is more than a measure of how much people eat or fail to eat. Malnutrition results from inadequate intake of protein, calories, and micronutrients and is characterized by frequent infections and disease. As a consequence the physical function of an individual is impaired to the point where he or she can no longer maintain natural bodily capacities such as growth, pregnancy, lactation, learning abilities, physical work, and resisting and recovering from disease.
- *Undernourishment* is more specifically used to describe the status of people whose food intake does not include enough calories (energy) to meet minimum physiological needs for an active life.
- *Starvation* is an extreme form of these conditions, characterized by a "state of exhaustion of the body caused by lack of food."[5] When starvation is accompanied by increasing mortality on an epidemic scale, we usually describe it as a *famine*.

These conditions are difficult to measure, so indicators have been devised that simply measure the height or weight of an adult or child and compare it with the same measurement in a reference population of well-nourished and healthy people. Three such measurements—stunting, wasting, and underweight—are commonly used. In practice they are relatively accurate reflections of hunger (Box 2.1).

The Food We Need

Determining the amount of food we need is complicated.

First, the sources of calories are very different. A *food calorie* or *kilocalorie* (kcal) is defined as the energy needed to increase the temperature of 1 kilogram (kg) of water

Box 2.1   Practical Measures of Hunger[6]

- *Stunting* or low height for age

  Caused by long-term insufficient nutrient intake and frequent infections. It generally occurs before age two, and the effects are largely irreversible. These include delayed motor development, impaired cognitive function, and poor school performance.

- *Wasting* or low weight for height

  Is a strong predictor of mortality among children under five. It is usually the result of acute significant food shortage and/or disease and reflects a recent and severe process that has led to substantial weight loss, usually associated with starvation and/or disease. Wasting is often used to assess the severity of emergencies because it is strongly related to mortality.

- *Underweight* or weight-for-age of a child

  It is estimated that the deaths of 3.7 million children aged less than five are associated with the underweight status of the children themselves or of their mothers.[7]

  Most famine reported in the media is the consequence of disasters such as drought, floods, war, or earthquakes, but disaster-related famine accounts only for 8 percent of hunger-related deaths; 92 percent result from chronic and recurring hunger. Child wasting tends to be a consequence of disasters because it measures intermittent, short-term, acute undernourishment. Chronic hunger is best captured by measures of child stunting.[8]

---

by 1°C and is used as a common measure of the energy that cells acquire from food through digestive breakdown and oxidation. There are several principal sources of food energy.[9]

- *Carbohydrates* are the main source of energy and hence play a key role in basic metabolism.[10] Each gram of carbohydrate provides about two kcals, and the World Health Organization (WHO) guideline is that carbohydrates should provide 55–75 percent of total energy.
- *Fats*, comprising saturated fats, monounsaturated fats, and polyunsaturated fats, are the second major dietary energy source. Some fats are also essential for proper growth and development in early life. Each gram provides about nine kcals. Dietary guidelines, backed up by scientific studies, generally advise that a diet high in saturated fats will lead to increased blood cholesterol levels and a higher risk of heart disease.[11] Ideally fats should provide 15–30 percent of total energy.

- *Proteins*, comprised of amino acids, are vital for both maintenance and growth.[12] The amount and quality of protein intake promotes good muscle and bone growth and overall optimal human health.[13] Some amino acids cannot be synthesized in the body and are, therefore, important constituents in the diet. Meat, fish, eggs, and dairy proteins show a greater digestibility than plant sources, such as pulses and nuts.[14] Ideally protein should provide 10–15 percent of total energy.

Second, individuals differ greatly in the calories they need.

Energy requirements vary with an individual's metabolism, physical activity, and growth needs as well as with the requirements for pregnancy and lactation. Important determinants are age, gender, body size, and composition. Adult men tend to require more calories than women. Children have much lower requirements, increasing with age up to a peak around adolescence. Adults over age sixty also have lower requirements. Within each range, the need is determined by body size and activity level (Table 2.1).[15]

From these needs it is possible to use a country's population composition broken down by age, gender, body size, activity level, and pregnancy status to work out its overall needs. The Food and Agriculture Organization of the United Nations (FAO) does this by determining for each country the minimum dietary energy requirement (MDER). This is the amount of energy needed for light activity and to maintain a minimum acceptable weight for attained height. The numbers of hungry are then those who get less than this amount.[17]

In some countries such as the Democratic Republic of Congo (DRC) and Eritrea the average food supply is less than the MDER, and the percentage suffering from undernourishment is very high (Figure 2.1). In other countries the average food supply is higher than the MDER, and the greater the difference, the lower the percentage undernourished. While many argue that hunger is not determined by how

Table 2.1   Energy needs in calories per day

Age	Men/boys	Women/girls
0–12 months	518–775 (beginning and end of range)	464–712
1–18 years	948–3410	865–2503
18–30 years	2100–4500 (depending on weight and activity)	1650–3850
30–60 years	2100–4200	1750–3400
Over 60 years	1700–3600	1550–3150

Source: FAO/WHO/UNU[16]

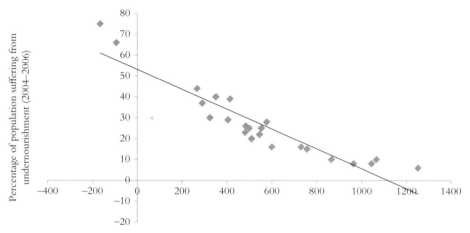

Average food supply (Kcal/capita/year) minus Minimum Daily Energy Requirements (2004–2006)

Figure 2.1 National proportion undernourished in relation to the gap between the Average Food Supply and the Minimum Daily Energy Requirement for 23 developing countries.[18]

much food is available but rather by how well it is distributed, these results indicate that the overall supply of food is a critical factor.

Not Just Calories

Poor nutrition is not just about a lack of calories. If a cereal diet supplies enough calories, it usually provides enough protein. Since, at least for the poor, cereals are the most important source of calories, there is a tendency for adequate nutrition to be equated with sufficient cereal production and consumption. Nevertheless, vegetables, fruit, livestock, and fishery products are also important sources of proteins, as well as vitamins and minerals. It is total food supply, measured in terms of quality as well as quantity, that determines whether people are adequately fed.

It is generally accepted that a well-balanced diet is essential, but for many years there has been a lively and, at times, acrimonious debate over where the balance lies and what components are most deficient. The focus of nutrition interventions has evolved from countering protein deficiency, to concern over protein-energy deficiency, and now to the prevention and treatment of micronutrient deficiencies.

In 1935, a Jamaican pediatrician, Dr. Cicely Williams, described a set of symptoms that she termed kwashiorkor that appeared to be related to lack of protein in the diet.[19] Children suffered from accumulations of liquid under the skin (edema), retardation in growth, wasted muscles, and some psychomotor damage.[20] The condition was especially prevalent in Africa (the name is of Ghanaian origin), in part because of the low protein content of the staple starchy foods.

Although there were no reliable data regarding the prevalence of kwashiorkor, the subsequent conventional wisdom was that protein deficiency was the main nutrition problem in the world, a view reinforced by a UN Protein Advisory Group that promoted the use of protein-rich food mixtures and by the shipment of vast quantities of dried skimmed milk to developing countries.[21] The terms *kwashiorkor* and *malnutrition* were often used interchangeably. Then, in 1963, Derrick Jelliffe, professor of Paediatrics and Child Health at Makerere Medical School in Uganda, pointed out that there were many other childhood conditions in addition to kwashiorkor. The main problem in all of these, he concluded, was a combination of calorie and protein deficiency.[22] The controversy intensified in the 1970s with the publication of an article on "the great protein fiasco" by Professor Donald McLaren, a nutritionist working at the American University of Beirut, in which he claimed the role of protein deficiency as a cause of child malnutrition had been overstated.[23] Others agreed that protein intakes seemed to be adequate in most countries, and the emphasis subsequently shifted to improving the health of women and children through growth monitoring and interventions such as oral rehydration therapy, breastfeeding, and immunization.[24]

The First 1,000 Days

In a five-part series on nutrition issued by the *Lancet* in 2008, adequate nutrition was found to be critically important during the first 1,000 days of a child's life (i.e., from conception to two years of age). Undernutrition during this crucial window increases the risk of early mortality and has largely irreversible long-term effects on health and on cognitive and physical development (Box 2.2).[25]

Micronutrient Deficiencies

The latest phase of this debate, partly reflected in the Scaling Up Nutrition (SUN) initiative (see Box 2.2), has been a focus on micronutrient deficiencies. Although the consequences of severe deficiencies—especially of vitamin A, iron, zinc, and iodine—have long been known, there has been little understanding that milder forms could be harmful and are widespread.[29]

Box 2.2 Interventions in the First 1,000 Days

Approximately 195 million children in the world are stunted. This is a third of all children who are under 5 years old. Of these, 90 percent live in just thirty-six countries, twenty-one of which are in Sub-Saharan Africa. In some African countries the proportion of children stunted is as high as 50 percent.

Undernutrition causes an estimated 3.5 million maternal and child deaths annually. As U.S. Secretary of State Hillary Clinton said at CARE's 2010 National Conference and Celebration, "These deaths are intolerable because they are preventable."

Direct nutrition interventions targeted at the 1,000-day window are cost-effective: the Copenhagen Consensus Panel, a panel of leading economists, found vitamin A and zinc supplementation to yield benefits worth $1 billion per year for an annual investment of $60 million.[26] In 2009, a World Bank study identified a package of thirteen interventions for the first 1,000 days, including promoting good feeding practices, providing micronutrients through supplementation or fortification, and therapeutic feeding. In return for implementing these interventions in the thirty-six highest burden countries (estimated to cost $11.8 billion annually) the lives of one million children annually would be saved. Despite this evidence, expenditure by major aid donors on nutrition totaled just $439 million between 2002 and 2007 (less than 1 percent of total bilateral development assistance).[27]

The Scaling Up Nutrition (SUN) framework, launched in April 2010, aims to set out a plan of action and to mobilize support for tackling child undernutrition. Several developing countries—including Bangladesh, Ghana, Malawi, Nepal, and Uganda—have indicated they wish to engage with the SUN agenda. The goal is that at least eight countries will start to receive intensive support for SUN by 2012.[28]

- *Vitamin A.* About 130 million preschool children suffer from vitamin A deficiency. As has been well established, lack of this vitamin can cause eye damage. Half a million children become partially or totally blind each year, and many subsequently die. As more recent research also has shown, lack of vitamin A has an even more serious and pervasive effect, apparently reducing the ability of a child's immune system to cope with infection. Trials in several developing countries where children under five were given vitamin A supplementation resulted in a 25 percent reduction in mortality from measles and diarrhea.[30]
- *Iron.* Lack of minerals in the diet can have equally severe effects. Iron deficiency is common in the developing world, affecting as many as a billion people. Iron deficiency is the leading cause of anemia (a reduction in the number of red blood cells below normal levels), which afflicts over 30 percent of women of childbearing age (15–49 years old) and 42 percent of all pregnant women. In Africa the figures are 47 percent and 57 percent, respectively. Iron-deficient

anemic women tend to produce stillborn or underweight children and are more likely to die in childbirth.[31] In Africa, 68 percent of preschool children are anemic. Globally, 136,000 women and children die annually due to iron-deficiency anemia.[32]

- *Zinc.* Zinc is also a key mineral in the diet. A lack of zinc increases the severity of diarrhea, pneumonia, and possibly malaria and also causes stunting. There is a high risk of low zinc intake in the developing countries; estimates for zinc deficiency are 70 percent of the population in Southeast Asia, 95 percent in South Asia, and 68 percent for Sub-Saharan Africa.[33]

- *Iodine.* Finally, lack of iodine is the single greatest cause of preventable mental retardation: severe deficiencies of this mineral cause cretinism, stillbirth, and miscarriage; mild deficiency can significantly affect the learning ability of populations. Over a billion people in the world suffer from iodine deficiency, and 38 million babies born each year (30 percent of the world's newborns) are at risk of brain damage.[34]

Lindsay Allen, a nutritionist at the University of California, Davis, maintains that "it has become recognized by the nutrition community that micronutrient malnutrition is very widespread, and is probably the main nutritional problem in the world."[35] Micronutrient deficiency can be rectified relatively cheaply and easily by supplementation or fortification. More fundamentally we have learned that improved nutrition is not simply a function of increased production of calories, or even of total food production. Nutritionists have played a key role in emphasizing the quality of food produced and consumed, but equally agricultural scientists have a vital role in ensuring that a diversity of food production is encouraged and maintained. As Allen argues this will require "much more communication . . . among the nutrition, agriculture and development communities."[36]

Malnutrition and Ill Health

For many hungry people, undernutrition and malnutrition lead to death, not necessarily through starvation (although this may happen in famine situations) but because a poor diet reduces the capacity to fight disease.[37] Diarrhea, measles, respiratory infections, and malaria are common in many parts of the developing world. Well-fed people can fight them off; the malnourished, especially children, will succumb. Undernutrition has been estimated to be an underlying cause in 35 percent of all under-five child deaths.[38]

How Is Hunger Calculated?

The FAO calculates the number of people who are hungry each year. It is a fairly complicated, essentially statistical, calculation that makes a number of key assumptions (Box 2.3).

Box 2.3 Estimation of the Proportion Hungry[39]

First, the FAO assumes that hunger is distributed in a population according to a log-normal distribution (Figure 2.2). At the upper end of the distribution are people who have plenty to eat (and may indeed be obese); at the bottom are those who are hungry.

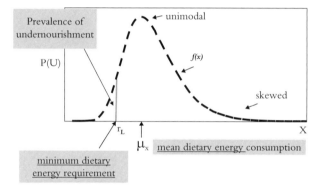

Figure 2.2 Theoretical distribution of dietary energy consumption, $f(x)$. ($P(U)$ is the proportion of undernourished in total population, (x) is the dietary energy consumption, and μ_x its mean. r_L is a cut-off point reflecting the minimum dietary energy requirement).[40]

There are four basic steps in calculating the proportion of a population that is hungry.

1. Calculate the mean for a particular country by deriving a *food balance*, which is based on the country's food production and imports, less such items as exports, animal feeds, and processing waste. The dietary energy supply per person is then found by aggregating all food commodities and converting them into energy values (kilocalories). This is then converted to a kcal/person/day figure, which is used as the *mean energy consumption* for that country (equivalent to the average supply represented by μ_x in Figure 2.2 above).

(continued)

Box 2.3 (*continued*)

2. Calculate the variation from the mean, which is assumed to be a function of income. The per person daily income for each household is calculated using household survey data and the households assigned to each decile for a given area.
3. Calculate the country's *minimum dietary energy requirement* (MDER or r_L) based on the sum of the requirements for long-term good health for different geographic and socioeconomic groups, using survey data on attained heights, by sex and age for light physical activity. There is also an additional pregnancy allowance, which is calculated using the country's birth rate.
4. Finally, the area under the curve between 0 and r_L gives the proportion undernourished.

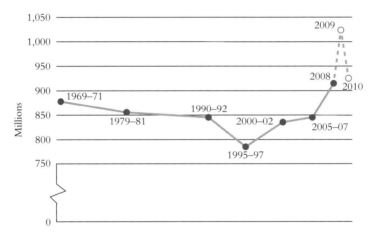

Figure 2.3   Numbers of undernourished in the world.[42]

Each year, the resulting hunger numbers are updated in the FAO annual report *The State of Food Insecurity in the World* (SOFI), which covers 106 countries.[41] 2010 figures indicate that hunger has been growing for the past decade after a steady decline in the previous three decades, probably because population growth is once again outstripping food production (Figure 2.3). In 2009, it reached a peak of over a billion people, the largest number in the history of the planet. Subsequent estimates have not been published by FAO since it is revising its methodology, but the true figure probably lies between eight hundred million and one billion.

The increase between 2007 and 2008 was caused largely by the food price spike. After 2009 it fell again, mainly due to declining international and domestic food prices and a more favorable economic climate in developing countries. Nevertheless, despite this increase in absolute numbers, the proportion of hungry people has continued to fall, from nearly 35 percent in 1969–1971 to just over 15 percent in 2011. The considerable advances in the Chinese economy has been a major factor in the decline, producing a fall from 15 percent undernourished in 1990 to 10 percent in 2004–2006.[43]

The Global Hunger Index

The FAO's approach to calculating hunger has received considerable criticism.[44] This is not surprising given the important role that numbers play in the popular advocacy for more attention to the problems of world hunger. A somewhat similar approach developed by the U.S. Department of Agriculture (USDA) is based on simulation models using prices, national accounts, and production equations, but it often differs radically from the FAO figures. While FAO estimated 1,020 million undernourished people in 2009, the USDA estimate was only 833 million people.[45]

Both of these approaches depend on the quality of data which is highly variable. But they have the advantage of presenting a quantified, and fairly rigorous, assessment of the number of people who are *expected to be hungry* given a certain level of availability of calories. However, they do not measure *actual hunger or its consequences*. Moreover, the FAO measurement does not show the variation within countries, nor the actual consumption of food (where a large proportion of the food is grown or captured at the household level and not traded, consumption is often underestimated). The FAO measurement also assumes calorie availability is a reliable guide to hunger.

In an attempt to overcome these limitations, a relatively new index—called the Global Hunger Index (GHI)—has been developed that combines availability with some of the consequences of food insecurity. The GHI sums, with equal weighting, the following three percentages:

1. the FAO proportion of the population undernourished,
2. the prevalence of underweight children under the age of five, and
3. the mortality rate of children under the age of five.

In effect the GHI combines malnutrition with the outcomes of malnutrition in a single index.[46] It ranks countries on a 100-point scale, with 0 being the best score (i.e., no hunger) and 100 the worst, although neither extreme is reached in practice.

- 0 representing no hunger
- less than 4.9 reflects "low" hunger

- between 5 and 9.9 reflects "moderate" hunger
- between 10 and 19.9 indicates a "serious" problem
- between 20 and 29.9 is "alarming"
- 30 or higher is "extremely alarming"

The results are published each year by a consortium of the International Food Policy Research Institute, Welthungerhilfe, and Concern Worldwide. However, because the measures of the consequences of malnutrition are only recorded periodically, changes in the index are only significant over a longer time span. The FAO index provides a better year to year assessment.

The harsh reality revealed by the FAO and Global Hunger indices is the existence of great inequality. In 2000 the United Nations agreed on a set of very ambitious Millennium Development Goals (MDGs).[47] The first of these aimed at halving the proportion of people suffering from poverty and hunger by 2015 (using 1990 as a baseline). Sub-Saharan Africa is worst performing (Plate 2).[48] There the GHI has fallen by only 13 percent since 1990, and the region is significantly off-track to achieve the MDG.

The countries with "extremely alarming" 2010 GHI scores are Burundi, Chad, the Democratic Republic of Congo (DRC) and Eritrea. In the case of the DRC, the GHI has increased by more than 50 percent since 1990. Most of these countries are, or have been, characterized by years of conflict and political instability. In others, particularly in southern Africa, the high prevalence of HIV and AIDS has played a role in creating a high score. South Asia has done better, the GHI falling by 25 percent since 1990, but India, Bangladesh, and Pakistan still have GHI levels above 20.

Overall the situation is grim. However, there are some regions and countries where significant progress has been made. Most successful have been Latin America and Southeast Asia, where the index has fallen by 40 percent since 1990. And in Asia and Sub-Saharan Africa, two countries—China and Ghana—have gone against the prevailing trend. China's GHI has fallen from 11.6 in 1990 to 5.7 in 2009 and Ghana from 23.5 to 11.5.[49] A fundamental question is why some countries have done better than others.

Household Surveys

Both the FAO index and the GHI are useful as advocacy tools but are less helpful in prioritizing need and targeting or delivering resources. They are also too crude to permit monitoring of progress or measuring the success of past interventions.[50]

A more practical and useful approach is to conduct detailed nutrition and food security surveys. These more comprehensive studies try to not only find out what the nutrition status is at the household level but also what might happen in the future.

They try to identify: Who are the food-insecure and vulnerable people? How many are there? Where do they live? Why are they food-insecure? What is the appropriate assistance to reduce vulnerability and food insecurity?[51]

Because these are conducted at the household level, they are costly in staff time and may not be representative of the country as a whole, but they do provide clear targets as well as very useful insights (Box 2.4).

Box 2.4 A Comprehensive Food Security and Vulnerability Analysis (CFSVA) and Nutrition Survey for Rwanda

A CFSVA was conducted by the World Food Programme (WFP) (through the National Institute of Statistics of Rwanda and with help from other partners), first in 2006 and then in 2009.[52] 5,400 households were visited from a stratified sample of sixteen districts or groups of districts. During the survey, food consumption data was collected at the

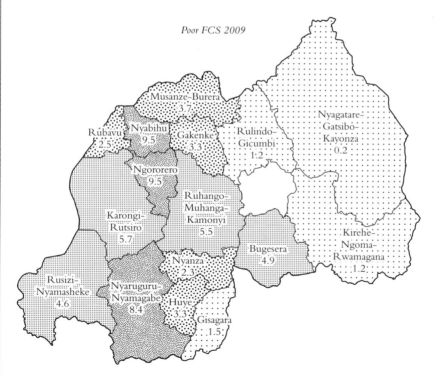

Figure 2.4 Proportion of population by district in Rwanda who are food insecure.[53]

(continued)

Box 2.4 (*continued*)

household level and used to obtain a snapshot of households' access to food. This information was transformed into a Food Consumption Score (FCS), and households were categorized as those with *poor* (food insecure), *borderline* (moderately food insecure), or *acceptable* consumption (food secure).

The three districts (or groups of districts) with the highest proportion of households reporting poor food security were located along the Crete of the Nile line that runs from North to South in Rwanda. Together these constitute 14 percent of the total population, but they account for 42 percent of all the households with a poor FCS (Figure 2.4).

The analysis also revealed that 21 percent of the poor FCS households are female headed, and 22 percent have a chronically ill member. 36 percent of poor FCS households also tend to cultivate less than 0.1 ha of agricultural land and have low diversity of agricultural production and ownership of livestock. Overall, there was a significant association between the FCS and the wealth index.

Encouragingly, the survey revealed a significant improvement in food security since 2006.

Since 2004 the World Food Programme (WFP) has completed more than thirty-five baseline surveys worldwide, most recently in Liberia, Tanzania, Mozambique, Pakistan, and the Republic of Congo.[54]

Access and Utilization

As is evident from the Rwanda analysis (see Box 2.4), food insecurity is not only a function of not producing enough food but is also linked to poverty and to a lack of utilization of, and access to, food.

In 1981, Nobel Prize–winning economist Amartya Sen published his classic study of the Great Bengal Famine that occurred in 1943 under British rule (Box 2.5).[55] Sen's analysis recognized the importance of access to food, as opposed to its supply.

Sen's study of this and other subsequent famines in Ethiopia, the Sahel, and Bangladesh led him to develop the concept of *entitlement*.[58] Put simply, people are entitled to food because they have produced it themselves—as farmers or sharecroppers—or they have earned the money to buy it or they can receive food as part of an exchange with their neighbor or kin, or under a government system of benefit.

Box 2.5 The Great Bengal Famine of 1943[56]

In 1943, Bengal's rice crop had been hit by a cyclone, had suffered from flooding and a disease outbreak, and the Japanese occupation had cut off supplies from Burma. At the time, and subsequently, the famine was attributed to the shortage of food these events created. But Amartya Sen's study suggested this was only part of the story.

The overall shortage of food grains was not that much lower than in previous years, for example 1941, when there had not been a famine. In Sen's view the evidence suggests it was not a "remarkable" shortfall. More important, in the rural areas where the famine hit hardest, was the failure of laborers' wages to keep pace with the inflation induced by the war economy. Well before the famine, the price of rice had doubled, but wages had risen little (Figure 2.5). A major cause of the famine was the inability of the rural poor to purchase the rice they needed. They migrated in large numbers to Calcutta, where relief measures proved inadequate despite the availability of sufficient food. In all, some three million people died.

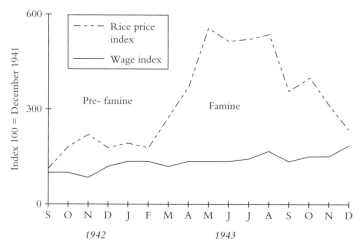

Figure 2.5 Wages and rice prices in a district of West Bengal.[57]

This emphasis on *access* to food rather than food production produced a sea-change in thinking about food security.[59] FAO, in a 1983 reappraisal of the concept of *food security*, stressed "physical and economic access to . . . basic food."[60]

Perhaps not surprisingly, assessments of food insecurity based on surveys of households and individuals routinely generate higher estimates of insecurity than more aggregated statistics. Such household surveys capture subjective measures of adequacy, risk exposure, and sociocultural acceptability, in addition to objective dietary, economic, and health indicators.[61] They can also be used to measure past consumption patterns using food expenditure and dietary diversity measures and how people cope with lack of food.[62]

Christopher Barrett, an economist at Cornell University, summarizes the reality of hunger as consisting of three "pillars"—availability, access, and utilization—that are "inherently hierarchical, with availability necessary but not sufficient to ensure access, which is, in turn, necessary but not sufficient for effective utilization."[63] As the recent price fluctuations have underlined, stability is a critical fourth pillar. Availability is relatively well captured using existing methods. However, household surveys, while illuminating, tend to be only partial in their coverage, both in space and time. Barrett argues for a global network of sentinel sites using a standardized core survey protocol for regular, repeated household- and individual-level monitoring. This would permit tracking of multiple food security indicators with targetable individual, household, and community characteristics across continents. It would also facilitate strict monitoring and evaluation of the impacts of various policy and project interventions.[64]

Hunger and Poverty

Not surprisingly, hunger and poverty go hand in hand. According to World Bank estimates, some 1.4 billion people—that is, a quarter of the developing world's population—were in poverty (defined as living on less than $1.25 a day) as of 2005 (Plate 3). This figure is estimated from household surveys that determine household per capita expenditure on consumption, including both cash spending and imputed values for consumption from a household's own production.[65]

Most of South Asia's 600 million poor are in India, where they compose over 40 percent of the population. Just over half of Sub-Saharan Africa's population is also in poverty. About two-thirds of East Asia's poor live in China, although they amount to less than 16 percent of the population there.[66]

Poor people have few or no assets, are unemployed, or earn less than a living wage and thus cannot produce or buy the food they need. As the food price spike of 2007–2008 brought home, even the poorest people are dependent on purchased food. They spend a much higher proportion of their income on food than wealthier people, making them particularly vulnerable to price increases. A five-person household living in Bangladesh living on under $1 a day per person typically spends its $5 as follows:

- $3.00 on food
- $0.50 on household energy
- $1.50 on other purchases

A 50 percent increase in food and energy prices, as occurred during 2007–2008 means there is virtually nothing left over for other expenditures.[67] In practice, people in this situation eat less: They have one less meal a day, women may reduce their food intake, or children may drop out of school. Not surprisingly the numbers of chronically hungry people in the world rose by 100,000 million following the food price spike.

Rural Poverty

To the casual observer, poverty seems to be worse in the cities, but, in reality, the urban poor fare better.[68] Although the cost of living may be low in rural areas, there are fewer opportunities to make a living. At the extreme, the urban poor can at least beg or steal. To quote one statistic, the incidence of malnutrition is three times higher in the mountainous Sierra of Peru than in the capital, Lima. In many countries in Sub-Saharan Africa and Asia, most of the poor are rural poor; in Malawi and Indonesia, for example, the proportion is over 70 percent.[69] Some live in rural areas with high agricultural potential and high population densities such as the Gangetic Plain of India and the island of Java. But the majority of rural poor live where the agricultural potential is low and natural resources are lacking, such as the Andean highlands and the Sahel.

One of the biggest concentrations of rural poverty is in northern India. In three states—Bihar, Madhya Pradesh, and Uttar Pradesh—rural poverty rates are some of the highest in the world: 41 percent, 31 percent, and 21 percent, respectively. These are accompanied by high rates of infant mortality and female illiteracy. Such rural populations are very isolated; the proportions of habitations without roads for the three states are 58 percent, 62 percent, and 43 percent. Not surprisingly, economic growth rates in the three states were only 2.5 percent in the 1990s (compared with 5.5 percent in states such as Gujarat and Maharashtra). One consequence was that the gap between India's rich and poor states widened in the 1990s. While parts of urban India compete for global business in software engineering and biomedical research, parts of rural India have poverty rates comparable to failed states and child malnutrition rates higher than any country in the world.[70]

Poverty and Land

The poor are typically landless or have very small farms, often too small for subsistence without appropriate inputs. Moreover, the size of landholdings is falling, with the fastest decline in Africa.[71] Average farm holdings for the continents of Asia and Sub-Saharan Africa (excluding South Africa) are, at the 2000 census, 1 ha and 1.5 ha, respectively (which equals around 2.5 and 3.7 acres).[72]

In Bangladesh about 60 percent of the rural population own less than half an acre and of those over 50 percent are poor.[73] When they do own land, it is often unproductive and is rarely irrigated. They find themselves unable to improve their land because of lack of income and access to credit. Despite using up to 80 percent of their income to obtain food and a similar proportion of working time for its production and preparation, they remain undernourished.[74] Nevertheless, figures from Bangladesh show that the landless and smallest landholders can get out of poverty. Between 2000 and 2005 poverty fell across all land-owning classes as a result of a broad-based poverty reduction program. The number of people in poverty fell most significantly (by 38 percent), however, among those owning medium to large (i.e., over 2.5 acres) farms than those who were landless (11 percent), indicating the importance of landownership in enabling the poor to climb out of poverty.[75]

Poverty by Ethnicity and Gender

Poverty is common among ethnic and other minorities. Indigenous ethnic groups are often isolated from mainstream society, and sometimes isolated geographically, contributing to their poor status. Over 15 percent of the Chinese ethnic minority population in 2002 were poor in comparison with just 9 percent of the ethnic majority.[76] In Latin America, too, the likelihood of being poor is greater for indigenous groups; Ecuadorian and Mexican indigenous peoples are 16 percent and 30 percent, respectively, more likely to be poor than ethnic majorities.[77] In Vietnam, despite an overall decline in the proportion of people living in poverty (from 58 percent in 1993 to 16 percent in 2006), the decline has been less impressive for the fifty-two ethnic minority groups that predominantly inhabit remote and mountainous regions.[78]

The Scheduled Tribes (ST) and Scheduled Castes (SC) of India make up 9 percent and 20 percent of the population, respectively, and 14 percent and 27 percent of the poor.[79] Such groups have been historically underrepresented in the political system, but since 1953 when political representation was adopted as part of the Indian Constitution, government seats have been reserved for ST and SC peoples in proportions equal to their representation within the population. The impact of such legislation has been an increase in targeted redistribution toward these groups and a net reduction of poverty.[80]

Women comprise a large number of the poorest and most oppressed worldwide, and some of the poorest households are headed by women (see the Rwanda food security analysis in Box 2.4). They often shoulder a disproportionate share of the workload. Women in the hill districts of Nepal work around sixteen hours a day, compared with the nine to ten hours men work. Many women are hungry as well as overworked, so creating a vicious circle of discrimination, poverty, and hunger.[81]

City of Joy

Such statistics provide an overview of the reality of millions of people's lives, but a full picture of the deprivation and misery they suffer can be obtained only from close acquaintance. Dominique Lapierre in his fictional book *City of Joy* describes, in the opening chapter, the circumstances that drive a West Bengal family to migrate to the slums of Calcutta (Box 2.6).[82]

Box 2.6   The Journey of a West Bengal Family to the Slums of Calcutta

Thirty-two-year-old Hasari Pal lives with his wife, Aloka, and three children, his parents, two younger brothers, and their families—in all, sixteen people. His father, Prodip, had once owned over three hectares of good rice land, but a large landowner, by bribing the judge in a court action, had wrested the land from him. The family was reduced to owning less than half a hectare, share cropping another plot and growing vegetables, tending a few fruit trees, and caring for a buffalo, two cows, and two goats on the land around their house. They survived until, one year, pests destroyed the rice crop. The land was mortgaged to the village moneylender, and one brother became an agricultural laborer.

For two more years they struggled on; then a storm brought down all the mangoes and coconuts. They had to sell the buffalo and one of the cows. Hasari's youngest brother began to cough blood, and money was needed for the doctor's fees and medicine. In the same year, his youngest sister married. The dowry and the ceremony cost the family 2,000 rupees (about $200). They spent their savings and pawned the family jewelry.

Then came the final blow: The monsoon failed, the young rice crop withered in the field, and the village wells dried up. The household had no more assets, nothing they could mortgage or pawn, and so Hasari Pal, Aloka, and their children set off, with trepidation, to find a new life in Calcutta.

When Hasari Pal and his family arrive in Calcutta, they find a place to camp on the pavement. Hasari gives blood for a few rupees and purchases some bananas, enough to keep his family fed for a few days. Then, by chance, he meets up with someone from his home village, a rickshaw puller who persuades the owner of the rickshaws to let Hasari work in the place of a puller who has just died. It is a hard life, and Hasari eventually dies from the strain of the work, but he has saved enough for a dowry for his daughter, and his family survives. The opportunity is there, for some at least, to slowly progress from the pavement to the slum and to the beginnings of a decent livelihood.

The Role of Technology

As evident from the story of Hasari Pal and his family (see Box 2.6), poor people live complicated and "risky" lives. As I described earlier in this chapter, food security depends on availability, access, utilization and, stability of food supply. Each on its own is not sufficient; moreover, poor households also need income to provide for health care and education costs and allow them to purchase more food. For this to happen they need highly productive technologies that will maximize their production given their land, labor, knowledge, skills, and other assets.

Amartya Sen's conclusion from his study of the Bengal famine that "Starvation is the characteristic of some people not *having* enough food to eat. It is not the characteristic of there not *being* enough food to eat,"[83] has been often quoted by activist groups arguing that there is plenty of food in the world; we do not need more, according to this perspective, just better distribution. Other activists have used the quotation to argue against developing new, and possibly hazardous, agricultural technologies. These are somewhat simplistic arguments, not least because they ignore logistical, market, and political realities (I will return to this later in the book).

Significantly, Sen in a later book added access to technology as one of the entitlements of the poor. He defined three overarching entitlements by which individuals can establish ownership and command of food:

1. *Endowment* (labor power, wealth, land, and other resources)
2. *Production possibilities* (the ability to earn money to buy food, or to grow your own food)
3. *Exchange conditions* (the ability to buy and sell goods at fair prices)

Sen argues that technology is part of the production entitlement: "This is where technology comes in: available technology determines the production possibilities, which are influenced by available knowledge as well as the ability of the people to marshal that knowledge and to make actual use of it."[84]

As we shall see in Chapter 3, it is this entitlement that the pioneers of the Green Revolution set out to provide.

3

The Green Revolution

The Japanese have made the dwarfing of wheat an art. The wheat stalk seldom grows longer than 50 to 60 centimeters. The head is short but heavy. No matter how much manure is used, the plant will not grow taller; rather the length of the wheat head is increased. Even on the richest soils, the wheat plants never fall down.

U.S. adviser to the Meiji government, 1873[1]

The Green Revolution was the product of two genes—one in wheat in Japan and the other in rice in China—that in the hands of visionary scientists and administrators transformed agriculture, not only in the developing countries but throughout the world. In retrospect it is easy to see the revolution, for revolution it was, as a series of obvious logical steps. But the journey was somewhat circuitous. Roger Thurow and Scott Kilman in their book *Enough* say, "From the beginning, the Green Revolution was the unintended outcome of unlikely work by determined individuals."[2]

In this chapter I address the following questions:

- How did the Green Revolution come about?
- What were its successes?
- What were its limitations?

The long circuitous route started with a short journey, in 1940, by Henry Wallace, the new vice president of the United States. Before taking up office he asked President Franklin D. Roosevelt if he could go on a trip to Mexico, ostensibly to practice his Spanish but also to build closer ties to the new Mexican president, Manuel Ávila Camacho. Wallace also happened to be a plant breeder who had been inspired as a young boy by George Washington Carver, the first black student at Iowa State University who went on to become a distinguished agricultural scientist, promoting peanuts, soybeans, and sweet potatoes for poor farmers in the south who sought to diversify out of cotton. Wallace had become very successful at breeding hybrid corn (maize) varieties—so successful that he and his colleagues had founded what would become the giant Pioneer Hi-Bred corn seed company.[3]

41

First Steps in Mexico

Wallace drove round rural Mexico, going into the fields to talk to the farmers, noting that despite the recent land reforms they were still desperately poor.[4] The soils lacked nutrients and were heavily eroded, the maize and bean yields only a third of those of the United States. While many thought the situation was hopeless, Wallace did not. His own experience as a maize breeder had taught him that science could make a big difference. On his return to the United States he called on Raymond Fosdick, the president of the Rockefeller Foundation in New York. Fosdick took little persuading. Although the Foundation had been focusing on health issues for many years, most notably hookworm and yellow fever, the staff were becoming increasingly convinced that poor nutrition lay behind much of the ill health in the developing countries. The Rockefeller Foundation's involvement in agricultural development was a momentous decision; to this day the Foundation has been a leading funder and innovator in the field (I was very proud to inherit such a distinguished history of achievement when I became president in 1998).

To undertake a more detailed assessment of what was needed, Fosdick sent to Mexico three distinguished agricultural scientists, who combined the core disciplines that, in hindsight, were to become relevant to the Green Revolution: Paul Mangelsdorf, a plant breeder; Richard Bradfield, an agronomist; and E. C. Stakman, a plant pathologist. Their subsequent report described what was needed to get agriculture moving in Mexico and convinced the Foundation of the feasibility of a major scientific program. A joint venture, the Office of Special Studies, was established by the Mexican Ministry of Agriculture and the Rockefeller Foundation in 1943.[5] The Office was headed by George Harrar, a plant pathologist, with Norman Borlaug, another plant pathologist, Edwin Wellhausen, a maize breeder, and William Colwell, a soil scientist. Eventually the office grew to twenty-one U.S. and a hundred Mexican scientists, mostly working at an experiment station at Chapingo on the rain-fed central plateau. Its goal was to improve the yields of the basic food crops: maize, wheat, and beans.[6]

The research program concentrated first on maize, the mainstay of the Mexican diet, consumed in the form of the thin, flat, unleavened bread called *tortilla*. Mexican agronomists had already discovered that most strains of maize grown in the United States were not well adapted to Mexican conditions, so the program set out to try and duplicate the U.S. achievement of breeding high-yielding, hybrid maizes but using indigenous varieties as a basis. Success came quickly. In 1948 1,400 tons of seed of the improved maize varieties were planted. The new seed, good weather that season, and the ready availability of fertilizer resulted in a record harvest, and, for the first time since the revolution of 1910, Mexico had no need of imports. By the 1960s over a third of Mexico's maize land was being planted to new high-yielding varieties, and maize yields were averaging over 1,000 kg per hectare (kg/ha). Total production had increased from two to six million tons.

Rust Resistant Wheat

Although maize was the main staple crop, Mexico was importing about a quarter of a million tons of wheat per year. Yields were very poor: less than 800 kg/ha, even though most of the wheat land was irrigated. "Most varieties were a hodge-podge of many different types, tall and short, bearded and beardless, early ripening and late ripening. Fields usually ripened so unevenly that it was impossible to harvest them at one time without losing too much over-ripe grain or including too much under-ripe grain in the harvest."[7] In northern and central Mexico, the soils had lost most of their fertility. On the newer, well-irrigated lands in the Pacific Northwest the soil was generally fertile enough to produce high yields, but stem rust was very destructive. Epidemics in three consecutive years, 1939–1941, in Sonora had caused many farmers to reduce their wheat land or stop growing the crop altogether.

The wheat program, under the direction of Norman Borlaug, began by testing over seven hundred native and imported wheat varieties for rust resistance.[8] By crossing some of the imported varieties that were more resistant to rust and higher yielding with the Mexican varieties Borlaug produced, in 1949, four rust resistant varieties each adapted to a particular ecological region of Mexico. By 1951 the new varieties were being grown on 70 percent of the total wheat land and, five years later, Mexico was producing over a million tons of wheat, with an average national yield of 1,300 kg/ha. Imports of foreign wheat were no longer required.

Starting on the Green Revolution

These results illustrated what science in the form of modern plant breeding could do for developing countries' agriculture. The next step was to focus on improving wheat yields three- or fourfold, an achievement that became the core of the Green Revolution. The history of the Green Revolution is well known but worth recounting here as a reminder of the power and limitations of innovative technology and of the crucial importance of the economic, social, and institutional environment to the success of agricultural development.

A careful analysis of the trends in agricultural productivity in a variety of countries, both developed and developing, suggests there is a point in history when yields begin to take off.[9] While agricultural production remains based on traditional practices, with few or no outside inputs, yield gains are modest, on the order of 1 percent or less per annum. Production grows, barely perceptibly and largely as a result of increases in cropped area, until a critical threshold is reached when a transition to a new basis for production occurs. Much improved varieties are introduced, farmers turn to inorganic fertilizers and synthetic pesticides, invest in irrigation and drainage, and adopt a range of other new technologies, including the purchase of agricultural machinery. In most developed countries the takeoffs occurred soon after the end of World War II. The Green Revolution, begun in Mexico, aimed to replicate this yield takeoff in the developing countries.

Short-Strawed Wheat

Borlaug and his colleagues realized that the next step in the wheat program was to improve yields through greater use of fertilizers. Experiments on properly irrigated soils showed that 140 kg/ha of nitrogen raised yields more than fourfold. Even on rain-fed soils, yields more than doubled and the addition of phosphate produced five- or sixfold increases. But the traditional varieties had a tendency to grow tall and *lodge*; that is, they fell down under the weight of the luxuriant green growth produced by the extra nutrient uptake, making the grain liable to rot and be less harvestable. The breeders also recognized the need for nonshattering varieties that would hold the wheat grains until they were ripe enough for mechanized harvesting and threshing. Wheats with better milling and baking quality were also required, now that Mexico was relying solely on its own grain and not blending with the stronger, imported varieties.

As the quote at the beginning of this chapter testifies, short-strawed wheat varieties had long existed in Japan. In 1935 the Japanese had produced a new dwarf, *Norin 10,* by crossing one of their traditional dwarfs with Mediterranean and Russian varieties imported from the United States.[10] This was spotted by a U.S. agricultural officer working in Japan in 1946, and seeds were sent to Washington State University. Initially the crosses with U.S. wheats resulted in sterile offspring, but several years later Orville Vogel, of the U.S. Department of Agriculture (USDA), produced a fertile cross resulting in a new semidwarf variety called *Gaines* that yielded a world record of over 14 tons/ha.

Norman Borlaug heard of Vogel's work and in 1953 obtained some of his breeding material to cross with traditional Mexican varieties. To do this he developed a system of shuttle breeding (Box 3.1).

Shuttle breeding not only speeded up the breeding program, but the new varieties were highly responsive to fertilizer applications (Figure 3.1) and insensitive to day-length. This meant separate breeding programs for each geographic region were

Box 3.1   Shuttle Breeding[11]

Norman Borlaug began breeding wheat in the summer months at Chapingo in the central highlands of Mexico. He proposed immediately taking the seed from the first crop north to a lowland station in the Yaqui Valley of Sonora. This went against the conventional wisdom that seeds required a rest period after harvesting and that crops had to be bred under conditions similar to where they would be grown. George Harrar vetoed the plan, causing Borlaug to resign. Elvin Stakman resolved the stand-off, Borlaug withdrew his resignation, and shuttle breeding began in 1945.

The results were spectacular. In Borlaug's words: "As it worked out, in the north, we were planting when the days were getting shorter, at low elevation and high temperature. Then we'd take the seed from the best plants south and plant it at high elevation, when days were getting longer and there was lots of rain. Soon we had varieties that fit the whole range of conditions. That wasn't supposed to happen by the books."

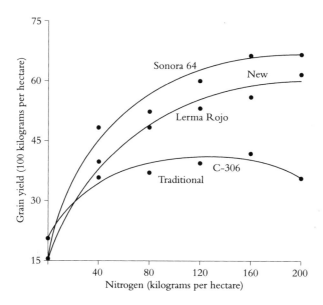

Figure 3.1   Responses of new and traditional wheats to fertilizers.[12]

not needed. Providing the temperature was above a certain minimum and there was sufficient water, the new varieties would grow almost anywhere.

By 1966 the new varieties were yielding 7 tons/ha. A decade later, virtually all of Mexico's wheat land was planted with these varieties and the country's average yield was close to 3 tons/ha, having quadrupled since 1950. By 1985 total production had increased to 5.5 million tons from around 1.4 million tons in 1961.

Dwarf Wheat to Asia

The next stage of this momentous revolution was an epic journey by Norman Borlaug and his colleagues transporting 250 tons of the new wheat seed to India and Pakistan (Box 3.2).

Subsequently the new wheat seeds were exported at a price only slightly higher than the world market price. In 1967–1968, Pakistan imported enough new seeds to plant over 400,000 hectares. Yield increases occurring after Pakistan and India's yield takeoffs in 1967 were on average around 50 kg/ha per year (Figure 3.2).

A major impetus for the rapid uptake were failures of the monsoon in 1966 and 1967. The United States, which was the only country carrying food reserves of any size, had responded by shipping one fifth of its grain crop to India.[15] This degree of dependence was recognized by both sides as being risky and undesirable. In some instances, food aid agreements had permitted countries to put off serious agricultural development plans and the large volume of food received had depressed prices,

Box 3.2 Seeds to India—as Told by Norman Borlaug to Paul Underwood[13]

In 1965, India and Mexico decided to import 250 tons of the new Mexican wheats. . . . They placed the order too late to ship it out of Guaymas Port in Mexico. So we had to ship from Los Angeles to catch the last freighter that would get it in time to Pakistan and India for planting in the middle of October. And then everything went wrong. I thought I had the border fixed so that these 35 big trucks—and we didn't have much money for this operation—were held up at the Mexican border for two days. And I had to pay them to hold the freighter. Then, we were held up for another day on the American side. Pure bureaucracy. Finally, my Mexican colleague that was supervising this, he called and he said, "They're on their way to the Los Angeles port." But they didn't get very far. It was the day of the Watts riot. The National Guard was out there. Stopped it. We had to because Watts was on the route to the port from Mexico. Finally, it loaded.

I went home to bed. I hadn't slept for two nights, on the telephone. I woke up, whatever it was, 18, 20 hours later, turned on the radio, and the war is on between Pakistan and India. Seed on the same freighter. It was transhipped to Singapore which meant that it arrived six weeks too late for planting. There wasn't time to check the germination like we ordinarily would have done to calibrate how much seed to plant. We planted—I was in Pakistan, and I saw it was miserable. And so I had decided to double the seeding rate. . . . We did a lot of praying and put on more fertilizer and handled the water. And by the time, shortly before Christmas, I left. I had visited both Pakistan and India, and I could see this beautiful harvest and the enthusiasm of farmers *"everywhere!"*

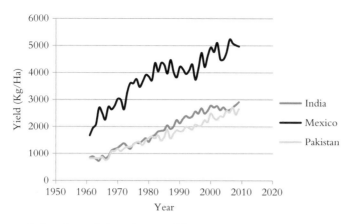

Figure 3.2 Growth in wheat yields in Mexico, India, and Pakistan.[14]

reducing the incentive for farmers to produce more. But the monsoon failures helped to wipe out the world's surpluses and prices rose dramatically (e.g., rice going from $120/ton to over $200/ton in 1967). Faced with a food crisis of this magnitude, a number of developing country leaders quickly recognized the potential of the new varieties.

Such was Prime Minister Indira Ghandi's enthusiasm that she ripped up a flower bed in front of her residence and planted the new wheats.[16] President Ayub Khan of Pakistan also took a personal and active part in the promotion of the Green Revolution. Both leaders were able to take some of the credit for its success.[17]

Short-Strawed Rices

Over much of Asia the staple diet was, and still is, rice. Unlike wheat, rice is mostly grown by smallholders (on farms of less than 3 ha) for home consumption. In the 1950s, over 90 percent of the rice produced in the world was grown in Asia; national yields averaged between 800 and 1,900 kg/ha.[18]

Early in the twentieth century, Japan, China, and Taiwan had developed many new rice varieties, which had led to significant increases in production, but the major impact on Asian rice yields came in the 1960s as a result of the skillful efforts of two groups of breeders, working in ignorance of each other, in the Philippines and China.

The benefits of scientists being able to pursue clear goals in multidisciplinary teams had convinced those involved in the Mexican program of the need to create purpose built research institutes that would attract scientists of the highest caliber. In 1961, the International Rice Research Institute (IRRI) was established at Los

Baños in the Philippine province of Luzon, as a joint venture between the Philippine government and the Ford and Rockefeller foundations. This was the first of a family of new research institutes equipped with first-class laboratories and adjoining experimental plots located on fertile irrigated land.[19] Excellent living conditions and international salaries added to the ability to attract the best scientists from around the world.

The Ford Foundation became a partner partly because its community development program in India, started in 1951, had underlined the importance of agricultural research. The program, largely run by social scientists, had assumed that improved technology was readily available and needed only a program of vigorous education for it to be implemented. However, the village extension workers often proved inexperienced in agriculture and, more important, encouragement of increased fertilizer use turned out to be ineffective because the traditional cereal varieties lodged. Forrest Hill, an agricultural economist and vice president of the Foundation, concluded the Foundation had "got the cart before the horse." What was needed, he believed, was innovative agricultural research to support the extension work.

"A Miracle Rice"

The first director of IRRI, Robert Chandler, assembled a team of rice experts drawn from Asia and the United States.[20] Experience in Mexico with the wheat program and the knowledge already obtained from breeding programs in India provided a blueprint for the new rices. A large collection of rice types was quickly amassed at Los Baños, and, of the crosses made in 1962, a particularly promising combination was between the tall, vigorous variety, *Peta*, from Indonesia and *Dee-geo-woo-gen*, a short, stiff-strawed variety from Taiwan, which contained a single recessive gene for dwarfing. In 1966, the new variety, named IR8, was released for commercial planting in the Philippines. It was immediately successful and, amid considerable publicity, was dubbed the "miracle rice."

IR8 combined the seedling vigor of *Peta* with the short straw of *Dee-geo-woo-gen*. It was, like the new wheats bred in Mexico, highly responsive to fertilizer and insensitive to photoperiod, maturing in 130 days. Under irrigation it yielded 9 tons/ha in the dry season and, on the IRRI farm, when continuously cultivated with a rapid turnaround between crops, produced average annual yields of over 20 tons/ha.

There was an immediate impact on Philippine rice production. By 1970 half of the Philippines' rice land was planted to the new varieties, and the yield takeoff had occurred. The Philippines became self-sufficient in rice production in 1968 and

1969. A decade later, 75 percent of the rice land was planted to the new varieties, and average yields were over 2 tons/ha and rising at nearly 70 kg/ha per year. The reelection of President Ferdinand Marcos in 1969 owed much to the attainment of self-sufficiency in rice production.

The Chinese Revolutions

In 1949, after the Chinese Communist Party came to power, 300 million tenant farmers were given land and other production assets, so ending the feudal system that had predominated for centuries. Farmers were quick to respond; by 1952 grain production had reached a record 164 million tons.[21] However, these gains were soon reversed. In the late 1950s the second five-year plan, known as the Great Leap Forward, was implemented (Box 3.3).

Box 3.3   The Chinese Great Leap Forward

The Chinese second five year plan aimed to utilize agriculture as a basis for industrial transformation. The leadership introduced a policy of collectivization and mass mobilization of rural labor. By 1958 the production units had grown to an average size of more than 2,500 households per unit. Labor was devoted to irrigation, flood control, and land reclamation, as well as small-scale industrial production. Rice breeding consisted of using large-scale collections of landraces (local varieties, largely arising through a combination of natural and human selection) for identification, evaluation, and selection of elite varieties.

The consequences were disastrous. In the years 1959 to 1961 some 45 million people died of malnutrition and starvation. The so-called Great Agricultural Crisis has been analyzed by Dennis Tao Yang, an economist at Virginia Tech and the Chinese University of Hong Kong, utilizing the entitlement concepts developed by Amartya Sen in his study of the Bengal Famine.[22] Although adverse weather was partly to blame, the root cause was the diversion of massive amounts of agricultural resources to industry and the significant increase in grain procurement from the peasants. This led to malnutrition, which in turn reduced their capacity to work in the fields, so lowering production still further.

However, it was more than economic mismanagement. The famine also has been analyzed by Frank Dikötter, a historian of China at the School of Oriental and African Studies (SOAS) at London University, using local archives that have recently become available.[23] This study reveals the devastating effect of setting highly unrealistic targets for agriculture, irrigation works, and local industrialization and enforcing these with rewards and extreme punishments that went from the very top downward.

49

The government responded to the crisis following the Great Leap Forward with a partial retreat from collectivization and by putting greater emphasis on developing modern agricultural technologies and inputs. A program of modern plant breeding was initiated at the Academy of Agricultural Sciences in the province of Guangdong. In many respects it was very similar to the work at IRRI. Although this was only known much later, the Chinese and Philippine teams were using breeding material containing the same dwarfing gene in the Taiwanese variety *Dee-geo-woo-gen*, which had probably originated in southern China. In 1959, the Chinese produced their first successful cross, similar in many respects to IR8, and known as *Guang-chai-ai*. It was rapidly taken up in the province of Guangdong and in Jiangsu, Hunan, and Fujian. By 1965, a year prior to the release of IR8, it was being grown on 3.3 million hectares.[24] By 1970 average yields had increased to 3.4 tons/ha.

Packages of Inputs

To begin with, the new varieties were distributed somewhat haphazardly. When IR8 was first released, farmers who turned up at IRRI could have 2 kg of seed free, providing they left their name and address. The seed spread within a few months to over two thirds of the provinces of the Philippines; however, it soon became apparent that a more organized approach to the supply of the seeds and other inputs was needed. Seed began to be distributed through Philippine government agencies and a newly formed Seed Growers Association, composed of private farmers who both grew and marketed the new seed. Standard Oil of New Jersey (ESSO) also established 400 agro-service centers in the Philippines as marketing outlets not only for fertilizer but for seed, pesticides, and farm implements.[25]

In Latin America, two Mexican farm advisers working in El Salvador had hit on the idea of putting together in one package all the basic inputs a farmer would need to try out a new variety on a small patch of ground. The idea quickly spread to other countries and was tried on a massive scale in the Philippines where a typical package contained 0.9 kg of IR8 seed, 19 kg of fertilizer, and 2.7 kg of insecticide. The packages were produced by governments and also sold by fertilizer companies.[26]

Donors provided logistic support for the Green Revolution. The U.S. Agency for International Development (USAID) began to fund fertilizer shipments in the 1960s and to finance rural infrastructure—farm to market roads, irrigation projects, and rural electrification. It also funded a large force of technical assistance experts. Contracts were signed with U.S. land-grant colleges to assist institution building in education, research, and extension and to create agricultural universities in the land-grant tradition. One of the most successful partnerships was between

Cornell University and the University of the Philippines Los Baños, next to the IRRI campus.

The global uptake of the new varieties was rapid and widespread, especially where the soils were good and there was adequate water supply. By the 1980s the Green Revolution and new Chinese varieties of wheat, maize and rice were dominating the grain lands of the developing world.[27]

Teething Problems

Inevitably, there were "teething problems" in the early years. Governments were unprepared for the rapid rise in production. Storage, transport, and marketing systems were sometimes overwhelmed. In India the schools had to be closed so they could be used to store the grain. A huge rice harvest in Kalimantan, on the island of Borneo, went to waste because there was insufficient transport to get it to the centers of demand in Java. These were the problems of success, however, and were quickly overcome.

Another early problem was the poor acceptability of some of the new varieties. Even poor, undernourished people retain a pride in eating good-quality grain. There was preference in India and Pakistan for chapatis made from the traditional white grains rather than the reddish grains of the new varieties, but this problem was soon solved by breeding for white grain color.[28] The grain of the first IRRI rices, such as IR8, was also rejected because it tended to harden excessively after cooking, and farmers received a lower price than for the traditional grains. Quality improved with later varieties, but many farmers continued to prefer the traditional taste. In Indonesia, the growing of traditional rice varieties was prohibited, a ban sometimes enforced by the destruction of crops in the fields. Farmers responded by cultivating the traditional rices in their home gardens or on the margins of fields growing the new varieties.

More Serious Obstacles

Other problems were more persistent and less amenable to simple solutions. They often required changes in government policies and the creation of new agencies. Inevitably the solutions tended to create yet further problems. An example was the credit needs of the new technology. Adoption of the high-yielding packages was expensive: In Bangladesh the cost of the necessary inputs was 60 percent more than for the traditional varieties. Small, subsistence farmers, often tenants or sharecroppers, could only afford the new packages if they borrowed from local money lenders, invariably at high rates of interest. In the Philippines, where the majority of farmers are tenants, cash was borrowed from the landlords at rates of 60–90 percent per annum, often producing a permanent state of indebtedness. The government response was to set up an Agricultural Guarantee Loan Fund, established at the Central Bank. This, in turn, supported many private rural banks that loaned without collateral and at reasonable interest rates. This new, more favorable credit system was an important factor in determining the very rapid uptake of the new varieties in the Philippines.[29]

However, there were drawbacks. Under the traditional feudal system, the tenants provided personal services for the landlord, gathering fuel or lending a hand with house repairs. In return, landlords helped with rice or money at times of economic hardship, providing a degree of protection against the outside world.[30] With the advent of the Green Revolution, this relationship became more commercial. The inputs had to be paid for, and in bad years there was no longer any latitude. Credit repayments under the government scheme were due on schedule and defaulters were punished. The banks were less accommodating than the landlords.

Environmental Effects

In the 1960s, when the Green Revolution was beginning to make its impact, little thought was given to environmental consequences. They were deemed either insignificant or, at least, capable of being easily redressed at a future date, once the main task of feeding the world was accomplished. I remember visiting the Ford Foundation office in New Delhi in 1968 to pursue the feasibility of an environment program in India and being told, "We are not interested in saving birds but in feeding people" (twenty years later I became the head of that office and oversaw a number of environmental programs).

There was also a strongly held view, one still commonly voiced, that a healthy, productive agriculture would necessarily benefit the environment. Good agronomy was good environmental management. It is a point with some force. Traditional agriculture is usually informed by ecological wisdom (see Chapter 6). Modern technologies can be environmentally sensitive, but only if they are designed and used with the benefit of modern ecological knowledge.

As is also frequently argued modern, intensive farming, in both the developed and developing countries, also benefits the environment by relieving the pressure on natural habitats. Norman Borlaug claimed the Green Revolution saved 1.8 billion hectares of land. In 1950, the world's grain output from 600 million hectares was nearly 700 million tons. Forty years or so later, the world's farmers harvested 1.9 billion tons—a 170 percent increase—but from about the same amount of land.[31]

Too often, however, the technologies accompanying the Green Revolution turned out to have adverse environmental effects. The heavy use of pesticides caused severe problems (see Chapter 12). There was growing human morbidity and mortality while, at the same time, pest populations became resistant and escaped from natural control. High levels of subsidy led to mis- or overuse of pesticides and fertilizers. By the mid-1980s the subsidies were 68 percent of the world price for fertilizers, 40 percent for pesticides, and nearly 90 percent for water. Among the consequences were contamination of water sources, an increase in pesticide-resistant rice pests, and a decrease in the natural enemies that control them. In the intensively farmed lands of both the developed and developing countries, heavy fertilizer applications produced nitrate levels in drinking water that approached or exceeded permitted levels.

The contribution of agriculture to global pollution has grown, with potentially serious consequences. Land has become degraded, forests and biodiversity lost, grazing land and fisheries overexploited. Moreover, as we have discovered in recent years, agriculture is both a victim and a culprit of climate change (Chapters 15 and 16) (Figure 3.3).

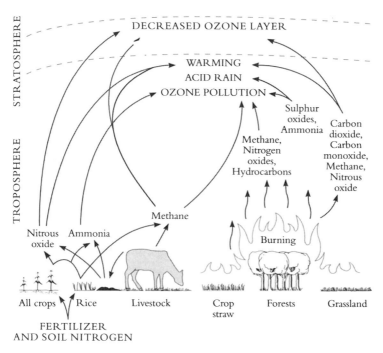

Figure 3.3 Global pollution caused by agriculture.[32]

Who Benefited?

The most controversial effect of the Green Revolution was its impact on the lives of the poor.[33] In the opinion of many commentators, the new technologies were inherently biased in favor of the rich, benefiting large farmers at the expense of small, and landlords at the expense of tenants. These critics saw little evidence of the benefits "trickling down" from the rich to the poor and argued that, without appropriate institutional reforms, efforts to introduce the new technologies were a waste of money. According to others, including University of Sussex economists Michael Lipton and Richard Longhurst, the technologies, despite their evident drawbacks, were essential if production was to be increased and labor demand maintained in the face of rapid population growth and little room for expanded cultivation. They posed the question: "In the absence of the Green Revolution, would not the situation have been a great deal worse?"[34] The answer to this question appears to be yes.

Nevertheless, three specific questions have to be answered:

- Did producing cheaper and more widely available grain reduce undernutrition and malnutrition?
- Did the Green Revolution benefit poor farmers, increasing their household incomes and consumption through greater production?
- Was there a positive effect, direct or indirect, on the employment and incomes of rural labor?

The answers are complex and, inevitably, depend on geographic, social, and political circumstances.[35]

Hunger and Poverty

In general, there has been a significant reduction in poverty and hunger in those countries directly affected by the Green Revolution, apparently because of cheaper food (Figure 3.4).

But this needs to be qualified. As Andrew Dorward, an agricultural economist at SOAS in London, points out, these global food price indices are based on weights (e.g., the U.S. consumer price index) derived from the expenditure patterns of richer countries or richer income groups within countries, and do not reflect the experience of low-income groups.[37] This is why the food price spikes of 2007–2008 and 2010–2011 had such a devastating effect on the developing countries.

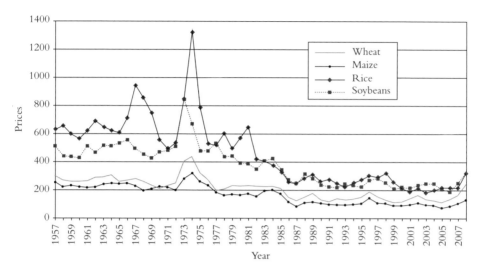

Figure 3.4 Declining prices of food commodities since the 1960s (constant U.S. dollars, based on FAO data).[36]

Even within countries the impact has been uneven. Among the urban poor, the incidence and severity of undernutrition has declined, particularly in China; this is also the case among the rural poor who live in the Green Revolution lands (where the new varieties predominated), but probably not elsewhere. Small farmers and the landless gained higher incomes in the Green Revolution lands, with, in some situations, rising real wages. Yet in many of the rural areas largely untouched by the Green Revolution, both poor farmers and the landless suffered losses in real income and increased hunger.[38] However, there is still considerable controversy over whether the Green Revolution decreased or increased inequality. A review of over three hundred studies found that for 80 percent of the studies inequality had worsened.[39]

In East and Southeast Asia both numbers and the proportion undernourished have declined dramatically over the past thirty years. By contrast, in South Asia and Sub-Saharan Africa, although the Green Revolution contributed to declining proportions, the numbers of undernourished have risen in each decade since 1970, largely due to population growth (Figure 3.5). Indeed the incidence would have fallen much faster in the Green Revolution countries were it not for the high rates of population growth.

Most significantly the higher yields meant that even though cost per ha increased, the cost per unit of grain harvested went down. As a consequence global food prices declined for the next four decades.

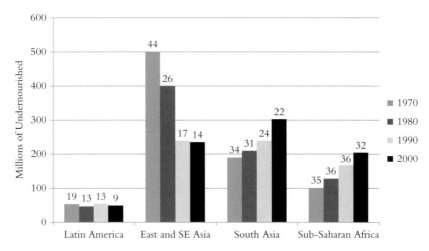

Figure 3.5 Changes in numbers and proportion of chronically undernourished in developing countries.[40] Percent undernourished above each column.

In addition some of the most visible and significant improvements in nutrition came from using the increased availability of food supplies to support highly targeted interventions.[41] For example, food stamps were provided in Jamaica, through registered health centers, to well-defined groups such as pregnant and lactating women and children under five. Another approach was to target certain disadvantaged regions, as in the Philippines. There, subsidized rice and cooking oil were made available to selected poor villages through local retailers.

The Benefits to Farmers

The next question is: How did farmers benefit from the new technologies? Most poor farms are small—less than a hectare of arable land—and, not surprisingly, poor farmers were initially reluctant to try the new technologies.[42] They were often tenants or sharecroppers, with poor security and mostly engaged in subsistence farming with no surpluses to be sold and hence no income to invest. Credit and inputs, such as fertilizers, were relatively more expensive for them because of the high transaction costs. The prospect of taking up new technologies appeared as a costly and highly risky undertaking, despite the promise of much higher yields. But the situation did begin to change. Cheap credit and subsidized inputs became available. Landlords, experiencing the benefits on their own land, encouraged their tenants to follow suit. Small farmers began to innovate, albeit step by step and at a slower rate than large farmers.[43]

Farm size, though, was only one factor in determining uptake of the new technologies. Usually, the topography of the land, the quality of the soil, and, most critical, the access to irrigation were more important than the amount of land available.[44] During the Green Revolution the adopters had slightly less land than the nonadopters in Mexico, but their land was of better quality.[45] In India, adoption was strongly correlated with water supply. Where irrigation was well developed, for example in the Punjab and Haryana, adoption was 100 percent for all farm sizes, but in the eastern states of India where water control is poor, adoption was low—less than 50 percent—even on irrigated land. In India, and elsewhere in the developing world, the new varieties had less impact on the less favored lands (i.e., unirrigated, with problem soils and topographies).

Increasing Employment

The third question is: What was the impact of the new technologies on employment and the incomes of rural labor? Many of the characteristics of the new Green Revolution cereal varieties and their associated technologies had the potential to increase rural employment. Extra labor was required to meet the higher standards and densities of sowing and planting (e.g., in the preparation of seed beds for the new rices), to ensure precise water control, to weed, to apply the greater amounts of fertilizer and pesticide, and to gather and thresh the larger harvests. But the biggest impact came from the widespread growth in double- and triple cropping on irrigated land, made possible by the shorter growing time of the new varieties. In India, it was soon realized that because the new wheats could be planted in December, thirty days later than the traditional wheats, and still be harvested at the usual time in the spring, it was possible to insert a rice crop beforehand. Large areas of land stretching from the Punjab to West Bengal adopted rice-wheat rotations.[46] Continuous rice cropping also became possible in some locations, and in the Philippines one ingenious farmer developed a rice garden in which rice was being planted every few days, creating an intensive mosaic of cultivation.

But, countering these trends for increased labor demand, the Green Revolution also provided strong incentives for mechanization.[47] Although, in theory, mechanization should be most common where labor is scarce and land abundant, in practice the form of mechanization has been the more important factor.[48] Various forms of mechanization have very different consequences for production and employment. Using machines to spray insecticides can increase yields and, generally, does not displace labor. Mechanical irrigation devices, such as water pumps, produce higher yields, and may or may not be labor displacing. By contrast, the spraying of herbicides and the introduction of tractors and two-wheeled power tillers to prepare land and of machines to harvest, thresh, and mill grain produce relatively little direct effect on yields and are strongly labor displacing.[49]

In Asia, the introduction of mechanical hullers to replace the hand-milling of rice significantly decreased the need for labor. In Bangladesh, rice is traditionally milled using a foot-operated pestle and mortar, called a *dheki*.[50] It provided much-needed employment, particularly for poor landless women, often being the only source of income for divorcees and widows in a culture where women need to observe the seclusion of purdah. A mechanical mill reduces the labor input from 270 to 5 hours per ton. By the 1980s over 40 percent of rice in Bangladesh was being mechanically milled, and the annual introduction of 700 new mills was replacing the work of 100,000 to 140,000 women a year.

For the Green Revolution lands as a whole, real wage rates remained the same, or rose only a little.[51] Long-term studies do show reductions in inequality from the Green Revolution, but this may not be the same as reducing poverty.[52] It is very difficult to generalize. As Peter Hazell, an agricultural economist, then at the International Food Policy Research Institute (IFPRI), points out, "[because] many of the rural poor are simultaneously farmers, paid agricultural workers, net buyers of food, and earn nonfarm sources of income, the net impact of the Green Revolution on their poverty status can be complex, with households experiencing gains in some directions and losses in others."[53] Yet for the smallest farms and landless agricultural workers, the gains can be too small to lift them out of poverty.[54] The poor can benefit in other ways. Perhaps most important, the study by Mahubab Hossain and his colleagues at IRRI showed the new high-yielding rices in Bangladesh helped stabilize employment earnings, reduce food prices, and moderate seasonal fluctuations to help the poor better cope with natural disasters.[55]

The Key Factors in the Green Revolution

According to some critics of the Green Revolution, the growth in production owed little to the new varieties and was primarily due to agricultural expansion, arable cultivation moving onto increasingly marginal lands. But the evidence is otherwise: although area growth was important in the 1950s, the subsequent gains were largely from increasing yield per hectare. Nevertheless, the yield growth was not solely attributable to the new varieties. They were necessary but not sufficient alone for success. Their potential could only be realized if they were supplied with high quantities of fertilizer and provided with optimal supplies of water. As was soon apparent, the new varieties yielded better than the traditional at any level of fertilizer application, although without fertilizer they sometimes did worse on poor soils.[56]

Because the new varieties were more exacting in their requirements, good irrigation, by providing a controlled environment for growth, became crucial. Most developing countries have a dry and a wet growing season. Potentially, dry season yields are 50 to 100 percent greater than in the wet, but the lack of rainfall in the dry season and the high evapotranspiration rates resulting from absence of cloud cover makes the crops liable to water stress. Without adequate irrigation, yields tend to be low and variable, whatever the level of fertilizer application. With irrigation and heavy fertilizer application, some of the highest cereal yields in the world were attained. In South and Southeast Asia the irrigated area grew at 2.2 percent per annum; by 1980 a third of the rice area was irrigated, producing a rapid increase in the dry season rice crop.

Just how much of the increased cereal production was due to the availability of the new varieties, how much to increased fertilizer use, and how much to the growth of irrigation is a matter of argument. Bob Herdt and Celia Capule, two agricultural economists at IRRI, carried out an analysis of the eight countries (Bangladesh, Burma, China, India, Indonesia, Philippines, Sri Lanka, and Thailand) responsible for 85 percent of the Asian rice crop. They showed the three factors had a roughly equal contribution. Of the extra 117 million tons produced between 1965 and 1980, 27 million tons was attributable to the new varieties, 29 million tons to increased fertilizer use, and 34 million tons to irrigation (the remainder was due to "residual factors" such as infrastructure and extension).[57]

Total Factor Productivity

The enabling environment—encompassing all those "residual" factors that are additional to the contribution of the inputs of new varieties, fertilizers, and irrigation—is also important. One way of estimating this is to calculate *total factor productivity* (TFP).[58] This is the ratio of total output growth to total input growth. It compares the increase in total output with the increase in total inputs (such as land, labor, fertilizers) and measures the difference. In effect it estimates the importance of other factors in the enabling environment such as investments in research, extension, human capital, and infrastructure (Box 3.4).

Box 3.4   Total Factor Productivity in India, 1956–1987[59]

In 1995, Mark Rosegrant of IFPRI and Robert Evenson, an economist at Yale, conducted an illuminating assessment of the importance of TFP before, during, and after the Green Revolution in India. They analyzed the productivity growth in five major and fourteen minor crops between 1956 and 1987 and showed that TFP contributed about half of the total growth crop in outputs. The most rapid growth was during the Green Revolution period of 1967–1976 (Table 3.1).

Table 3.1   Annual growth rates in Indian crop outputs, inputs, and total factor productivity, in percent (three-year moving average)

| | Period | | | |
Measure of growth	1957–67	1967–76	1976–86	1957–86
Crop output	2.18	2.68	2.07	2.25
Crop input	1.08	1.28	1	1.11
Total factor productivity	1.1	1.39	1.05	1.13

Source: Rosegrant and Evenson[60]

The same study revealed the contribution of different components of TFP. Public research and extension was very important, explaining over half of the growth in TFP, while private sector R&D contributed 25 percent. Foreign agribusiness firms provided farm machinery and chemicals, while Indian firms facilitated the foreign investments and complemented public sector research. Markets (used as an indicator of rural infrastructure) along with increases in irrigation (over and above the input of water) accounted for 10 percent. Irrigation in TFP creates an improved environment for crop technology.

The new, imported, high-yielding varieties (HYVs) were important during the Green Revolution, but afterward Indian R&D was the key source of new forms of HYV. Interestingly literacy became significant only after the Green Revolution when efficiency in input use became important as compared with input and crop variety promotion. Post–Green Revolution technologies tend to be highly complex, knowledge-intensive, and location specific, requiring more skills and knowledge.

Rosegrant and Evenson concluded: "India has achieved significant total factor productivity growth and . . . this growth enabled the economy to increase food production even though India began the period with high population densities and limited potential for cropland expansion as a source of output growth."

The Green Revolution Summarized

The Green Revolution succeeded because it focused on three interrelated actions:

- Breeding programs for staple cereals to produce early maturing, day-length insensitive, and high-yielding dwarf varieties
- The organization and distribution of packages of high-payoff inputs, such as fertilizers, pesticides, and water regulation
- Implementation of these technical innovations in the most favorable agroclimatic regions and for those classes of farmers with the best expectations of realizing the potential yields

In many ways it was a triumph for technology rather than science. The dwarfing genes had been known for decades in China and Japan and most of the breeding techniques were well established. What made the difference was the combination of investment in first class research centers and in the organization of delivery of the inputs necessary to make the science productive.

However, it was not simply a product of the technology. The enabling environment was crucial. Governments made substantial investments in agricultural research, in ensuring farmers had access to credit and inputs, and in getting markets to work efficiently. The favored countries benefited from governments willing and able to make and direct the necessary investments, In aggregate, Asian governments consistently spent about 15 percent of their annual total budget on agriculture during the Green Revolution. This compares with a meager 5 percent in Africa.

Uptake of the technologies was fastest and most dramatic in those lands such as Sonora in Mexico, Luzon in the Philippines, the lowlands of Java, and the Punjab of India and Pakistan where irrigation was already well developed and where large farmers, often larger than in the rest of the country, had good access to credit, had a greater propensity to take risks, and were likely to be rapid adopters. These lands scattered through Asia and Latin America became the so-called Green Revolution lands.

The success is undisputed, but it was a revolution with the following serious limitations:

- Its impact on the poor was less than expected.
- It did not reduce, and in some cases it encouraged, natural resource degradation and environmental problems.
- Its geographic impact was localized.
- There were eventually signs of diminishing returns.

William Gaud, the administrator of USAID, first coined the name Green Revolution.[61] At the time it was an appropriate description of a momentous event. Then, the image it conveyed was of a world covered with luxuriant and productive crops—the green swathes of young wheat and rice fields. It was truly also a revolution, in the scale of the transformation it achieved, although it did not go far enough.

The world has changed significantly and in many respects since the 1960s. Food production has greatly increased—global production of cereals has risen from 900 million tons in 1960 to 2,500 million tons in 2009, faster than the rise in global population over the same period from 3.1 billion to 6.9 billion—and, until recently food prices have fallen. In theory, as I discussed in Chapter 2, there is enough to go round, yet a billion people are chronically hungry, repeated food price spikes add to the numbers of hungry, and feeding the world is complicated by changing food consumption patterns, the growth of biofuels, declining availability and quality of water, declining yield growth in some crops, and the threats from climate change.

In the rest of the book I describe a theory of change that can address this multitude of problems in a way that continues to increase productivity but in a manner that is stable, resilient, and equitable. I call this a "Doubly Green Revolution."

4

The Political Economy of Food Security

> I was strongly inclined to do away forever with distributions of grain, because
> through dependence on them agriculture was neglected.
>
> The Emperor Augustus of Rome[1]

Hunger has always been political. The practice of agriculture evolved from hunting
and gathering in several locations around the world, ten to fifteen thousand years
ago, and it is likely that there were disputes between hunter-gatherer communities
over access to different parts of the landscape.[2] Agriculture stimulated the creation
of settled communities, and undoubtedly this intensified conflict—over the best arable land and for rights to water and grazing. Disputes would have been acute when
harvests were poor.

Populations grew and the first states were organized around the provision of food,
and especially the control of, and access to, water for irrigation. Some of the earliest
civilizations, for example, in ancient Egypt, Mesopotamia, India, China, and pre-
Columbian Mexico and Peru, were *hydraulic*; that is, they relied on piped or chan-
neled water for production of food and other crops. The anthropologist Karl Wittfogel, in his book *Oriental Despotism,* argued that such large-scale irrigation creates
societies that depend on forced labor, large bureaucracies, and armies, leading to
hierarchical, centralized, and despotic rule.[3] According to various critics, his thesis
is too simplistic, and they have taken a more systems approach to the rise of civilizations.[4] However, such hydraulic civilizations are clearly among the first, well-
developed "political economies."

Food security was a major preoccupation in the ancient world. Hunger was
common, and the specter of famine hung over rulers and subjects alike. In the
Roman Republic and the Empire that followed, the task of feeding the growing
population of Rome was a political, economic, and logistic challenge (Box 4.1).

The Emperor Augustus, quoted at the beginning of this chapter, reflects on the
central dilemma of the political economy of food security: How much reliance should
be placed on households to provide for themselves, how much on the market, and

Box 4.1   Food Security in Ancient Rome[5]

Although Rome was not the earliest state to concern itself with food security, the Romans have left us many firsthand accounts in books and letters that make it possible to reconstruct how food crises were tackled. For example, Rome experienced grain shortages and wild price fluctuations in 58 to 56 BC. On one occasion crowds besieged the Senate and threatened to burn the senators alive, apparently encouraged by the tribune Clodius who had passed a law increasing the number of people entitled to subsidized grain. The price of grain eventually fell when Cicero proposed Pompey be put in charge of the grain supply.

The population of Rome mushroomed and by the end of the period before Christ numbered about a million people, with between 200,000 and 400,000 tons of grain being imported each year. As Rome acquired more overseas territories, the centers of grain production progressively moved to where it was easier and cheaper to cultivate cereals—from Italy to Gaul (southern France), to Baetica and Tarraconensis (Spain), to North Africa, and eventually under the Emperor Augustus to Egypt. Most of the trade, storage, and marketing was in private hands, although the state acquired grain as a form of taxation and distributed this, partly subsidized—at one point to as many as 20 percent of the population. Prefects of the grain supply were responsible to the state for ensuring an adequate supply and its distribution.

Many of the senators in the Republic and the subsequent emperors were conservative in their approach to public welfare and disapproved of the subsidies. Not surprisingly, however, these were popular and persisted for several hundred years.

to what extent should governments intervene? These are questions that have not gone away.

The Nature of Political Economy

Political economy—the study of the economies of states (polities)—had its origins in the seventeenth century. In simple terms, political economy recognized that economics was shaped by politics, political and social institutions, and ideas—and vice versa. Leading exponents in the eighteenth and nineteenth centuries included David Ricardo, Karl Marx, Thomas Malthus, John Stuart Mill, and Adam Smith, whose *Wealth of Nations* published in 1776 during the British agricultural revolution was the classic text.

Political economics came to be replaced by the more focused discipline of economics, which tackles such issues as the optimization of an individual's utility or profit and the determination of price by supply and demand, with little reference to political or institutional aspects. Today, *political economy* is defined in many ways, but fundamentally it has gone back to its roots and is about the interrelationships between economics, politics, institutions, and governance. In its broadest sense it includes how these relate to peoples and cultures and how all these factors affect nation-states and their interrelations.[6] Jean Drèze and Amartya Sen in their introduction to a key modern work, *The Political Economy of Hunger,* refer to the title as "an explicit reminder of the need to adopt a broad perspective to understand better the causation of hunger and the remedial actions that are needed."[7]

In this chapter I discuss these questions:

- What is the nature of food security?
- How is food security affected by price volatility?
- What are the potential costs and benefits of trade liberalization?
- What is the role of food aid?

Food Security Today

In many respects, little has changed since Roman times. Then, as now, natural disasters are not the sole cause of food crises. Harvest failures often play a role, but equally important are warfare, piracy, speculation, and political mismanagement. As I write these words, there is major hunger in the Horn of Africa affecting about ten million people in Somalia, Ethiopia, and Kenya. It is partly a result of the worst drought in sixty years, but made far worse because of the civil war, lawlessness, and political interference in Somalia.

Definitions of Food Security

This diversity of causes is part of the reason why the definition of *food security* has been subject to so much debate (see Chapter 1).[8] In the 1970s, the dominant cause seemed to be lack of food supply. The good harvests of the 1960s, largely a consequence of the Green Revolution, had led to falling prices. In response, the United States took land out of production and together with Canada ran down its grain reserves. Then, in 1972, there were major harvest failures across the globe; by 1974 grain prices had doubled, and there was only enough grain in store to feed the world's population for three and a half weeks.[9] Food security was back on the political agenda and defined by the World Food Conference in terms of "availability at all times of adequate world supplies of basic food-stuffs."[10]

Despite the seriousness of the situation, this emphasis on food supply as the first priority soon came under challenge, first from nutritionists and nutrition planners, and then from the work of Amartya Sen, which I discussed in Chapter 2.[11] Sen's studies of the Great Bengal Famine of 1943 had led him to recognize the importance of access to food, as opposed to its supply, and changed the way we think about food security. Influenced by Sen, the World Bank adopted the following definition of food security in 1986: "Food security is access by all people at all times to enough food for an active, healthy life."[12]

This definition was subsequently elaborated by the Food and Agriculture Organization (FAO) in 2001: "Food security [is] a situation that exists when all people, at all times, have physical, social and economic access to sufficient, safe and nutritious food that meets their dietary needs and food preferences for an active and healthy life."[13]

These definitions are implicitly multidimensional and form the basis of the modern political economy of food security. In an ideal universe, based on a simple economic model—what Sen calls an exercise in "*instant economics*"—the market can ensure everyone has the food they need.[14] If prices of a staple crop rise, farmers plant more of that crop in the next season, and prices should fall. But there is inevitably a time lag and farmers may overproduce, creating a glut with low prices reducing production next year. Supply may meet demand but in a jerky, unstable fashion. Farmers may also not be able to produce more because of lack of access to land or inputs, to higher-yielding seeds, or to the credit to buy them. The input and output markets may not be efficient and the infrastructure (e.g., storage, roads) may be poor. Although market forces may work to solve these problems, they may be weak or slow because of a lack of potential financial investments. The private sector may choose to invest where the returns are greater or less risky. Not surprisingly, developing country farmers are slow to respond and following a food price spike, prices remain high.

Nevertheless, although food markets are often highly distorted, as I describe below, they remain the core mechanism whereby supply meets demand.[15]

Food Self-Sufficiency

Whenever a food crisis occurs there are political campaigns for countries to become food self-sufficient. Often the first reaction to rapidly rising prices is to impose export bans, as happened in 2008 and 2010. But these tend to make the situation worse: There may be a temporary respite in the exporting countries, yet elsewhere prices will rise further, and the fall in domestic prices in the exporting countries may discourage their farmers from increased planting.

Nevertheless, there are benefits in countries trying to achieve a greater degree of food self-sufficiency. Benin, Burkina Faso, Niger, and Senegal in West Africa are major rice economies, importing an average of almost 1.6 million tons of rice per year between 2001 and 2007, at a value of around $435 million.[16] The New Rices for Africa (NERICAs; see Chapter 9, Box 9.4) could provide much greater domestic production. Senegal, through its Grand Agricultural Offensive for Food and Abundance (GAOFA) initiative achieved over 50 percent rice self-sufficiency in 2009.[17] But such achievements will be beneficial in the long run only if the prices of the locally grown rice compete with those of the imported rice from Asia where highly efficient irrigated production keeps prices low.

Greater self-sufficiency could result in higher incomes for farmers and reduce levels of rural poverty. But this will depend on how it is done. How much can domestic production be increased? Where will the increases come from? Will the inputs need subsidizing and for how long? How resilient will increased production be in the face of climate change? Some of these questions are raised by the Malawi fertilizer and seed subsidy program that has been analyzed by Andrew Dorward of the School of Oriental and African Studies in London and Ephraim Chirwa at the University of Malawi (Box 4.2).

If the cost of self-sufficiency is high, higher than the costs of importing foods, the added costs will fall on consumers and taxpayers. In economic terms, resource allocation will be inefficient; the same resources invested in a different crop or sector could provide much higher returns. The desire for self-sufficiency is understandable and the political pressure may be intense, but the costs can be high and food security may remain elusive.[22] Wider considerations may be decisive. In the Malawi case because diversification out of maize and stimulation of the rural economy are being stifled by the longer-term problems of input affordability and lock-in to low productivity, Dorward and Chirwa argue there is a justification for sustained and consistent subsidy and other investment over many years.[23]

Food Sovereignty

An extreme version of food self-sufficiency is the concept of food sovereignty, which first emerged in 1996, principally as a response to the arguments around agriculture that had played out in the Uruguay round of the world trade negotiations.[24] A key originator of the concept was the self-styled international farming and peasant movement La Via Campesina, an organization created in 1992 that coordinates member groups across the world.

Box 4.2   Impact of the Malawi Subsidy Program[18]

Malawian smallholder agriculture is characterized by thousands of very poor farmers dependent on low input maize production on small landholdings that are nitrogen deprived. Production is not normally sufficient for farmers to meet annual consumption needs, and they depend on casual laboring and other income-earning opportunities to finance the purchase of the balance of their needs. 60 percent of farmers are net buyers of maize.

Malawian farmers are locked into a vicious circle of low maize productivity; the unstable maize prices depress wider labor and agricultural productivity and inhibit the growth of the nonfarm economy (Figure 4.1).

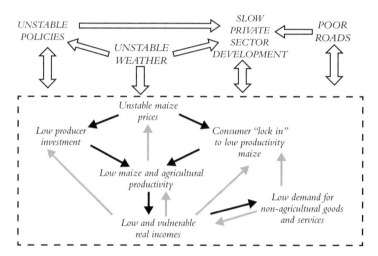

Figure 4.1   A vicious circle creating a low productivity maize production trap in Malawi (gray lines represent feedback loops).[19]

In principle a relatively simple way out of the trap is to increase the use of inorganic fertilizer and sowing of hybrid and open-pollinated maize varieties. But for more than half of the smallholder population, commercial fertilizer purchases in adequate quantities are unaffordable. In the early 2000s food insecurity increased: There were national food shortages due to poor harvest and late and expensive government-funded imports leading to large increases in maize prices.

In 2005–2006, the government began a program of subsidizing both fertilizers and maize seeds using coupons or vouchers (building on a pilot starter pack program begun in 1998).[20] The aim was to achieve food self-sufficiency and increased income of resource-poor households through greater food and cash crop production. By

(continued)

2008–2009 over 60 percent of farmers were receiving subsidized fertilizers. The cost rose from 6 percent of the government's national budget in 2005–2006, to 9 percent in 2007–2008, and to 16 percent in 2008–2009 although was expected to fall back in subsequent years because of a limitation on the amount of fertilizer to be subsidized. The increment in total production of maize grew from 975,000 tons in 2005–2006 to over 2 million tons in 2008–2009. As a result there has been greater household food security, higher real wages, wider economic growth, and poverty reduction.

For the 2011–2012 season, however, economic crisis has limited the amount of fertilizer imports and some 200,000 farmers will not be allocated vouchers to receive subsidized fertilizer and maize seed, without which many will not be able to afford these inputs. Although the program is considered successful overall, it is a heavy fiscal burden to the government and donors ($152.3 million in 2010–2011). Concerns over the program's effect on the private sector, corruption, and cost mean its long-term sustainability is uncertain.[21]

In April 1996, the Second International Conference of La Via Campesina was held in Tlaxcala, Mexico, and the first definition of *food sovereignty* was agreed on:

> Food sovereignty is the right of each nation to maintain and develop its own capacity to produce its basic foods respecting cultural and productive diversity. We have the right to produce our own food in our own territory. Food sovereignty is a precondition to genuine food security.[25]

This movement was borne out of a strong opposition to the dominant neoliberalism and the increasing globalization underlying the world trade negotiations. Many of its principles are highly radical. Most observers would argue that the strong streak of autarky that runs through the advocacy—complete food self-sufficiency—is unrealistic and inherently highly risky. Ignoring comparative advantage by growing crops and raising animals in places to which they are not well suited is likely to be costly, increasing food prices and decreasing food security. Nevertheless much that the movement advocates is consistent with the themes in this book.

Price Volatility

A key concern of both food security and sovereignty is the stability of food prices. The less stable the prices the less accessible and affordable the food, both for importing nations, at one level, and for individual households at the other.

In principle, the more open, efficient, and transparent the markets the greater should be the price stability.[26] But we have seen extreme volatility in global prices in recent years, and there is a general agreement this cannot be dampened, at least in the short run, by leaving it to the market. At present the situation is made worse by panic responses, in particular by the resort to export bans. The problem requires some form of systematic and concerted action by the international community.

Joachim von Braun, former director general of International Food Policy Research Institute (IFPRI) and now at the University of Bonn, believes food markets are now closely connected to the speculative activities in financial markets and hence should be subject to the kind of regulation that is applied to the banking and financial system. He also proposes that new regulatory processes should be accompanied by a three-tiered grain reserves policy:

1. The creation of a small, independent physical reserve at the World Food Program exclusively for emergency response and humanitarian assistance
2. A modest physical reserve shared by nations at the regional or global level
3. A virtual reserve where each country would commit to supplying to a fund, if needed, for intervention in grain markets (especially maize and wheat) [27]

Public reserves can help during food crises in three ways: (1) as domestic price stabilization tools; (2) as a source of emergency food for humanitarian aid during crises; and (3) as sources for food distribution programs. These recommendations, which were initially regarded as somewhat controversial, are being implemented by some countries such as Kenya and Nigeria, at least at the national level; Kenya tripled its grain reserve in 2011, and Nigeria has adopted a policy that 15 percent of the total annual grain harvest should be held in reserve.[28]

Protectionism

In Chapter 2 I showed that the proportion of a population who are chronically hungry depends, in part, on a country's food supply. Producing more food domestically will increase the supply but so will increasing imports.

Over the past thirty years there has been a massive expansion of world trade in food and other agricultural products; the value for global exports and imports has risen from around $1.3 trillion in 1978 to $16 trillion in 2008. In monetary terms, cereals are now second only to petroleum in value in international trade.[29] Some 276 million tons of cereals are traded across the world, of which nearly half is coarse grains.[30]

However this trade is highly distorted because of the levels of subsidy provided to their farmers by developed countries. These subsidies either stimulate excessive production, resulting in cheap exports, or they block imports. In the words of

Robert Paarlberg, a political scientist at Wellesley College: "[the subsidies] cause too much food to be produced in regions not well suited to farming, such as alpine countries in Europe, desert lands in the American southwest, or the municipal suburbs of Japan, and too little to be produced in the developing countries of the tropics where agricultural potential is often far more bountiful."[31]

Although agriculture accounts for less than 10 percent of global merchandise trade, it remains one of the most protected sectors, with tariffs on agricultural products being an average of 62 percent, exceptionally high in comparison with tariffs on, say, industrial goods (an average of 4 percent). Agriculture is the only sector where quantitative restrictions and export subsidies are allowed. According to Organisation for Economic Co-operation and Development (OECD) estimates, support to its agricultural producers, much of which is trade distorting, is $250 billion per year. Of this, $100 billion is direct subsidies to farmers, totaling 30 percent of gross farm receipts.[32] The rest includes tariffs and other restrictions on imports that disadvantage developing country farmers.[33] 38 percent of the distortions are due to the European Union (EU) and 16 percent to the United States.[34]

In economic theory a liberal trading system, free of subsidies, can lead to a more efficient allocation of resources because countries are producing those goods they have a comparative advantage at producing and importing those goods they are ill-equipped to produce. Under free and unfettered trade, food prices should be lower than they might otherwise be. In the history of agriculture, from the time of the Roman Empire onward, there have been repeated attempts to achieve free trade and its potentially manifold benefits. Nevertheless, for what appear valid political reasons, countries frequently retreat into protectionism, despite signing up to the free trade ideal. This occurred most recently during the food price spikes of 2007–2008 and 2010–2011 when countries ranging from India to Russia imposed export bans on grain.

In practice, there are three kinds of protection that countries can adopt: market access (e.g., import tariffs), export support (e.g., export subsidies), and domestic support (e.g., direct payments to farmers [Box 4.3]).

High Tariffs

Certain very high import tariffs, known as *megatariffs*, are imposed by the United States and EU on goods that could compete with domestically produced goods (e.g., in the US on sugar, peanuts, and dairy products and in the EU on grains, sugar, and dairy products). Two of the biggest distortions occur in the trade of sugar and cotton. In both cases producers in developing countries are not receiving the prices for

Box 4.3   Forms of Agricultural Protection[35]

1. *Market access:* The most distorting are tariffs (in effect taxes) that are added to imported goods, mostly by developed countries, to protect domestic producers from being forced out of the market by global competition. The U.S. Department of Agriculture (USDA) estimates tariffs account for 52 percent of world agricultural price distortions.[36]

2. *Export support:* These are commonly in the form of export subsidies, which pay developed country farmers a premium on top of the world price exports. This results in increased domestic production and prices but depresses world prices, so adversely affecting developing country farmers growing export crops. The USDA estimates export subsidies account for 13 percent of world agricultural price distortions, with the EU accounting for 90 percent of these subsidies.[37]

3. *Domestic support:* These involve direct payments to farmers, either developed or developing. In one form of domestic support the government sets a fixed price at which it buys agricultural goods, regardless of world prices. This is the most distortionary form and is referred to as an *amber box policy*. Farmers are encouraged to maintain normal levels of production despite low global prices. The USDA estimates domestic support as a whole accounts for 31 percent of world price distortions.[38]

These three groups are often interrelated. Thus, when domestic price support causes farmers to produce more than the market requires, export subsidies are used to compensate farmers for disposing of their surpluses in world markets.[39]

their produce that they would be entitled to under a free trade process. Since the Doha trade round began in 2001 (see "Doha Round" section below) nearly $32 billion has been given in subsidies to cotton farmers in the United States and the EU, with a further $16 billion to farmers in China and India. The main losers are the ten million West African cotton farmers in Benin, Burkina Faso, Chad, and Mali (the so-called Cotton-4).[40] Cotton accounts for between 5 percent and 10 percent of their gross domestic product (GDP), and it is calculated that elimination of the subsidies would increase world prices by 6 to 14 percent.

Tariffs also tend to be high on processed goods, which discourage developing countries from pursuing the processing of raw goods as a means of adding value. Overall, the global cost of current trade policies and domestic support for agriculture in developed countries is expected to reach between $70 billion and $200 billion by 2015, with more than 90 percent of this cost attributed to market access restriction and tariffs on agricultural production. A large proportion of the cost is being borne by developing countries because their exports are lower in amount and value than they should be and because they are less able to pay for the food imports they require.[41]

Trade Liberalization

Agriculture typically constitutes a far greater percentage of GDP and employment in developing countries than in developed countries, employing over 50 percent of developing countries' populations in comparison with 9 percent in the developed world. Developed countries also have a smaller percentage of exports as agricultural products. For all developing countries the proportion is an average of 27 percent of total exports, but for countries such as Burundi, Sudan, Ethiopia, Uganda, and Paraguay agriculture contributes over 80 percent of exports.[42]

Trade liberalization should therefore increase the export opportunities for developing country farmers and the prices they receive. The World Bank estimates full liberalization would on average increase the prices of international agricultural commodities (crops for human food and animal feed as well as livestock products) by over 5 percent, of cotton by over 20 percent, and of oil seeds by 15 percent. It would also increase developing countries' share of global agricultural exports by an estimated 9 percent, with Latin America and Sub-Saharan Africa being the largest beneficiaries.[43] Such increased incomes should attract further investment, driving economic growth in rural regions leading to greater opportunities for rural business (i.e., a virtuous circle). Consumers should also benefit from increased incomes through greater purchasing power as well as reductions in tariffs placed on domestic food products.[44] Liberalization is also thought to stabilize food prices: between the 1980s and 2000s, rice prices experienced reduced seasonal highs and lows in Bangladesh, Nepal, and Sri Lanka, countries that had adopted liberal policies.

In practice, there are actual and potential downsides to liberalization. Using so-called computable general equilibrium (CGE) models, the USDA has estimated that of an increase in net global GDP of $31 billion per year, 92 percent will go to developed countries and a mere 8 percent to developing country producers. Only by taking into account the longer-term changes in investment and productivity that trade liberalization can bring, will developing countries significantly benefit.[45]

In the short term, net importing countries will be harmed as will those whose preferential access to industrialized country markets is diminished (e.g., banana and sugar exporters in the Caribbean and Africa who have guaranteed access in the EU). In some cases uneven liberalization leads led to an influx of cheap foreign imports, while export tariffs on such goods as fruits, vegetable, cocoa, and sugar remain high.[46]

Finally, a major concern is that liberalization will fail to help rural subsistence farmers unconnected to markets. These farmers, a large proportion of the world's poor, are limited by a lack of an *enabling environment* (e.g., transportation and communication, infrastructure, credit systems, mechanization, and access to inputs) allowing them to respond to market changes, new investments, and potential productivity gains (I introduced this concept in Chapter 3 and will discuss it more fully in Chapter 8). Instead liberalization will most likely favor medium to large producers and could make some people even more food insecure.[47]

Negotiating Freer Trade

Despite these reservations there has been a long, tortuous, and often bitter battle to achieve greater trade liberalization. In 1949, the General Agreement on Tariffs and Trade (GATT) was set up, largely in response to an initiative by the United States and United Kingdom, to oversee trade negotiations, and this has been continued by its

Box 4.4   The Results of the Uruguay Round

Under the Agreement on Agriculture in the Uruguay Round, programmed reductions in domestic support, in export subsidies, and in tariffs and other limitations on market access were expected to benefit farmers in the developed and developing countries alike. By reducing the protection afforded developed country farmers, there would be less food exported at prices that undercut developing country farmers. And because import barriers were to be reduced, there would be growing demand for all agricultural products (staples and nonstaples), leading to increased agricultural production in both developed and developing countries.

In practice, however, the agreement has been only moderately successful. Overall levels of protection, as a percentage of agricultural output, have remained relatively constant. The Producer Support Estimate (PSE) calculated by the OECD aggregates tariffs, export subsidies, domestic support, and other means of protection into a single value. As a percentage of total agricultural output in the United States, the European Union, and for the OECD as a whole, the PSE has not decreased appreciably since the beginning of the Uruguay Round, largely because, in the words of Thomas Beierle of the World Resources Institute, "the United States and the European Union, in particular, have been able to meet their reduction commitments on market access, export support, and domestic support without actually reforming their agricultural policies."[49]

Nevertheless there have been some positive outcomes for the developing countries; namely, the reduction of import duties on tropical products and the inclusion of "special and differential treatment" that provides longer phase-in periods for developing countries and no commitment to reduce protection for least developed countries.[50]

successor, the World Trade Organization (WTO), founded in 1995. These two bodies have overseen a sequence of rounds of negotiations on trade, the most recently concluded being the so-called Uruguay Round that finished in 1994 after eight years of negotiations and involving 123 countries (Box 4.4).[48]

The Doha Round

A new round of trade negotiations began in November 2001 at the WTO Fourth Ministerial Conference in Doha, Qatar. Its aim has been variously described as "tackling unfinished business" or "overcoming the impasse over agriculture." Unlike previous rounds, there is a much stronger developing country voice. Prior to the start, a quarter of demands came from developing countries and 133 developing countries signed a document containing extensive joint demands. In the Uruguay and previous rounds, most of the negotiating power was exerted by the developed countries and in particular by the so-called Quad (US, Canada, Europe, and Japan). The Doha Round should have a different political dynamic because 80 percent of the 142 WTO members are developing countries.[51]

What has become clear in the negotiations is that each WTO member must see benefits in some areas if they are to make sacrifices in others. For developing countries this will be gains from globalization. For the United States and EU, gains in manufacturing, services, or investment or in new rules for intellectual property, labor standards, and environmental protection may balance the freer access to rich countries' agricultural markets by developing countries.

However, for much of the past decade the talks have been stalled.[52] The negotiations are riven by arguments between different power blocks and even within individual groupings; for example, between Brazil, India, and China there are strong disagreements. Despite being in the majority the developing countries have had little influence. Alan Beattie, an economist at the *Financial Times,* analyzes the history of free trade and points out that, once installed, protectionist agreements are very difficult to uninstall.[53] This is especially true where a single, politically powerful lobby representing a relatively small geographic area can do deals and team up with other similarly motivated lobbies. This is true of the cotton farmers of the southern U.S. states. There are only about 20,000 cotton farmers, but they receive $2–$3 billion in subsidies.[54] A draft agreement on cotton between the United States and Brazil has been drawn up but is dependent on satisfactory completion of the Doha Round. At the time of this writing, at the end of 2011, little progress has been made. In Beattie's words, "If warm words were edible, the Doha round could have fed the world for a decade."[55]

The Experience of Liberalization

Despite the failure of Doha, developing countries have benefited from liberalization measures partly derived from the Uruguay Round and partly through other processes. A study by the World Bank of seventeen developing countries in the late 1980s showed that, for the most part, tax and trade policies of the previous twenty-five years had harmed their farmers.[56] However, a 2010 analysis of the effects of liberalization since the mid-1980s reveals a substantial amount of policy reform and some clear benefits. There has been an increase in the developing countries' share of the world's primary agricultural exports from 43 to 55 percent and of the share of farm output from 58 to 62 percent.[57] For developing countries as a group, net farm income is estimated to be nearly 5 percent higher than it would have been without the reforms of the past quarter century, which is more than 10 times the proportional gain for nonagriculture.

According to the World Bank, policy reforms since the early 1980s have improved "global economic welfare" by $233 billion per year. Moreover, removing the distortions remaining as of 2004 would add another $168 billion per year.[58]

Liberalization in Madagascar

In practice it is the experience of individual countries that counts. Liberalization in Madagascar began in 1983, and, while both imports and exports have grown considerably since then, the benefits have been uneven (Box 4.5). Prior to trade liberalization, state intervention in the agricultural sector was high. The government had attempted to stabilize agricultural prices, but this led to reduced incentives for

Box 4.5   Liberalization in Madagascar

Maize exports

While trade liberalization was taking place in Madagascar, the EU established a set of policies, called *Posei*, under which the island of Réunion, an overseas territory of France in the Indian Ocean east of Madagascar, was granted permission to give tax breaks on imported cereals as a way to promote livestock production. This new market provided a huge opportunity for Madagascar, but one that was not ultimately realized. Why not? First, Réunion established storage and other infrastructure to handle large shipments of maize, but much of the maize coming from Madagascar was divided into smaller, more frequent shipments. Second, due to these smaller shipments, handling and transport costs from Madagascar were relatively high. Finally, Madagascar could

(continued)

---

not meet the supply levels required by Réunion's livestock producers. Maize is now largely imported by Réunion from Argentina and France.[61]

Contract farming of high-value vegetables

The Malagasy government supports growth in export production through the development of an export processing zone (EPZ) starting in 1989. Due to the benefits received by exporting firms operating within the EPZ, output grew by approximately 20 percent annually between 1997 and 2001. For example, Lecofruit, which in the early 1990s produced relatively small quantities of gherkins, in partnership with the French company S.E.G.M.A. Maille, diversified into snow peas, French beans, asparagus, and mini-vegetables for European markets. The EU's Everything but Arms (EBA) agreement of 2001 granted Madagascar (as one of the least developed countries) duty-free and quota-free access to the EU for certain goods such as fruits and vegetables.

Lecofruit purchases vegetable products from almost 10,000 contract farmers, who receive a guaranteed and fixed price for their produce, seeds and small quantities of chemical inputs, and some technical assistance in return for their compliance with product quality and hygiene standards and the employment practices of the supermarkets in Europe. The benefits include increased farmer incomes, reduced seasonality of incomes, reduced risk, and improved soil fertility, which has had a knock-on effect on rice production (rice production was shown to be much higher in contract farmer's plots than in noncontract plots).[62]

export and domestic production. Yields in Madagascar were consistently low (yields of rice, the country's primary crop, are among the world's lowest), reflecting poor water management, a lack of nutrient replenishment, and limited adoption of improved technologies.[59]

On the one hand, trade liberalization increased the value of exports and imports as a share of GDP from less than 30 percent in 1986 to over 70 percent in 2004. On the other, commodities that were typically strong exports have significantly declined (e.g., from over 60 percent to just 1 percent of total agricultural exports in the case of coffee). This is due in part to the emergence of new competitors such as Vietnam as well as to the degradation of Madagascar's plantations.[60]

The lesson from Madagascar and elsewhere is that successful liberalization depends on an appropriate enabling environment, which I discuss further in Chapter 8.

Food Aid

For many countries in many situations, the benefits of trade liberalization are a long way off and hunger is unabated. Humanitarian aid, provided by governments and by charitable donations from many sources, targets those people who are suffering from natural or manmade disasters, and also populations where hunger is chronic and seemingly intractable.

Provision of humanitarian assistance is a relatively small component of overall aid but is often both critical and publicly very visible, particularly after an earthquake or cyclone or in times of conflict. Since 2000, humanitarian aid has accounted for approximately 8.3 percent of total overseas development assistance (ODA). This increased to 11.3 percent in 2005 due to the Southeast Asian tsunami and Kashmir earthquake. In 2008, it went up to 11.8 percent ($16.9 billion). Apart from shelter, water, and medical assistance, the main form that humanitarian aid takes is staple foodstuffs, and the key provider is the World Food Programme (WFP).

The modern form of food aid owes much to the Marshall Plan at the end of World War II when some $13.5 billion of aid (25 percent in the form of food, feed, and fertilizer) was provided to war-ravaged Europe by the United States and Canada.[63] The food was sold by the recipient countries for local currency so freeing their hard currency for other imports. Although in essence humanitarian, the Plan had mutual benefits for donors and recipients: The former had grain surpluses they needed to dispose of and farmers whose livelihoods required protection, while the recipient countries were able to feed their populations (125 million Europeans were averaging only 2,000 calories per day) and rebuild their industries.

The same principles continued to operate for many years in relation to developing countries, particularly under the United States' PL 480 (or "Food for Peace") program and the European Union's Instrument for Humanitarian Aid. These have tended to be highly politicized programs, although less so with the creation of the WFP in 1961. The amounts of aid were at their highest in the 1960s, prior to the Green Revolution, when approximately half of all developing country cereal imports were in the form of aid. There were subsequent, smaller peaks at the time of the oil crisis in the early 1970s and mid-1980s. Sub-Saharan Africa is now the largest recipient of food aid. While food aid to Asia has declined from about 4 million tons per year in 2001 to about 1.5 million tons in 2009, aid to Sub-Saharan Africa has averaged around 4 million tons over the same period.[64]

Today some 4 million tons is provided for emergencies, but the amounts under programs and projects to address chronic hunger have shrunk to less than 3 million tons.[65]

Much recent criticism of food aid has focused on the dangers of creating dependency and of undermining the functioning of local markets, thus reducing the incentives for farmers to produce more food. While food aid has provided a valuable market for U.S. grain surpluses, these have shrunk, and much maize is now grown for bioethanol. More recently the emphasis of the WFP has been on raising funds to source supplies in the markets of developing countries themselves (Box 4.6).[66]

For the future, food aid will be limited to countries such as those in Sub-Saharan Africa where the import requirements, although significant for the country, are small with respect to overall world trade. Much is still to be learned if relief is to be speedy and result in a sustainable level of security. Two examples illustrate the scale of the problems the WFP encounters (Box 4.7).

There had been hope that famines were a phenomenon of the past. But the combination of climate change, poor governance, natural disasters, and lack of preparedness are conspiring to make famines more severe, if not more frequent. As I mentioned at the beginning of the chapter, the 2011 food crisis declared in the Horn of Africa was triggered by a record drought but made infinitely worse by civil war and political interference. In a recent report on humanitarian emergencies prepared by a panel on which I sat and chaired by Lord Paddy Ashdown, two of the key recommendations were (1) to invest more time and money in anticipating emergencies and (2) to help countries build more resilience into their development.[80]

Box 4.6   Purchase for Progress

The World Food Programme (WFP), a major staple food buyer in developing countries, established the Purchase for Progress (P4P) initiative in 2008 as a means of procuring more food from smallholder farmers. The food is predominantly purchased via contracts with farmer organizations, small and medium traders, agrodealers, and NGOs that run development projects with smallholder farmers. P4P also invests, with the financial support of a range of donors, in capacity building, such as postharvest storage and processing, and supports emerging trading systems such as warehouse receipt systems (Chapter 8).

Since the launch of P4P's five year pilot program in 2009, P4P had by 2011 contracted with around 200 farmer organizations with another 660 being targeted; has provided training in management, farming, and postharvest techniques to over 65,000 farmers, traders, and warehouse operators; is working with over 160 parties and has purchased over 170,000 tons of food from twenty countries. The potential benefits of a program like P4P to local development and food security are enormous.[67] For one female farmer in Mali, Djèneba Coulibaly, her income increased immediately after P4P was established in her community in 2009 because she was able to sell her crops directly rather than through a middle man. She has also increased her farming area from 0.5 ha to 1 ha.[68]

Box 4.7 Recent Food Crises in Africa

Sudan[69]

The Sudan receives more food aid than anywhere else in the world yet exports vast amounts of domestically produced crops to other countries. The Darfur region, which lies in the west of Sudan, has long been inhabited by two groups of people: 60 percent subsistence farmers and 40 percent nomadic or seminomadic herders of livestock. Traditionally they have lived together in relative harmony, with the herders crossing the land of the subsistence farmers and using their wells.

Severe drought and subsequent famine in 1984–1985 caused the deaths of an estimated 95,000. The consequent shortages of food and water and the ensuing land degradation forced the herders to migrate southward, which increased competition for land and water between them and the subsistence farmers. The conflicts escalated when the Janjaweed militia, with government support, began to force the farmers from their homes and take possession of the wells.

Despite the WFP providing monthly food rations to an estimated two million people in Sudan, rising to three million in the lean season, the conflict that has destroyed the region's infrastructure has directly or indirectly affected over half of the six million people living there, with 2.7 million people displaced since the conflict began. Aid agencies continue to be obstructed from accessing many areas due to the fighting and political insecurity.[70]

Niger

In 2010, Niger faced its worst food crisis in years when a tenth of the population were in need of emergency food aid and half of the population were experiencing food shortages.[71] The country faced a shortfall of $50 million in aid, Niger's free health system was close to bankrupt, and 378,000 children were at risk of severe acute malnutrition with a further 1.2 million at risk of moderate malnutrition.[72]

When Niger's transitional, civilian government (following a military coup) called for help in April 2010, it cited the reasons behind the impending famine as being a poor 2009 harvest, which left the country with a grain shortfall of more than 400,000 tons.[73] Harvests in many parts of the country had looked good in 2009, but a severe locust invasion (the worst invasion in 15 years) stripped farmers of their entire crop, and droughts in 2010 made recovery near impossible.[74] Rising food prices (which have, in some areas, doubled), falling incomes, high grain price instability, climatic variability in the Sahel region, and limited ability to import grains are just some of the cited reasons behind Niger's continuing pattern of food crises.[75]

The WFP raised the alarm in December 2009, but donors were slow to provide funding and there were difficulties buying food in the region.[76] In July 2010, WFP stated it needed $371 million. A month later it was forced to scale back aid due to a lack of funding.[77] An operation to distribute 30,000 tons of grain was planned for August,

(continued)

including the distribution of basic rations to children between 6 and 23 months through-out the country. But due to a massive shortfall in donor funds, the WFP revised its plans and aid was distributed only to families with children under the age of 2 years. In southern Niger, where families survived on wild leaves and fruit for six months, food aid, provided by the government, was finally disbursed in May 2011.[78] However, sub-sequent drought and attacks by pests has reduced harvests, leaving half of Niger's farm-ing villages facing food shortages at the end of 2011.[79]

The Lens of Political Economics

This chapter has emphasized the importance of looking at food security through the lens of political economy. At its core, food security may be an economic prob-lem, but its solution is dominated by political, technological, institutional, and behav-ioral factors. The operation of market forces alone, whether within a country or on a global scale, will not create food security. Supply can meet increased demand but farmers in developing countries, especially smallholders, find it difficult to respond quickly to market signals. They need labor, inputs (seed and fertilizers), credit, insur-ance and access to a market that will purchase their harvest. If they are purchasers, they can benefit from lower prices only if they have access to a store that sells at a fair price. The benefits will flow only if there is an appropriate enabling environ-ment. In Chapter 5 I present a theory of change aimed at addressing these problems in the context of both food security and the wider objective of agricultural devel-opment.

Part II

5

A Doubly Green Revolution

> We require a revolution that is even more productive than the first Green Revolution, even more "Green" in terms of conserving natural resources and the environment and even more effective in reducing hunger and poverty.
>
> Gordon Conway, *The Doubly Green Revolution*[1]

The concept of a *theory of change* has become common currency in philanthropic organizations such as the Rockefeller Foundation and the Bill & Melinda Gates Foundation. In essence, it is a simple conceptual tool that tries to answer the question: Given a certain goal, what is the route that is most likely to get there? A much quoted (reputedly Chinese) proverb "Give a man a fish and you feed him for a day. Teach a man to fish and you feed him for a lifetime," sums up a theory of change, not just for a particular objective but for development in general. Teaching skills, suggests the proverb, is a better route than charitable donations. This was also a message given by Frederick Gates, a Baptist clergyman and educator, to John D. Rockefeller when he was setting up his philanthropic endeavors at the end of the nineteenth century.[2]

I proposed a theory of change (although I did not refer to it as such) for agricultural and rural development in the first edition of this book, published in 1997.[3] What the developing countries need, I argued, is "a Doubly Green Revolution." This sequel has been written because I believe the theory of change embodied by this concept is still as relevant today, in some respects even more so. As I emphasized in Chapter 1, we are beset by periodic food price spikes, chronic hunger is now afflicting about a billion people, and the drivers of the chronic crisis—which range from increasing populations and per capita incomes to the effects of climate change—are intensifying. The many challenges and problems are formidable and can be tackled and solved only by approaches that, in combination, add up to a revolution.

In this chapter I begin by setting out the central hypothesis of a theory of change for agricultural development in the developing countries, which I refer to as the achievement of a Doubly Green Revolution. I then describe in some detail four of the key components: higher productivity, improved stability, more resilience, and greater equitability.

The Concept of Development

Development is a concept that, like food security, has numerous definitions and practical interpretations. It is primarily a term used by developed, industrialized countries to describe the path whereby so-called developing or less-developed countries get to be more like themselves (i.e., the developed countries). Economists define *development* as economic development and equate it with economic growth, often abbreviated simply to growth. The focus is on creating new wealth and includes raising a country's gross domestic product (GDP), increasing personal incomes, investing in the process of industrialization, and generating technological progress.

The developed countries, such as the United States, Japan, and the countries of Western Europe, have all experienced industrial revolutions accompanied by technological advances in many fields. Today, there are several so-called emerging nations such as China, Brazil, and India, as well as smaller, fast-developing countries, including the "eastern tigers" such as South Korea and Vietnam, which seem to be reasonably successful in following the same pathway.

This pathway to development brings about increased material wealth, reduces poverty, hunger, and the burden of disease, and improves educational opportunities. Yet there is a general consensus that development is more than this; to these primarily material freedoms should be added the freedoms, not just from hunger or illness, but also from tyranny, discrimination, and repression, that are so clearly articulated in Amartya Sen's book *Development as Freedom*.[4] Developed countries are also assumed to function both efficiently and fairly. Moreover, this model often implicitly assumes the new wealth will be relatively equally shared; that is, the poor will benefit from the trickle down of the benefits acquired by the better off.

In practice, development is not all of these things, or at least not all at once. In the United Kingdom, where the industrial revolution first took hold, wealth increased and the poor benefited to some extent through a trickle-down process. However, most of the freedoms took longer to realize—through the Reform Acts of the nineteenth century, universal education, and the creation of the Welfare State in the twentieth century.[5] For the United States, some of the freedoms were won by gaining independence and in the subsequent civil war. Industrial growth increased material wealth, and subsequent struggles increased rights for the poor. A key question today is whether the development process can be speeded up in developing countries, by some combination of outside aid, foreign investment, and the universal distribution of modern technology.

The Resource Curse

In recent years, it has become fashionable to argue that the less-developed countries should pursue rapid industrialization, in many cases on the back of the exploitation of rich mineral, oil, and other resources. Around one third of the growth in GDP in Africa between 2000 and 2008 (at nearly 5 percent per annum) has come from these resources and from the associated government spending they generate. Africa is projected to continue to profit from the rising global demand for natural resources given that the continent contains 10 percent of the world's oil reserves, 40 percent of its gold, and 80 to 90 percent of the chromium and platinum metals.[6]

But so far, the experience has not been accompanied by much trickle down and indeed many African countries, particularly oil-producing countries, have suffered from conflict and have reversed progress toward poverty reduction and greater social freedoms.[7] Although in the short term the export of natural resources may increase wealth, it tends to lead to a distortion of the economy, characterized by a lack of investment in such sectors as agriculture and manufacturing. Furthermore an economy highly reliant on nonrenewable natural resources may not be sustainable. Such resources are part of a country's overall capital stock, and for a country to be on a long-lasting development path, it must save enough to cover their depletion. This frequently does not happen.

Agricultural Development

The alternative, or at least complementary, route is to focus on agriculture as a developmental pathway. Agriculture, unlike oil or minerals or even many forest products, is a renewable natural resource. By using a few natural inputs—such as water and recycled organic waste—it is possible to continue to produce modest crop yields on the same piece of land for many hundreds, if not thousands, of years (see Chapter 13). Moreover, while there have been conflicts over ownership of arable or grazing land and for control of lucrative export crops such as rubber, tea, or cocoa, they have not been as severe as those involving oil or minerals. Significantly, as Paul Collier and Benedikt Goderis of Oxford University note, there is little evidence of the resource curse applying to agriculture. Agricultural commodities can be produced in many different locations, unlike oil and minerals, so reducing the source of conflict.[8]

Agricultural development has been key in the early stages of the development of the industrialized countries and, indeed, of the newly emerging economies. The United Kingdom began its development journey with a major agricultural revolution in the eighteenth century, and more recently the eastern tigers, such as South Korea and Vietnam, have similarly invested heavily in agriculture as part of the development process. These experiences amply demonstrate the power of agriculture as an engine for economic development. Increased production and employment in agriculture and natural resources can generate, directly or indirectly, considerable employment, income, and growth in the rest of the economy.

Not surprisingly very few countries have experienced rapid economic growth without preceding or accompanying growth in agriculture.[9] As Michael Lipton, an economist at the University of Sussex, notes: "No country has achieved mass dollar poverty reduction without prior investment in agriculture."[10] In the early stages of development, it is often the lead export sector and foreign exchange earner since much of the manufacturing is agriculturally related.

Agriculture dominates the least-developed countries; it typically accounts for over 80 percent of the labor force and 50 percent of GDP. While this may be seen as a sign of "backwardness," it is a potential strength because any individual small improvement, say in a crop yield, can be multiplied throughout the agricultural economy. Productive innovations can lead to fast and inclusive growth. This was true of the Green Revolution where the new short-strawed wheat and rice varieties were taken up on a large scale. It has also been true of more recent agricultural transformations; for example, in Malawi where government subsidies for fertilizers and modern seeds greatly increased production (see Box 4.2 in Chapter 4).

Analysis of data from forty-two countries during the period 1981 to 2003 shows that a 1 percent gain in GDP originating from agriculture generates a 6 percent increase in overall expenditure of the poorest 10 percent of the population; in contrast, a 1 percent gain in GDP originating from nonagricultural sectors creates zero growth.[11]

The Virtuous Circle

In effect there is a virtuous circle generated by agricultural development. As agriculture develops—resulting in greater yields for both subsistence and cash crops—farmers become more prosperous, and the rural poor, whether landless or on smallholdings, also benefit through wage labor. Chronic hunger decreases. The rural economy also grows—through the creation of small rural businesses—providing more employment and improved rural facilities, especially schools and health clinics. Roads and markets develop so that the rural economy connects to the urban economy and to the growing industrial sector. Free trade provides opportunities for greater imports and exports. High-value agricultural exports (e.g., coffee, cocoa and cotton) in particular can accelerate agricultural development, further intensifying the virtuous circle.

Another Green Revolution?

For the virtuous circle to work, productivity has to grow. So how do we achieve this? Some have argued that we need a repeat of the Green Revolution; that is, introduction of new technologies similar to those of the semi-dwarf cereal varieties that will deliver a quantum leap in yields and production. For a number of reasons this is unlikely to be an effective strategy. The ideal environments for the Green Revolution varieties are already fully exploited. The poor and hungry live today in very different circumstances. Moreover, the technologies of the first Green Revolution were developed on experiment stations favored with fertile soils, well-controlled water sources, and other factors suitable for high production. There was little perception of the complexity and diversity of farmers' physical environments, let alone the diversity of their economic and social environments.

The Green Revolution, it is often said, "passed Africa by," either because some of Africa's principal staple crops, such as sorghum, millet, and cassava, were not targeted or because the environment—physical, socioeconomic, and political—was not conducive. Yet Africa was not alone; for example, even though the Green Revolution was successful in India, much of that country, too, has been passed by. At the end of the 1980s growth in Indian agricultural GDP per annum had increased from 1.4 percent, at the beginning of the Green Revolution, to 4 percent, but it has subsequently slowed to around 2 percent over the 1990s, with rain-fed areas being hardest hit.[12] From 2000 growth slowed even further. Yields stagnated, and by 2004 per capita food grains production was back to 1970s levels.[13]

Indian farmers are increasingly vulnerable to declining water tables, aging, and poorly managed transportation and irrigation infrastructure as well as to unstable markets. The size of farms is also decreasing: in 1960–1961, the percentage of cultivated area operated by farms over 4 hectares (ha) in size was about 60 percent, but this had fallen to 35 percent by 2002–2003, and the proportion of farms under 1 ha in size has risen from about 40 percent to nearly 70 percent in the same period.[14]

Thus, in both South Asia and Sub-Saharan Africa the challenge today is to develop interventions appropriate for relatively small farmers in more diverse, poorly endowed, risk-prone environments. This will require a variety of locally adapted interventions targeted on specific needs.

A Doubly Green Revolution

We require a new revolution that does not simply reflect the successes of the first. It must not only benefit the poor more directly but also be applicable under highly diverse conditions and be environmentally sustainable. In effect, we require a Doubly Green Revolution—a revolution that is even more productive than the first Green Revolution, even more "green" in terms of conserving natural resources and the environment, and even more effective in reducing hunger and poverty.[15]

The route to this revolution is an agriculture that possesses the following characteristics:

- *Highly productive*—By 2050 we will need to have doubled food production and in an efficient manner.
- *Stable*—Agriculture must be less affected by the vagaries of the weather and the market.
- *Resilient*—It must be resistant to, or tolerant of, stress or shocks, especially those generated by climate change (see Chapter 15).
- *Equitable*—We have to provide accessible food and incomes, not just to the better off, but to the poor and hungry.

Moreover it has to be:

- *Sustainable*—We must achieve a pattern of equitable growth that lasts from generation to generation and ensure we do not undermine the environmental and natural resource base on which agriculture depends.

This, in summary, is the theory of change.

Smallholders as the Agents of Change

At the core of the Doubly Green Revolution are smallholders, comprising the vast majority of farmers in the developing world. Very approximately, there are 400 to 500 million small farms—that is, less than 2 ha in size—in the world. Assuming a farm family size of about five, this implies that over two billion people are dependent on smallholdings for their livelihoods—about a third of the world's population.

Nevertheless, being a smallholder is a relative term. Having less than 2 ha is usually regarded as a smallholding in a developing country; 80 percent of all African farms (33 million farms) fall in this category. But in many parts of Latin America a small, predominantly subsistence farm is 10 ha. Bangladeshis would regard this as a large commercial farm. Clearly "smallness" is a relative term depending on the resources of the holding, not only the land but the labor, skills, finances, and technology available. Irrigated rice farmers, for example, can produce high yields with two to three crops a year, consuming and selling their harvest even though their farms may be much less than 2 ha in size.

In all continents, farms of less than 1 ha and with few resources are usually unable to produce a surplus for sale and cannot provide enough work, food or income for the family. Such "marginal" farms in India compose 62 percent of all holdings and occupy 17 percent of the farmed land.[16] As Prabhu Pingali of the Bill & Melinda Gates Foundation reminds us, however, the Chinese agricultural revolution was brought about by very small smallholders, with on average only a *mu* of land, just 1/15th of a hectare.[17]

As has long been recognized, smallholders are in many respects highly efficient.[18] Small farms produce more per hectare than large farms; many studies have shown there is an inverse relationship between farm size and production per unit of land (however, Paul Collier and Stefan Dercon of Oxford University point out this may be true of small versus large farms [i.e., 1 ha versus 10 ha] but not of small versus very large farms [i.e., 1 ha versus hundreds or thousands of ha).[19] In the developing countries where labor is relatively cheap and capital relatively expensive, there are few economies of scale. Household labor is the key to smallholder production— usually a family with long experience of the local environment and knowledge of what works and what does not. Because it is "on the spot," the labor is readily available and motivated. Most important it is flexible, able to respond immediately to the vagaries of the farming calendar and adaptable to the frequent crises affecting the farm, whether they be pest and disease outbreaks, droughts or floods, or slumps in market prices.

Home Gardens

You can see this most clearly in the home garden of a smallholding, a traditional system of agriculture that goes back to the origins of domestication (probably the first wheats were cultivated when the farmer, most likely a woman, brought seed back from the wild stands of wheat to sow on the midden by the dwelling) (Box 5.1).[20]

For ecologists like me, such gardens are a delight. But as Paul Collier warns, we must be careful not to romanticize.[23] The high labor input may be capable of producing a large and varied harvest, but the returns to the labor are small. Labor productivity for home gardens is low and typically insufficient to bring people out of poverty.

Box 5.1   The Javanese Home Garden[21]

Home or kitchen gardens are particularly well developed on the island of Java in Indonesia, where they are called *pekarangan*. The immediately noticeable characteristic is their great diversity relative to their size: they usually take up little more than half a hectare around the farmer's house. Yet, in one Javanese home garden fifty-six different species of useful plants were found, some for food, others as condiments and spices, some for medicine, and others as feed for the livestock—a cow and a goat, some chickens or ducks, and fish in the garden pond. Much is for household consumption, but some is bartered with neighbors, and some is sold.

The plants are grown in intricate relationships with one another. So dense is the planting that to the casual observer the garden seems like a miniature forest. But it is not a natural ecosystem; it is the product of intimate knowledge and daily care and attention, usually by the woman of the household (Figure 5.1).

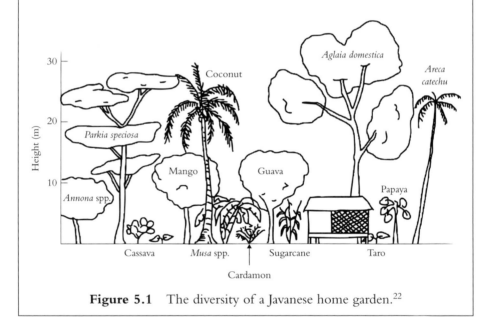

Figure 5.1   The diversity of a Javanese home garden.[22]

Increasing Productivity

As the home garden illustrates, productivity can be measured in several different ways (Box 5.2).

Achieving a growth in total global production is important since it will ensure there is enough "to go round." But as I have argued, this is a necessary but not sufficient condition for agricultural development. For poor farmers and the landless it

Box 5.2   Measures of Productivity

In general terms, *productivity* is the net increment in valued product per unit of resource (land, labor, energy, or capital). It is commonly measured as annual yield or net income per hectare or man-hour or unit of energy or investment. In this book, I have used various measures of productivity, reflecting different combinations of output and input.

Most agronomists refer to *yield*, expressed as kilograms of grain, tubers, or meat per hectare, or per kilogram of nitrogen, or per hour of human effort. Nutritionists may be more interested in the output of calories, proteins, or vitamins, while economists may measure the monetary value of the agricultural production at the market, usually expressed as income less expenditure; that is, as net income or profit.

We also frequently measure total production of agricultural goods and services per household or per region or nation.

Labor productivity expressly measures the return (production or income) per unit of labor; that is, per worker, whether a farmer or a farm laborer.

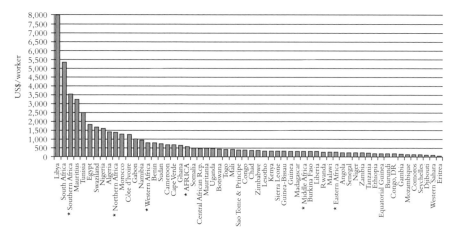

Figure 5.2   Labor productivity in African agriculture, 2003–2005, US$ per worker.[25]

is, in many respects, labor productivity that is the most crucial factor because it determines whether they escape poverty. Steve Wiggins of the Overseas Development Institute has estimated that labor productivity needs to be over $700 per worker per year to get the worker and their dependents out of poverty.[24] For half the countries of Africa, the current average is less than $350 (Figure 5.2).

In many respects this challenge lies at the heart of the Doubly Green Revolution. How can we increase the returns so that smallholder farming becomes a route out of poverty? It is not enough to help smallholders achieve subsistence, even if it is sustainable. They also need incomes—not least to pay medical bills and for schooling, as well as to pay for food when harvests are poor. The goal of the Doubly Green Revolution is not just sustainable existence but sustainable development.

Some analysts argue this can occur only through large-scale farming. The obvious advantages lie in the economies of scale relating to the costs of transactions off the farm—procuring inputs, obtaining credit, and marketing; for example, the costs of meeting standards that satisfy buyers from processors and supermarket chains.[26]

But the experience of large farms, especially in Sub-Saharan Africa is mixed. There have been spectacular failures, especially where inappropriate mechanization has led to severe soil erosion as in the ill-fated Groundnut Scheme in Tanganyika (now mainland Tanzania) in the 1940s or the export vegetable cultivation in Senegal in the 1970s. Large farms in Africa require experienced management, otherwise the costs and the environment can conspire to bring about failure.

Nevertheless large farms can be appropriate and can be successful when large capital investments are necessary; for example, to support large processing plants that in turn require large-scale production. Examples are farms devoted to high-value and specialist crops such as fruit, vegetables, flowers, and intensive pigs and poultry and for export crops such as sisal, sugar, oil palm, tea, rubber, and coffee.[27] Large commercial farms can be close to the frontiers of technology, finance, and logistics and hence globally competitive.[28] This has been the Latin American experience, with farms of more than 10,000 ha growing soybeans in the Brazilian Cerrado and sugar estates of over 300,000 ha in Southern Brazil, many focused on ethanol production. In Malaysia, large-scale contract farming under FELDA (Federal Land Development Authority) as well as an emphasis on rural and agricultural development have enabled diversification into, and expansion, of oil palm production. Between 1960 and 2001 the land area cropped with oil palm grew from 55 ha to 3.4 million ha (54 percent of total agricultural land use). Alongside this growth, the incidence of poverty among smallholder farmers declined from around 68 percent in 1970 to 21 percent in 1990.[29] These are impressive developments, and large-scale farming will have an increasingly significant role in achieving food security. Nevertheless, smallholders, because of their sheer numbers and the total land area they occupy, will have to play the dominant development role at least for several decades to come. The question is: Can we significantly increase their labor productivity?

In practice the way forward lies not in *either* small *or* large farms, but in making a deliberate choice of *both* small *and* large farms. They both have a role to play in future agricultural development. In the words of Prabhu Pingali: "Under the right circumstances smallholders can be just as productive, just as innovative, just as competitive, and just as risk-taking as larger farms."[30]

Increasing Yields

For most environments, the only way to increase labor productivity is to raise yields. The Green Revolution resulted in dramatic increases in the yields of the main cereal crops. These translated into rising average national yields, especially in South Asia and China, but not in Sub-Saharan Africa. Cereal yields there have remained stubbornly stuck at an average of little over 1 ton per ha, about the level of yield under the Roman Empire in Europe, two thousand years ago (Figure 5.3).

The yield increases in South Asia and China have partly come from plant breeding (see Chapter 9).[32] But there is now evidence of declining yield growth for some crops (see Figure 1.8 in Chapter 1).[33] At the national level some crops have hit plateaus, while others continue to increase in yield (Figure 5.4). Rice yields, for example, grow where irrigation expands, as in Indonesia, or remain low as in Thailand where 75 percent of the rice is rain-fed and on poor-quality soils. However, where all rice is produced with irrigation, as in South Korea, the yields increased rapidly until 1980 but thereafter have hit a plateau at 80 percent of the yield potential.

The yield potential of rice has changed little over the past forty years. There are many definitions of *yield potential* but probably the simplest and most useful in practice is "the yield of an adapted crop variety or hybrid when grown under favourable conditions without growth limitations from water, nutrients, pests, or diseases."[35] The *yield gap* is then the gap between this yield and the average farmers' yields.[36] In most irrigated wheat, rice, and maize systems, yields appear to be at or near 80 percent of yield potential (as for South Korean rice in Figure 5.4).[37] We urgently need new high-yielding varieties of staple crops (and of livestock breeds) that make more efficient use of sunlight, water, and nutrients and so increase the yield potential, but

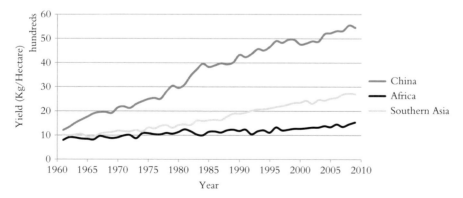

Figure 5.3 Average cereal yields have increased steadily since the Green Revolution in China and South Asia but remained stagnant in Sub-Saharan Africa.[31]

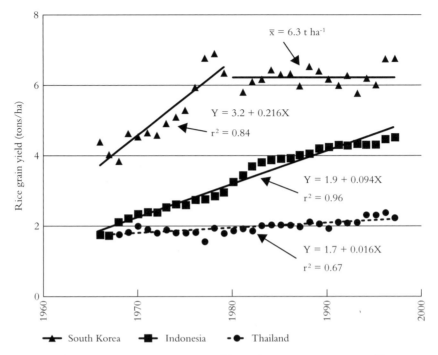

Figure 5.4   Average rice yields in three Asian countries.[34]

this will not be easy. At the same time we should be concentrating on decreasing the yield gaps for rain-fed crops. This will also not be easy, but the gains may be greater.

As the story of Mrs. Namarunda illustrates (see Chapter 1), we need improved soil fertility, increased access to water, better pest, disease, and weed control without environmental damage, access to microcredit and affordable inputs especially of fertilizers and seed, relevant information that is readily available and understandable, and training in improved skills. Some of these form part of the wider set of factors such as research, extension, infrastructure, markets, and education that are captured by the measure of *total factor productivity* (see Box 3.4 in Chapter 3); others relate to specific technologies, and I discuss these in later chapters.

Reducing Instability

Over time productivity, whether of product or income, may rise, fall, or remain constant. It will also fluctuate about this trend. The yield of a crop is likely to mirror the variability in the climate. Income may also fluctuate, not only reflecting changes in yield but also variations in the market price of inputs, such as labor, fertilizers, and pesticides and of the product itself. *Stability* is a measure of such variability in the face of these relatively minor and commonplace disturbing forces. It is most conveniently measured by the reciprocal of the coefficient of variation in productivity.

Except in highly controlled circumstances, crop production will vary with the weather, rainfall, temperature, and other deviations from the norm. Instability will also occur over different time horizons. Both production and prices tend to vary with the season. Typically, prices are low at harvest and high immediately before it as stocks run out, creating the "hunger season." Production also varies annually.

Various agronomic measures, such as better pest and disease control, will reduce such instability (these are discussed in Chapter 12). Fluctuations in rainfall may result in poor and erratic water supply, which can be solved by systems of small- or large-scale irrigation (see Chapter 14). Farmers can cope with a modest fluctuation in production. Different crops or types of livestock on a mixed farm may fluctuate asynchronously, so that the overall production remains stable. But this is true only for relatively large farmers (i.e., with a hectare or more of land). In the 1970s in Bangladesh some 40 percent of farmers had less than half a hectare of land and had to buy rice even in good years.[38] A poor rainfall year in these circumstances can be devastating. Average farm size in Bangladesh has declined to 0.68 ha, a level unlikely to sustain livelihoods.[39] The only alternative here is to develop nonfarm and off-farm sources of income to offset the consequences of climatic and other variability.

Building Resilience

Sometimes agricultural production is affected not by a normal, relatively small-scale variation but by an extreme event. The price spikes described in Chapter 1 are examples. Similarly, sudden increases in the prices of inputs—say seeds or fertilizers—can be a severe blow to farmers. Other kinds of shock include a sudden drought or flood or a major pest or disease outbreak. Tropical cyclones, earthquakes, tsunamis, and eruptions of civil conflict can be devastating, especially for small poor farmers. Sometimes the damage is caused not by a sudden shock but by an accumulated stress; for example, growing indebtedness or rising soil salinity on a farm (Box 5.3).[40]

For each stress or shock, there is an appropriate adaptation consisting of one or more countermeasures that serve to maintain the resilience. In practice there are many different kinds of countermeasure that can be deployed (Box 5.4).

I discuss various forms of biologically based resilience in Chapter 6, and in Chapter 15 I look in more detail at the ways agriculture can remain resilient in the face of climate change.

Box 5.3   The Nature of Resilience[41]

Resilience is the capacity of an agricultural system to withstand or recover from stresses and shocks.

A stress is defined as a regular, sometimes continuous, relatively small, and predictable disturbance (e.g., the effect of growing soil salinity or indebtedness). A shock, by contrast, is an irregular, infrequent, relatively large, and unpredictable disturbance, such as is caused by a rare drought, flood, or a new pest. Lack of resilience may be indicated by declining productivity, but equally, as experience suggests, collapse may come suddenly and without warning.

In Figure 5.5 agricultural development is illustrated as an increasing trend. Along comes a stress or shock, which in some circumstances can be fully resisted, such as when a dam or barrage prevents a flood. More often the development path is adversely affected, and growth falls; this is generally followed by recovery, which may be fast or slow. In some cases the disturbance is too great, and recovery may not fully occur. Agricultural development may resume along a less productive path or in extreme cases may collapse altogether. A resilient pathway is one that persists, despite the stresses and shocks, in its upward trajectory.

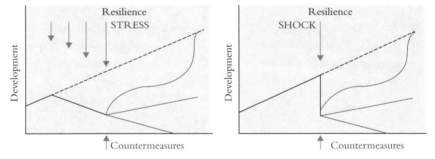

Figure 5.5   The patterns of resilience, showing the effects of stress (left) and shock (right).[42]

Note: In the first edition of this book I referred to this property as sustainability, but in practice sustainability has come to mean a much broader concept. Resilience, similar in definition to the concept of C. S. Holling, is more appropriate.[43]

Equitability

High, stable, and resilient agricultural production is an important measure of the success of an agricultural system's performance, but of equal importance is how fairly the produce is shared. An African village that has a high, stable yield of sorghum, using practices and varieties broadly resistant to pests and diseases, might be regarded

Box 5.4   Forms of Countermeasure against Stress or Shock

- *Institutional.* Zone land to protect against flooding; create government warning systems for cyclones.
- *Economic.* Create weather crop insurance to compensate for climatic extremes; develop microcredit schemes to help farmers develop alternative sources of income.
- *Infrastructure.* Build embankments and other structures to provide protection against floods.
- *Medical.* Administer livestock vaccines to protect against serious disease outbreaks.
- *Environmental.* Conserve mangrove shelterbelts and coastal forests to protect against sea level rise.
- *Agronomic.* Breed drought- and flood-resistant crop varieties; develop resilient cropping and livestock systems.
- *Livelihoods.* Improve income diversity; create rural-urban linkages.

as more successful than another village having lower, less stable, and resilient yields. However, it is not only the pattern of production that is important but also the pattern of consumption. Who benefits from the high, stable, and sustainable production? How is the harvested sorghum, or the income from the sorghum, distributed among the people of the village? Is it evenly shared or do some villagers benefit more than others?

Equitability describes this pattern of distribution of productivity. *Equitability* can be defined as a measure of how evenly the productivity of an agricultural system is distributed among its human beneficiaries. The more equitable the system the more evenly are the agricultural products—the food or the income or the resources—shared among the population of the farm, village, region, or nation. It can be represented by a statistical distribution or by a measure such as the Gini coefficient.[44]

Equitability, of course, is not a direct measure of lack of poverty. Everyone in a community can be equally poor or rich. However, when productivity rises, the more equitable the distribution of the gains the more likely is there to be a reduction in poverty. Equitability also refers to the way income and other benefits are shared by more traditionally disadvantaged members of a community—by women in particular, but also by ethnic minorities and lower castes.

Even when agricultural innovation and development stimulate economic growth, this does not necessarily lead directly to a reduction in poverty. Much depends on the nature of the innovations and how broad based is the agricultural development they generate. While the introduction of irrigation and new varieties can create employment and incomes, certain kinds of mechanization associated with agricultural intensification will destroy jobs (see Chapter 3). For equitable economic growth, agricultural innovation needs to be deliberately focused on increasing production while, at the same time, creating employment both in agriculture and in related, rural-based industry.

As we saw in Chapter 3, the impact of the Green Revolution on hunger and poverty was uneven. The rural and urban poor gained from low food prices, but in many situations the benefits went to the large farmers rather than small, and the landless continued to suffer. In many of the rural areas largely untouched by the Green Revolution, both poor farmers and the landless have suffered losses in real income and increased hunger. It has been only in the past twenty five years, with the development of new rain-fed rices and high-yielding sorghums and millets that the Green Revolution has moved outside the Green Revolution lands. One consequence has been a widening gap between those who own or rent a little land and those who are truly landless. The number of landless has risen rapidly, although it is difficult to get good estimates because census definitions are not always consistent.

The core of the Green Revolution technology—the new varieties—were potentially poverty alleviating. But this result occurred only if technological innovation was positioned within a broad-based agricultural and rural development. Technologies are by themselves not enough. As Michael Lipton puts it, "There is a limit to technical cures for social pathologies."[45] Too often the new technologies were injected into communities with rapidly growing populations already dominated by excessive inequalities where, in the absence of countervailing policies, the powerful and the better-off have acquired the major share of the benefits. As a consequence, over 20 percent of the developing world's population was still poor and hungry in the 1980s twenty years after the Green Revolution began. Today that proportion has dropped to nearly 14 percent, but the absolute numbers have increased to nearly one billion.[46]

The Example of North Arcot

There are very few long-term analyses of rural development for individual villages or districts, but an illuminating example is the history of North Arcot in southern India, documented by Peter Hazell of IFPRI and C. Ramasamy of Tamil Nadu Agricultural University and by Barbara Harriss-White of Wolfson College, Oxford, and S. Janakarajan.[47]

North Arcot is a rice-growing district in the state of Tamil Nadu in southern India, dependent on tank and tube well irrigation. Although it was not a prime Green Revolution area, and the overall benefits were not as great as in the Punjab, they were still significant: rice output grew by nearly 60 percent between the mid-1960s and late 1970s. Originally, most of the farms were small (average 1.2 ha) and one third of rural households consisted of landless laborers. Initially the larger farmers (about 2 ha) were the adopters. However, by the early 1980s over 90 percent of the land was planted to new varieties with no difference between large and small farms, largely due to improved access to credit and the release of new varieties better adapted to small farms with less reliable water.

The installation of irrigation pumping and threshing machines produced a fall in labor demand, with family labor increasing while hired labor fell by 25 percent per farm. This would have been worse if there had been an increase in use of tractors, which was probably inhibited by the small farm size. Nevertheless, wage rates and employment earnings rose, partly as a consequence of the growing rice economy. There were increased opportunities in dairying and in nonfarm activities—such as government employment programs and a growing silk weaving industry—and there was substantial migration to urban areas. As a result there was no increase in landlessness and little or no increase in size of farms.

All groups benefited from a growth in income, consumption expenditure, and in their diets; calories per adult equivalent rose, on average, among rice farmers from about 1,800 calories to over 3,000 in 1983–1984 and for landless laborers from 1,700 to over 2,500 calories (Figure 5.6).

In the next decade (from 1983–1984 to 1993–1994), however, progress in North Arcot was mixed:

- Rice remained the major crop (47 percent of cropped area), but production stagnated, largely due to competition and overuse of the tube wells and their subsequent abandonment.

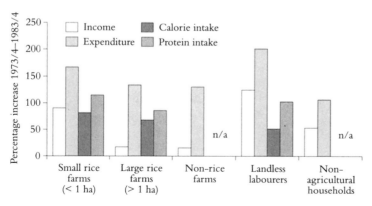

Figure 5.6   Increases in standard of living in North Arcot, southern India.[48]

- Crops were increasingly diversified and cropping intensity increased.
- The proportion of households cultivating less than 1 ha rose to over 70 percent, and the distribution of land became increasingly unequal.
- Agricultural employment declined. (By 1993–1995, nearly a quarter of the economically active population was engaged in nonfarm activity.)
- More women began working in agriculture, but this did not improve female earnings.

Summing up the North Arcot experience, Barbara Harriss-White and colleagues comment:

> The Green Revolution has succeeded in keeping rice production at medium-term constant. . . . This has been achieved at the cost of increasing instability of output, stagnant yields and plunder of the water table. . . . It is evidently now the non-agricultural economy which is providing development dynamism to the region.[49]

A Theory of Change

The North Arcot example well illustrates the concerns that helped to stimulate the thinking behind the Doubly Green Revolution. While the Green Revolution generated a significant increase in productivity and incomes in North Arcot that was relatively equitable, the subsequent history raises important issues of sustainability not least of the critical environmental resources. It also highlights the importance of considering agricultural development within the larger context of rural development and the linkages to urban communities. As I discuss in Chapter 6, this linkage—whether via migration as in North Arcot or through trade or remittances—is crucial to helping people escape poverty.

The Doubly Green Revolution is a theory of change, not a recipe; it provides the concepts that will guide progress toward both greater food security and better agricultural development. The theory stresses the importance of considering four objectives from the outset—increased productivity, greater stability, greater resilience, and higher equitability—and pursuing them together, not one by one, and in relation to the larger socioeconomic, political, and environmental landscape. As I argue in the next chapter, there are trade-offs between them, and it is where these trade-offs are at a minimum that we will find a sustainable agriculture.

6

Sustainable Intensification

Agri cultura "est scientia, quae sint in quoque agro serenda ac facienda, quo terra maximos perpetuo reddat fructus." [Agriculture "is a science, which teaches us what crops are to be planted in each kind of soil, and what operations are to be carried on, in order that the land may produce the highest yields in perpetuity."]

Marcus Terentius Varro, *Rerum Rusticarum*[1]

As I argued in Chapter 1 there is little likelihood of significantly more arable land becoming available; yet we know that we have to approximately double food production by 2050. For the past fifty years the only significant increase in arable land has been for crops such as oil palm and soybean on cleared rainforest or the Brazilian Cerrado. For the future the only solution to the food security problem is to get more production out of the existing land, but to do it in a way that is sustainable. This will depend on human ingenuity, in particular in harnessing the benefits of ecological processes.

In this chapter I address the following questions in relation to sustainable intensification:

- Why are ecological concepts and theory important?
- What are the roles of ecological processes in developing practical sustainable technologies?
- How does sustainable intensification relate to the conservation and utilization of biodiversity?
- What is the contribution of the farm household in developing sustainable livelihoods?

The Nature of Sustainability

Ecology has informed and underpinned agricultural production since the first faltering steps in domestication and cultivation. When someone (probably a woman) living ten thousand years ago in the Fertile Crescent (present day Iraq, Jordan, Syria, Lebanon, and Israel) carried seeds of wild wheats and barleys from the great natural cereal stands of the region and sowed them near her house, she initiated the process of domestication. She also began the process of crop cultivation, creating what were to become complex home gardens based on ecological processes that deliver a sustainable form of agriculture.

The concept of sustainability was not articulated for many thousand years later. When he wrote the words at the beginning of this chapter, Marcus Terentius Varro, a Roman landowner of the first century BC, was eighty years old and had recently remarried. *Rerum Rusticarum*, one of a number of Latin treatises on agriculture to survive to the present day, was written for his wife as a handbook of advice on how to run the estate he had purchased for her. In the passage, he places crops in their environment and in the phrase "quo terra maximos perpetuo reddat fructus" (which can be translated as *quo terra* [that the land], *reddat fructus* [yields], *maximos* [the highest], *perpetuo* [in perpetuity]) defines sustainability in a way that is extraordinarily clear, elegant, and succinct.

Unfortunately, Varro's original clarity of meaning has been lost. Sustainability has become a highly politicized term and, in the process, acquired a diversity of meanings.[2] Plant breeders, agronomists, and other agriculturalists interpret sustainability as maintaining high yields and production; sustainable agriculture can embrace any means to that end. But for environmentalists the means are crucial: sustainable agriculture is a way of providing sufficient food without degrading natural resources. For economists, it represents an efficient, long-term use of resources, and for sociologists and anthropologists it embodies an agriculture that preserves traditional values and institutions. It has become an all-embracing term. Almost anything that is perceived as "good" from the writer's perspective can fall under the umbrella of sustainable agriculture—organic farming, the small family farm, indigenous technical knowledge, biodiversity, integrated pest management, self-sufficiency, food sovereignty, recycling, and so on.

This diversity of interpretation is, nevertheless, to be welcomed as part of a process of gaining consensus for radical change. Popular interest in sustainability was first aroused by the 1987 report, *Our Common Future* (also known as the Brundtland Report), of the World Commission on Environment and Development chaired by the former prime minister of Norway, Gro Harlem Brundtland. The report drew the attention of politicians and the public, in both the developed and the developing world, to the threats to the world's survival posed by the way we treat our environment. The report's often quoted definition of *sustainable development*—"development that meets the needs of the present without compromising the ability of future generations to meet their own needs"—was welcomed as a spur for political action.[3]

But the subsequent debate has been confusing, as different interest groups have wrestled with the practical implications.

The Brundtland definition in the context of agriculture is valuable as a policy statement, but it is too abstract for the farmers, research scientists, and extension workers who are trying to design new agricultural systems and develop new agricultural practices. They need a definition that is scientific, open to hypothesis testing and experimentation, and practicable.

The Contribution of Ecology

Central to this debate, in theory and practice, lies the contribution of one of the great revolutions in modern biology, the emergence of ecology as a sophisticated discipline. The origins of ecology lie in the nineteenth century and are allied to the development of evolutionary theory. Indeed, Charles Darwin can be considered one of the first ecologists (Box 6.1).

Until the middle of this century, ecology remained largely descriptive, often little more than natural history. But in recent years, the development of powerful hypotheses, the influence of mathematical tools, and the design of appropriate laboratory and field experiments have transformed natural history into science.[5] Like most sciences, it has its pure and applied sides, and its applications are now influencing many aspects of renewable natural resource management, including agriculture.[6]

Box 6.1   The Kelp Beds Off Tierra del Fuego

In a passage in his *Voyage of the Beagle* Charles Darwin describes the rich communities supported by the giant kelp beds off Tierra del Fuego.

The leaves of the seaweed are covered with corallines, shells, and crustacea. "On shaking the great entangled roots, a pile of small fish, shells, cuttle fish, crabs of all orders, sea eggs, starfish, beautiful Holuthuriae, Planariae, and crawling nereidous animals of multitude of forms, all fall out together. . . . Amidst the leaves of this plant numerous species of fish live, which nowhere else could find food or shelter; with their destruction the many cormorants and other fishing birds, the otters, seals, and porpoises, would soon perish also; and lastly, the Fuegan . . . perhaps cease to exist."[4]

In many ways the kelp community was like a tropical rain forest, with the rich diversity of life and the complex web of food-chains supporting a very high level of productivity.

The Concept of Agroecosystems

Population, community, and ecosystem ecology have begun to provide a better understanding of the complex dynamics that arise within agriculture; for example, in crop populations, in multiple cropping systems and agroforestry, and in the management of rangelands. Some of the most fruitful work has come from collaboration between ecologists and agricultural scientists, often at the newer, so-called "ecoregional" International Agricultural Research Centers (IARCs), such as at the Centro Internacional de Agricultura Tropical (CIAT) in Colombia, but also at independent centers such as the Multiple Cropping Center at Chiang Mai University where I and a group of Thai agronomists and agricultural scientists worked in the 1980s. We were attempting to understand and improve on the complex agricultural systems that farmers had developed in the valleys of northern Thailand.[7] The concepts and definitions that follow are based on this work, which in turn was influenced by earlier ecological studies of the savanna ecosystems of Zimbabwe.[8]

It takes little effort to recognize agricultural systems, such as the rice fields of northern Thailand, as modified ecological systems (Figure 6.1). Each field is formed, from the natural environment, by building up a ridge of earth that defines its boundary. Inside, the great diversity of the original wildlife is reduced to a limited set of crop, pest, and weed species—although still retaining some of the natural elements, such as fish and predatory birds. The basic ecological processes—(1) competition between the rice and weeds, (2) herbivory of the rice by the pests, and (3) predation of the pests by their natural enemies (and of the fish by the predatory birds)—remain the same. But these ecological processes are now overlain and regulated by the agricultural processes of cultivation, input (with fertilizers), control (of water, pests, and diseases), and harvesting.[9]

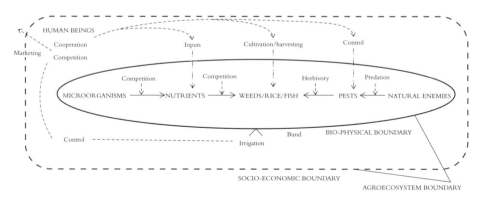

Figure 6.1   The rice field as an agroecosystem.[10]

However, this is only a partial picture of the transformation. The agricultural processes are, in turn, regulated by economic and social decisions. Rice farmers cooperate or compete with one another and market, exchange, or consume their produce. The resulting system is as much a socioeconomic system as it is an ecological system and has a socioeconomic boundary, although it is not as easy to define as the biophysical one of the earthen ridge. This new, complex, agro-socio-economic-ecological system, bounded in several dimensions, is called an agroecosystem. More formally, an *agroecosystem* is "an ecological and socio-economic system, comprising domesticated plants and/or animals and the people who husband them, intended for the purpose of producing food, fibre or other agricultural products."[11]

Agroecosystems defined in this way fall into a hierarchy. At the lowest level is the individual plant or animal and its immediate microenvironment. The next level is the crop or the herd or flock of animals, contained within a field or paddock. The fields combine to form a farming system that is managed by a farm household. And the hierarchy continues upward in a similar fashion. Agroecosystems also have distinctive properties that can be measured and used as indicators of performance.[12] These properties, introduced in Chapter 5, are productivity, stability, resilience, and equitability. They are relatively easy to define, but not as easy to measure (Figure 6.2).

Productivity measures how much an agroecosystem produces over time (e.g., tons of rice per hectare). The *stability* of production can also be measured: How does production vary from year to year? Productivity and stability can be either high or low. *Resilience* measures how well the productivity tolerates or recovers from stress or shock, and *equitability* measures how fairly the products of the agroecosystem are shared among the beneficiaries.

Trade-Offs

These four properties are essentially descriptive in nature, summarizing the status of the agroecosystem. But they can also be used in a normative fashion, as indicators of performance, and in this way can be employed to trace the historical evolution of an agroecosystem and to evaluate its potential, given different forms of land use or the introduction of new technologies. In agricultural development, there is almost inevitably some degree of trade-off between the different system properties. New forms of land use or new technologies may have the immediate effect of increasing productivity, but this is often at the expense of lowered values of one or more of the other properties. Agricultural development typically involves a progression of changes in the relative values of these properties, with successive phases of development producing different priorities.

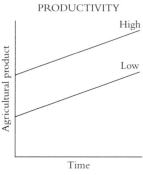

PRODUCTIVITY

Productivity: The output of valued
product per unit of resource input.

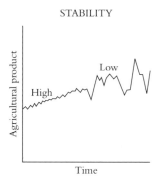

STABILITY

Stability: The constancy of productivity
in the surrounding environment (usually
measured from a time series by the
coefficient of variation in productivity).

RESILIENCE

Resilience: The ability of the
agroecosystem to maintain productivity
when subject to a stress or shock.

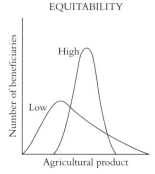

EQUITABILITY

Equitability: The eveness of distribution
of the agroecosystem among the human
beneficiaries, i.e. the level of equity that
is generated (a common measure is a
Lorentz curve).

Figure 6.2   Graphic representation of Agroecosystem Properties.[13]

In the Green Revolution, for example, there was a conscious choice to focus on productivity, implicitly at the expense of the other properties. As Randolph Barker, Robert Herdt, and Beth Rose of the International Rice Research Institute (IRRI) put it, "If the modest resources available to the international agricultural research system in the 1960s had been concentrated in the less favorable environments, it is likely that no major breakthrough would have been made."[14]

As I described in Chapter 3, the introduction of new short-stalked varieties of wheat and rice along with irrigation, packages of fertilizer, and pesticides dramatically increased yields. Overall food production in the developing countries kept pace with population growth, both more than doubling. However, the vast expanses of monocropped cereals required tight control to maintain their stability, and they were prone to pest and disease epidemics. The new rices, in particular, were attacked by devastating outbreaks of bacterial blight and brown planthopper.

At the same time, the Green Revolution agroecosystems have shown poor resilience in the face of environmental degradation. Now, there is also evidence of increasing production problems in those places where yield growth in the Green Revolution was most marked.[15] Of particular concern is the growing scarcity of water; in some of the most intensively cultivated districts of India, the groundwater table is falling by up to 2 meters a year.[16] A further factor is the cumulative effect of environmental degradation, partly caused by agriculture itself.[17] In India, fertilizers such as urea were heavily subsidized for over three decades leading to overuse and degradation of soil, declining yields, and rising levels of imports.[18]

Virtually all long-term cereal experiments in the developing countries exhibit marked downward trends in yields. This and other, albeit largely anecdotal, evidence from Luzon, Java, and Sonora suggest there are serious and growing threats to the productivity and resilience of the Green Revolution lands.[19] Equitability has also been low. The larger landowners have reaped most of the benefits, while the poor and landless have missed out.

A Sustainable Agriculture

The question is: Do the trade-offs have to be so severe? Can we find agricultural technologies and processes that generate high levels of all four properties, so minimizing the trade-offs? The small area overlapping high values in Figure 6.3 represents what we commonly refer to as *sustainable agriculture*. It is a key means of achieving sustainable intensification and, more broadly, the Doubly Green Revolution.

Sustainability, so defined, applies both vertically up and down the hierarchy of agroecosystems, and horizontally between different agroecosystems and between agroecosystems and natural ecosystems. A sustainable agriculture developed for a farm may not be sustainable in a larger landscape. For example, high productivity on a farm may depend on local water supplies and sources of organic matter that are not available elsewhere. Sustainability on the farm may reduce sustainability of neighboring agroecosystems (e.g., depriving them of water) or neighboring natural ecosystems (e.g., by removing organic matter so reducing their biodiversity). Most important is the assessment of sustainability over time—sustainability as durability. In the North Arcot example in Chapter 5, the initial sustainability declined in later decades. I will provide some examples of these effects in Chapters 12 to14.

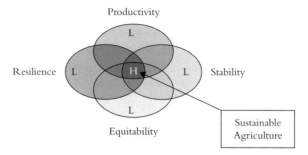

Figure 6.3   Trade-offs between livelihood properties (H=high, L=low).[20]

The complexity of the challenge in designing and implementing sustainable agro-ecosystems is daunting. Yet, I am optimistic, in part because of the potential of two key, emerging areas in the biological sciences: (1) *modern ecology*, with its sophisticated understanding of population and ecosystem processes and (2) *molecular and cellular biology*, which, with their associated technologies, are having far-reaching consequences on our ability to manipulate living organisms. I will discuss the role of ecology here, and in Chapter 9 turn to the complementary role of modern biotechnologies.

Ecology and Agriculture

Ecologists have long been fascinated by what makes ecological systems relatively stable and resilient.[21] There is a general supposition that the key factor is *diversity*—measured in its simplest form as species richness. Many theoretical and empirical studies have attempted to illuminate the relationships between diversity, on the one hand, and stability and resilience, on the other. The results have often been contradictory, sometimes even reversing the cause-and-effect relationships. For example, diverse systems may be relatively fragile and maintained only in stable environments. In unpredictable environments, the communities may be more resilient yet simpler. What is clear from these studies is that it is not the sheer variety of the species present that affects stability and resilience, but their nature, their function in the systems, and the relationships they have with one another.

It is also clear that "natural communities represent not random assemblages of species but rather collections of species that can coexist."[22] In many respects this is even truer of agroecosystems. The diversity of crop and livestock species and of their varieties and breeds are present because human beings have recognized they *can* coexist. Humans have also recognized that there are strong benefits in their coexistence. People do not live only on the calories provided by staple crops; they need sources of proteins, vitamins, minerals, and other nutrients. A diverse agroecosystem provides for a diverse and healthier diet. As farmers also recognize, different species can benefit each other. Trees and shrubs provide shade for herbs, legumes provide nitrogen, and livestock furnish manure. Mixtures of crops can also deter pests, and when a disaster strikes the farm—a drought or cyclone—the more diverse the farm, the more likely something will survive.

The home garden, described in Chapter 5, is a good example. Part of the reason for the minimal trade-off between the agroecosystem properties in the home garden is the deliberately in-built diversity that helps stabilize production, buffers against stress and shock, and contributes to a more valued level of production. But equally important is the intimate nature of the home garden. The close attention that is possible from family labor ensures a high degree of stability and resilience, and the link between the garden and the traditional culture leads to an equitable distribution of the diverse products.

A Doubly Green Revolution seeks to exploit these relationships, through a variety of ecologically based approaches building on the world's endowment of environmental services.

Ecosystem Services

Successful, sustainable agriculture depends on the sustainability of the natural environment and vice versa. While it is possible to create entirely artificial agroecosystems, isolated in controlled climatic chambers and provided with synthetic inputs, most of the world's farmers depend on the bounty of the natural environment and are subject to the many threats it presents.

In 2005, the Millennium Ecosystem Assessment (MA), which was set up under the auspices of the United Nations, produced a comprehensive report that assessed the state of each of the world's different ecosystems and the processes affecting them.[23] The report referred to "ecosystem services" to describe the elements of ecosystems and their processes that are of specific benefit to people. These benefits can range from the provision of water or plants and animals for food to less obvious contributions such as the insects that pollinate crops or the geochemical cycles that remove the pollutants we produce from the air and the water we consume.

The MA examined trends in ecosystem services over the past half century. "Severely degraded" services include capture fisheries, water supply, waste treatment and detoxification, water purification, natural hazard protection, regulation of air quality, regulation of regional and local climate, regulation of erosion, spiritual fulfillment, and aesthetic enjoyment. Two of these services—freshwater and capture fisheries—are being exploited at levels that cannot be sustained under current

111

demand, much less the demand of a growing population. Only four services—food production, livestock, crops, and aquaculture—have been enhanced in the past fifty years, in large part due to the power of modern agronomy and husbandry.

Agronomy and Husbandry

Agronomy and husbandry comprise the tools whereby modern ecology is applied to agriculture. They are used to recognize and identify relevant environmental services and exploit them in a productive and sustainable manner. Box 6.2 provides examples of modern agronomic practices.

Box 6.2   Agricultural Tools and Technologies with High Potential Sustainability[24]

Intercropping: the growing of two or more crops simultaneously on the same piece of land. Benefits arise because crops exploit different resources or mutually interact with one another. If one crop is a legume, it may provide nutrients for the other. The interactions may also serve to control pests and weeds.

Rotations: the growing of two or more crops in sequence on the same piece of land. Benefits are similar to those arising from intercropping.

Agroforestry: a form of intercropping in which annual herbaceous crops are grown interspersed with perennial trees or shrubs. The deeper-rooted trees can often exploit water and nutrients not available to the herbs. The trees may also provide shade and mulch, while the ground cover of herbs reduces weeds and prevents erosion.

Sylvo-pasture: similar to agroforestry, but combining trees with grassland and other fodder species on which livestock graze. The mixture of browse, grass, and herbs often supports mixed livestock.

Green manuring: the growing of legumes and other plants to fix nitrogen and then incorporating them in the soil for the following crop. Commonly used green manures are *Sesbania* and the fern *Azolla,* which contains nitrogen-fixing, blue-green algae.

Conservation tillage: systems of minimum tillage or no-tillage, in which the seed is placed directly in the soil with little or no preparatory cultivation. This reduces the amount of soil disturbance and so lessens run-off and loss of sediments and nutrients.

Biological control: the use of natural enemies, parasites or predators, to control pests. If the pest is exotic, these enemies may be imported from the pest's country of origin; if indigenous, various techniques are used to augment the numbers of the existing natural enemies.

Integrated pest management: the use of all appropriate techniques of controlling pests in an integrated manner that enhances rather than destroys natural controls. If pesticides are part of the program, they are used sparingly and selectively so as not to interfere with natural enemies.

These and other technologies will be discussed further in subsequent chapters. Here I will illustrate what is meant by sustainable technologies by discussing one example: intercropping.

The Dynamics of Intercropping

Intercropping is the practice of growing two or more crops together on a given piece of land. It does this by balancing two key ecological processes—competition, on the one hand, and commensalism or mutualism, on the other.[25] Typically, crop plants in a field will compete with one another, and the aim of the farmer is to plant them as close together as possible to utilize all the available land, but not so close that the yields are diminished by competition. If the crop plants are all of the same variety, they usually have been bred to minimize damaging competition. When different crop species or varieties are grown together, the competition may be fierce; trees grown in a maize field, for example, may shade out the crop. But this can be compensated for by one plant gaining benefits from the other (*commensalism*) or by both plants benefiting the other (*mutualism*). The tree may be a legume and provide nitrogen for the crop plant beneath, an example of a commensal relationship.

Intercropping has several advantages. It can provide a more efficient use of resources, such as soil nutrients and light that would not otherwise be utilized by a single crop; support or shade a companion crop; or host a great diversity of insects, bacteria, and other organisms that contribute to pest and disease control. Care must be taken, however, to plan the timing, spacing, and mixture of crops so that the balance between competition and commensalism or mutualism is enhancing.

Some of the basic intercropping strategies include:

- *Mixed cropping*: Crops are mixed, relatively randomly, in the available space.
- *Row cropping*: Crops are regularly arranged in alternate rows, or one crop is alternated with multiple rows of another crop.
- *Relay cropping*: crops are spaced apart temporally, so a fast-growing crop may be grown with a slow-growing crop, or one crop is planted much later than the second in the same available space.

There are limitless variations on these strategies, depending on the crops grown in an area and on factors such as pests, weeds, soil fertility, and water availability. One example is the so-called MBILI system (Box 6.3).

Box 6.3   The MBILI System of Kenya[26]

In western Kenya farmers traditionally row-crop maize with legumes to obtain yields from both of the crops as well as to gain soil fertility from the nitrogen-fixing legumes. Nitrogen is returned to the soil from the falling leaves and decomposing roots of the bean plants.

Researchers at the Sustainable Agriculture Centre for Research, Extension and Development in Africa (SACRED-Africa) in Kenya, noticed that the single rows were not allowing the legumes enough light, and that the second maize crop often failed due to insufficient late rains. They pioneered a new system, known as MBILI (Managing Beneficial Interactions in Legume Intercrops) and literally meaning "two" in Swahili. It consists of intercropping double rows of maize and legumes, allowing for better light and soil conditions within the understory legumes, while maintaining the same plant populations.

SACRED has also introduced the use of higher-value legumes, such as green gram and groundnut to improve profits, as well as faster-maturing maize varieties to reduce dependence on late rains. Yields from the system can be nearly 3 tons of maize and over 500 kg of legumes per ha.

The Utilization of Biodiversity

The ecological processes described above are outcomes of the interactions between the great diversity of living organisms—millions of species of plants, animals, and microorganisms—that live on this planet. Biological diversity underpins ecosystem services and, in turn, is maintained by these processes, and yet human actions have caused the species extinction rate to climb to 1,000 times higher than the prehistoric rate of extinction (as estimated from fossil records).[27] Conserving this rich biodiversity is of intrinsic value; it is part of our heritage. But it is also of direct and indirect utility to human prosperity and survival. Ecosystem services directly support over one billion people living in extreme poverty, and yet biodiversity and ecosystem service conservation often remain low priority in policy making because their values are not fully captured by traditional economic markets.[28]

Biodiversity also has a more immediate utility as a source of technologies—particularly for agriculture and health; much of this valuable, but poorly understood, biodiversity is found in developing countries—countries that may lack the legal frameworks or technological expertise with which to capture the benefits of such genetic resources.

Conserving Germplasm

The histories of modern plant breeding (described in Chapter 9) illustrate the crucial importance of maintaining the collection of genetic resources (germplasm) representative of the wide range of variation in crop plants and their relatives. The Food and Agriculture Organization (FAO) estimates there are globally some six million accessions stored in more than a thousand gene banks. Some of the biggest collections in the developing countries are in China and India, and there are major collections in most of the IARCs.[29] *Ex situ* collections (i.e., those preserved as seeds in storage rather than as *in situ* living plants in the field) now comprise over seven million accessions.

Support for *ex situ* collections is now overseen by the Global Crop Diversity Trust founded in 2004,[30] which supports germplasm collections around the world, and the Svalbard Global Seed Vault, established in 2008, provides a permanent storage facility for crop biodiversity below the permafrost on Spitsbergen, an island in the Norwegian archipelago of Svalbard, in the Arctic Ocean. The seed collections are duplicates of national collections that can be replaced if lost.

Equally there is a need to preserve crops, and their close relatives, *in situ;* that is, in the natural and agricultural habitats in which they have evolved and developed. The wild relatives of crops are often grown in areas especially threatened by human activity. Recent estimates show that 6 percent of wild relatives of cereal crops (e.g., wheat, maize, rice, sorghum) are under threat of extinction, as are 18 percent of legume species (the wild relatives of beans, peas, and lentils) and 13 percent of Solanaceae species (the plant family that includes potato, tomato, eggplant, and pepper).[31]

In 2004, the International Treaty on Plant Genetic Resources for Food and Agriculture came into force to help ensure that crops and their close relatives that may carry beneficial traits for future breeding are preserved *in situ*. The aim is to protect the rights of farmers.[32]

Habitat Conservation

The major threat to biodiversity conservation worldwide is habitat destruction, a significant cause of which is the extension of agriculture and commercial forestry into terrestrial natural habitats. Pollution from fertilizers and pesticide also has a major impact on both terrestrial and aquatic biodiversity. How do we slow this destruction and conserve biodiversity?

Part of the answer lies in creating systems of protection for natural ecosystems, in particular forests, where so much of the biodiversity lies. But to date, only 12 percent of the planet is under some form of protection: about eighteen million square kilometers (km²) of protected land and over three million km² of protected territorial waters. Even protected areas are often poorly managed and suffer from pollution and climate change, irresponsible tourism, infrastructure development, and increasing demands for their land and water resources.[33]

Despite these daunting facts, sustainable intensification can make a positive contribution. As I mentioned in Chapter 3 greater intensification of production on the same land can ease the pressures on nearby natural environments. More specifically, appropriate technologies of intensification that greatly reduce extraction of water from natural environments and also reduce the pollution caused by excessive fertilizer or pesticide use will help to conserve biodiversity.

Farm Households

Finally, sustainable intensification is also about people. It depends crucially on how farm households (farmers and their families) take their environment—physical, biological, social, economic, and political—and wrest from it a sustainable living.

Farm households are usually complex entities, consisting of different generations, men and women and their offspring, and often extending through brothers and sisters and their families to embrace a considerable number of people. While agriculture has sometimes been seen as essentially a male activity (and it is still common for agricultural extension workers to be male and to interact only with men), in recent years there has been growing recognition among the development community of the critical importance of women in farm households.

Agnes Quisumbing and her colleagues at IFPRI distinguish three contributions that women make to food security: food production, access to food, and nutrition security.[34] Women account for half the production of food in developing countries. In Sub-Saharan Africa, where women and men customarily farm separate plots, it is as high as 75 percent, with the men concentrating their efforts on cash crops.[35] African women are also responsible for 90 percent of the work involved in processing food. Men play a greater role in food production in Asia and Latin America, but women are still major contributors. They are mostly responsible for transplanting, weeding, and harvesting in rice production, and they are normally the cultivators of home gardens and vegetable plots.[36]

Access to food additional to that produced on the farm depends on some form of income. A study carried out in Rwanda by Joachim von Braun and colleagues at IFPRI showed a close relationship between women's income and calorie consumption in the household, even though women earned considerably less than men.[37] Lack of incomes for women not only results in lower overall household income, it crucially affects children's access to food since women typically spend a high proportion of their income on food and health care for children.[38]

Nutrition of children is also closely related to health care. Children are better nourished where breast feeding is practiced and attention is paid to providing nutritious food for weaned infants and to the maintenance of hygiene. This is predominantly, and usually exclusively, women's work; yet if they are active as farm workers or earning off-farm income, there are severe constraints on their time.[39] Studies in Africa and Asia have shown women to be spending less than an hour a day in direct child care.[40] As Jeffery Leonard of IFPRI points out, "[W]omen's multiple roles in poor households perpetually conflict with each other. For example, increased time spent on out-of-household chores or non-household employment can directly reduce the time women have for child-rearing and other household duties. Conversely, the time and energy devoted to just gathering fuel and water increasingly has been recognised as a major impediment to efforts to increase women's contributions to food production, household income, or family welfare."[41]

The Nature of Livelihoods

A *livelihood* can be simply defined as a means whereby members of a farm household secure a living, but this definition embraces a great diversity of enterprise. In the developed countries, a household's livelihood is typically the product of one or two adults working a set number of hours for an employer and, in return, receiving a set wage. For rural people in the developing countries, a livelihood can be constructed from a range of production and income-earning opportunities, on and off the farm. Farming may be at the core of a livelihood, but nonagricultural activities play a critical role, especially as farm households move out of poverty. Income can be earned in a myriad of ways, sometimes augmented by remittances from town-based, or even overseas, members of an extended family. Diversity in the livelihood on and off the farm is a strategy for enabling farm households to cope with challenging and risk-prone environments, and with social and political pressures. Periodic disasters happen frequently enough to make resilience a major objective.

In 1992, when we were both living in India, Robert Chambers of the Institute of Development Studies (IDS) and I wrote a paper that laid out the concept of sustainable livelihoods (Box 6.4).[42] Part of our reason for writing this paper was to convey a sense of the complexity of the lives of rural people in developing countries. We also wanted to explain why understanding this complexity within a conceptual framework is important, particularly if rural development is to do more good than harm.

Box 6.4   The Dynamics of a Rural Livelihood[43]

A *livelihood* comprises the capabilities, assets (including both material and social resources), and activities for a means of living. The dynamics are illustrated in Figure 6.4, which is based on the work of Jeremy Swift of IDS.

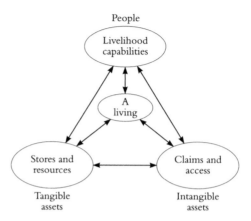

Figure 6.4   Components and flows in a livelihood.[44]

A livelihood may be constructed from one or—more commonly—a combination of these (with surprisingly few exceptions, developing country farmers do not rely exclusively on farming):

- Land on which crops or livestock are husbanded
- Natural resources—timber, fuelwood, wild plants, fish, and other wild animals—that may be harvested
- Opportunities for off-farm employment
- Skills employed on the farm in manufacture of handicrafts

Livelihoods are, in effect, agroecosystems at a higher level in the hierarchy. Thus they also embrace the same properties of productivity, stability, resilience, and equitability. For example, a livelihood is resilient when it can cope with and recover from stresses and shocks, and maintain or enhance its capabilities and assets, while not undermining the natural resource base. It is also equitable when the produce and income is apportioned among the members of the household to ensure women and children get their fair share.

Some idea of the range of rural livelihoods that exist in developing countries emerges from the 2009 Rwandan Comprehensive Food Security and Vulnerability

Analysis and Nutrition Survey described in Box 2.4 in Chapter 2 and discussed below in Box 6.5.

The power of livelihood analysis is its ability to identify the most vulnerable households and to gain some information about why they are vulnerable. This can be of value in targeting recipient households in emergencies. It can also determine the interventions that will help the most marginal households move out of poverty.

Box 6.5 Rural Livelihood Strategies in Rwanda[45]

For the 5,400 households surveyed, 33 percent of households reported one activity, 57 percent reported two, 10 percent reported three, and less than 1 percent reported a mix of four activities, giving an average of 1.8 activities per household. Of these activities, agriculture dominated, with 82 percent of households reporting agriculture as their main activity. This was followed by day labor (28), livestock production (23), and small trade (10).

From the data collected, a principal component and cluster analysis revealed nine categories, based on the similarity in mix and importance of activities (Figure 6.5).

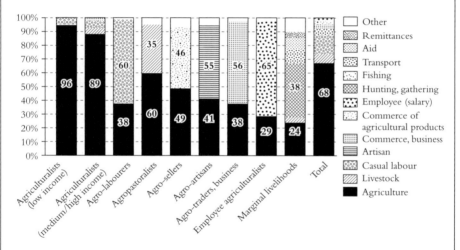

Figure 6.5 Activities contribution to livelihood (%) by livelihood group.[46]

The first two groups listed account for nearly half of the population of Rwanda. Agriculturalists, both low- and medium/high income, depend almost exclusively on agriculture for their livelihoods, with the lowest average number of income-generating activities.

Three groups (low-income agriculturalists, agro-laborers, and marginal livelihoods) were identified as being especially vulnerable. They account for 46 percent of the total

(continued)

Box 6.5 (*continued*)

> population and for a disproportionate amount of households in the lowest wealth quintile (76 percent) and in the poor consumption food score category (73 percent). A high proportion of the three groups had poor access to land (mostly less than 0.1 ha), fertilizers, and credit. Their households were also more frequently headed by a woman and/or an uneducated person, and they had a higher occurrence of stunting in children: 54.7 percent for agriculturalists (low income), 54.5 percent for agro-laborers, and 53.4 percent for marginal livelihoods.[47]

In and Out of Poverty

The structure and pattern of individual households and their livelihoods, as described above, determine whether people are able to escape from hunger and poverty. One of the most insightful set of studies of the nature of poverty in the developing countries and why people both escape from and fall into poverty was pioneered by Anirudh Krishna, a political scientist at Duke University.[48] He and his collaborators studied households in villages in India, Uganda, and Kenya based on the villagers' own perceptions and analysis of poverty. In Western Kenya, the researchers examined twenty villages in Siaya and Vihiga districts comprising 1,700 households.[49] The villagers were assembled in a series of meetings and asked to identify the following:

1. Which households in each village had managed successfully to escape from poverty (316; i.e., 19%) over the past twenty-five years and how many had fallen into abiding poverty (325; i.e. 19%). This relative lack of change reflects the general state of the Kenyan economy where there has been little improvement in GDP per capita in the past three decades and over half of the population remains poor.
2. How do households become poor? Poor health and health-related expenses were the reason overwhelmingly cited as responsible for households declining into poverty. Following these were heavy funeral expenses (often involving slaughter of the household's livestock), large family size, and land subdivision.[50]
3. How do poor households get out of poverty? What does an extremely poor household do with the first bit of money it acquires? What does the household do when a little more money flows in? This was repeated until a sequence of stages of the progress out of poverty was agreed on. The villagers identified fourteen such stages (Box 6.6).

Box 6.6   The Trajectory out of Poverty in Western Kenya[51]

- Acquire food
- Acquire clothing
- Repair house (primarily thatch roof)
- Primary education for children
- Purchase a chicken
- Purchase a sheep or goat

Beyond this point, households are no longer considered poor.

- Purchase local cattle
- Improvements to housing, furniture
- Secondary education for children
- Buy or lease land

Beyond this point, households are reconsidered relatively well-off.

- Purchase dairy cattle
- Buy land/plots
- Construct permanent houses
- Invest in a business

In India and Uganda, the stages were somewhat different. For example, in Rajasthan, India, the poverty cutoff was after the same first three stages as in Uganda, plus a fourth stage of repayment of private debt.

The critical element in helping households to escape from poverty was diversification of income by establishing links with the urban economy. Seventy-three percent of households who had escaped from poverty reported a member who had obtained a job, mostly in the private sector. In some cases they had established a craft or trade in a city, while a significant number (36%) had established a small business in the neighborhood of the village. Examples of the latter included retail shops; butcheries; selling agricultural products, fish, and paraffin; trading in timber, firewood, and charcoal; making shoes and bricks; weaving baskets; and brewing alcohol. Of the households who escaped poverty, 57 percent also diversified on-farm income through production of cash crops (e.g., staple cereals, tea, and sugar cane). Livestock acquisition also played a key role in the process (and loss of livestock was often critical in forcing households into poverty).

Significantly, acquisition of education, by itself, did not seem enough to escape from poverty. Social connections were important: Almost invariably, households that had made a successful entry into the urban economy, whether the informal or formal sector, possessed a privileged connection, such as an uncle or cousin who could take the new entrant under their wing.

Sustainable Livelihoods

In constructing and managing sustainable livelihoods, farm households balance the trade-offs between productivity, stability, resilience, and equitability, striving to create a situation where all of these are relatively high. This is not easy. Most smallholders live in heterogeneous, risk-prone environments. In the past two decades, we have gained a much better understanding of how poor people behave in the face of such adversity from studies of famines, including the experiences of aid workers attempting to mitigate famines while ensuring that immediate relief is tied to longer-term development.[52]

In a study by Susanna Davies, then at IDS, of the effects of drought on rural households in Mali in the 1970s, she describes a complex cycle of responses (Figure 6.6). In the first two years of the drought, food stocks are reduced, then surplus cattle are sold. Various coping strategies—such as collecting wild foods or temporary migration—are tried in the third year. In year 4, they may have to borrow food from their kin; however, in subsequent years, as the rainfall improves, they pay back the loans and reinvest in cattle.

Households will inherit and accumulate assets and will draw on these at times of adversity. But if, as has happened more recently in Mali, droughts are of longer

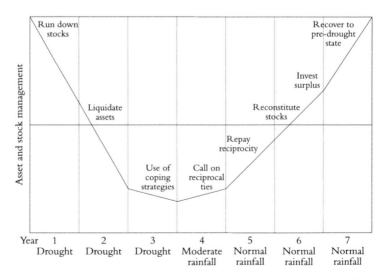

Figure 6.6 Responses of households to drought in Mali.[53]

duration and assets have been drawn down to the point at which households find it difficult to recover, then outside intervention is crucial to (1) donate food directly, (2) provide short-term credit, seed for sowing, or off-farm employment, and (3) encourage investment.[54]

Once basic survival is ensured, and conditions are secure, farmers will replace and improve the assets of their livelihoods. They may enhance resources by using their labor to construct terraces that will improve the stock of soil or by investing in a cart to take produce to market. Claims may be established by investing in a marriage or by giving presents. Access to information may be obtained by purchasing a radio. Capabilities can be increased by education or apprenticeship.

A common supposition is that basic human needs can be arranged in a strict hierarchy; people strive to satisfy one need before moving on to the next. In the hierarchy, food is assumed to be the need that has to be met first. But we now know that, except in a trivial sense, this is far too simplistic a model. All people, whether they be rich financiers or poor farmers, make both short- and long-term decisions: In the words of Susanna Davies, "They juggle between immediate consumption and future capacity to produce."[55]

The Rights to Land

Part of the clue to achieving a sustainable livelihood is the security of ownership or access to land. Security of access to land is usually a necessary, although not a sufficient, condition for investment in land improvement. Sometimes it may take the form of security of ownership or tenure, or it may be more complicated. Thus the long-term management of their livestock by pastoralists in Zimbabwe depends on a complex pattern of traditional rights that extend over a range of habitats, some of which may be rarely exercised.[56]

Without such rights, there is little incentive for a farmer to invest in improving the soil, water, and other resources on which agriculture depends. In the past, changes to the systems of land rights in Africa were not thought essential to increasing agricultural productivity for two main reasons: (1) land was abundant in Africa and (2) farmers rights to work the land as tenants based on customary systems of ownership were sufficient. Neither belief, however, is now valid.

In recent years the approach to the land rights issue has been to recognize customary tenure as equivalent to the freehold/leasehold culture by systems of law.[57] For example, in Eritrea in 1995, majority customary rights were transferred into lifetime rights under guaranteed government protection. Tanzania, Uganda, and Mozambique, too, recognize customary tenure, although in different systems, as legally owned.[58] An alternative approach has been to use local community organizations to help sort out tenure issues and to address the more difficult issue of inequality in land ownership.[59] Such systems may also reduce transaction costs associated with maintaining records and negotiating, contracting, and policing property rights, which in some cases may exceed the value of the land itself.[60]

The Challenges of Intensification

Better ownership of, or access to, land enables farm households to consciously examine the trade-offs between productivity, stability, resilience, and equitability and make informed decisions. There is plenty of evidence that small farmers can do this well. Given sufficient assets, they can produce sustainable agricultural systems that help to get them and their families out of poverty. But their decisions with regard to the larger environment are not necessarily wise or optimal. Livelihood activities can destroy or enhance the natural resource base. Livelihoods can contribute to desertification, deforestation, and soil erosion. They can also improve the environment by conserving wild environments, by harvesting natural resources on a sustainable basis, by planting trees, and by putting organic matter back into the soil. I discuss these issues further in later chapters.

In Chapter 7 I describe what kinds of technology are likely to be appropriate in the light of the needs of a sustainable agriculture.

7

Appropriate Technology

Innovation is the engine of social and economic development in general and agriculture in particular.

Calestous Juma, Professor of the Practice of International Development,
Harvard University[1]

In the first chapter I described the conditions under which a woman farmer in Kenya labors to feed herself and her children. She has to contend with numerous pests and diseases, has poor access to good seeds and fertilizers, and is afflicted by periodic drought. Yet, as I described, it does not have to be this way. Appropriate technologies and other interventions can provide answers; she can grow enough on her one hectare to feed her family and also grow some crops to provide cash for schooling and health care. With the right technologies she and her family can escape poverty and create productive and sustainable livelihoods.

Both . . . And . . .

There are, however, no magic bullets and no "one-size-fits all" solutions. We have to eschew dogmatic assertions that one form of intervention, especially one type of technology, is intrinsically best whatever the circumstances. Why?

First, because a Doubly Green Revolution embraces a range of outcomes. For example, we are seeking *both* increased production *and* a more sustainable environment, also a production that is *both* stable *and* resilient, and *both* these three goals *and* greater equitability.

Second, a Doubly Green Revolution must work in a wide diversity of environments: natural, social, and economic. We will need *both* microcredit *and* macroinvestment, *both* free-range *and* stall-fed livestock rearing, *both* biofuels *and* food crops, *both* water harvesting *and* large-scale irrigation, *both* organic *and* biotechnological solutions.

Our interventions, wherever and whenever they occur, must be *appropriate*. I discuss this concept more fully in this chapter; however, the *both . . . and . . .* theme runs throughout the book.

In this chapter I address the following questions:

- Where are the barriers to agricultural development and how can technology help to overcome them?
- What kinds of technologies are appropriate?
- How can developing countries develop indigenous scientific and technological capacity?

The Barriers

Developing country smallholders face considerable and, in some cases insurmountable, barriers to their farming. They suffer from lack of inputs, high fertilizer costs, inappropriate technologies, poor land tenure, lack of water, poor extension, poor markets, and poor infrastructure, to name a few of the barriers. It is a familiar list, but what is interesting about this list is how it changes from region to region, country to country, or even from one locality to another. It varies enormously depending on where you are in the developing world. The most important barriers in Mali are not the same as those in Malawi, and the barriers in western Kenya are a differently ordered set to those in eastern Kenya. Consequently, a crucial challenge is to identify the barriers to agricultural development from place to place, why they vary and, most important, how they can be turned into opportunities.

The barriers and opportunities also vary from crop to crop and indeed from one value chain to another; they are different for cassava and coffee, for example. I am using the term *value chain* here in a wider sense than is the conventional usage, to embrace the whole chain from the plant molecule at one end to the human molecule that is eventually affected in the man, woman, or child who consumes the food, at the other (Figure 7.1).

For example, in the case of the production of so-called Golden Rice, a new rice variety produced by genetic modification that provides consumers with vitamin A, the chain begins with a gene that codes for beta-carotene in a maize plant. This gene is then engineered into a rice plant and expresses beta-carotene, the precursor of vitamin A, in the rice grain. After the grain is harvested and ingested by humans, the beta-carotene is cleaved by an intestinal enzyme to produce molecules of vitamin A.[2]

For both place and value chain, the diagnostic challenges are to identify where precisely the best places to intervene are; what intervention is best value for money

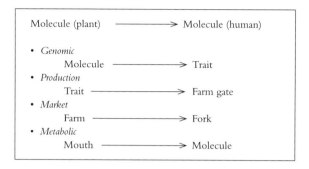

Figure 7.1 The long value chain.

or effort; what will have the most multiplier effect; and what is the most appropriate intervention.

The interventions may come in one of the following three principal forms:

- *technologies*, e.g., a new seed variety or agronomic system
- *financial investments,* e.g., a microcredit or crop insurance scheme
- *processes,* e.g., a system of quality control or of agricultural education

We need a new kind of diagnostic tool that will give us these answers from one place and value chain to another, which would point scientists, policy makers, and farmers in the right direction. In this chapter I focus on technologies. Other interventions are dealt with in Chapter 8 and elsewhere in the book.

What Is Appropriate?

In the twentieth century, there was a widespread view that increased agricultural production could be best effected by utilizing technologies from developed, industrialized countries, most of which have been produced over the past 150 years through application of modern physical, chemical, and biological knowledge.[3] These "conventional technologies" typically deliver desired products in a ready-to-use "packaged" form (e.g., a seed of a new crop variety, a bag of fertilizer, a synthetic pesticide, or a tractor). They were characteristic inputs in the Green Revolution. They are often global or at least regional in their appropriateness and are easy to market. While it is true that such technologies have frequently "worked," they have often been inappropriate, for one reason or another, to small farmers. Their benefit to the poorest communities in many circumstances has been limited. In some cases they have been unaffordable or required delivery systems, maintenance, or training to which poor farmers do not have adequate access. Conventional technologies may also have environmental side effects that are more severe or less manageable in developing countries than in the industrialized communities for which they were developed.

Partly for these reasons, there has been a greater interest, in the latter part of the twentieth century, in alternative technologies of various kinds. However, whatever the source and wherever the application, the important feature of a technology is that it is *locally appropriate*. By this I mean it has to be:

- effective
- readily accessible and affordable
- easy to use
- environmentally friendly
- serves a real need[4]

In terms of the agroecosystem properties described in Chapter 6, local appropriateness also implies seeking interventions that are productive, stable, resilient, and equitable.

In this chapter I first discuss the more selective and appropriate use of conventional technologies. I then examine the role of local traditional technologies, and how these have been combined with conventional technologies to produce intermediate technologies. Next I discuss the promise for emerging new platform technologies and how they too can be appropriate. Finally, I describe the role of agricultural research

Box 7.1   The Forms of Technology[5]

> *Conventional:* Technologies from industrialized countries developed through the application of modern physical, chemical, and biological knowledge, and delivered as products in a packaged form for a regional or global market; *e.g.. synthetic fertilizers and pesticides.*
>
> *Traditional:* Technologies that have been developed, usually over an extended period of time, by communities in developing countries to meet local needs; *e.g., home gardens, rainwater harvesting techniques.*
>
> *Intermediate:* Traditional technologies that have been improved in appropriate respects by their integration with modern conventional technologies; *e.g., intercropping, traditional treadle pump improved by engineers.*
>
> *New Platform:* New scientific "platforms" for innovation, based on advanced sciences, that have the potential to be developed simultaneously for the needs of the industrialized and the developing world; *e.g., genetically modified, drought- or pest-resistant crop varieties, mobile phones.*

and development and the nature of the innovation systems that developing countries need to create. I will base the discussion on the definitions in Box 7.1.

Selective Use of Conventional Technologies

Most conventional technologies were developed in the nineteenth and twentieth centuries, as products of large-scale industrial processes often reliant on petrochemicals. The development of cement and concrete in the nineteenth century made large-scale dams and irrigation systems possible. Widespread use of synthetic fertilizers was the result of the innovative production of ammonia from atmospheric nitrogen by Fritz Haber and Carl Bosch at the beginning of the twentieth century. Synthetic pesticides had their origins in the large-scale chemical processes developed before and during World War II.

Such technologies continue to be highly effective and relatively inexpensive in developed countries where they are widely used. But, as I argued earlier, they may not result in products appropriate to developing countries. For instance, the products are often costly in terms of energy and petrochemical inputs. Even where they can be produced cheaply on a large scale, transport costs make them expensive in developing countries, particularly in inland Sub-Saharan Africa. Conventional technologies may also have deleterious local environmental effects. Large-scale dam construction and impoundment can displace communities and destroy biodiversity, while synthetic fertilizers, if used to excess, can pollute water sources. Synthetic pesticides can be very destructive to wildlife and harmful to humans if they are not applied correctly. In each instance, the poor tend to suffer disproportionately from these unintended effects. Pesticides are also prone to the evolution of resistance.

Precision in Application

For these various reasons there has been a move, since the 1960s, to use conventional technologies in a more sparing and selective manner. The high cost of fertilizers has spurred more precise applications in the developed countries: using controlled-release fertilizers, fine-grained soil analyses, and variable application on individual fields employing tractors fitted with geographic information systems (GIS) devices.[6] This is not likely to be appropriate for smaller-scale farmers in developing countries, but even small farmers can substitute more selective placement of fertilizers instead of the widespread practice of broadcasting them.

One example is fertilizer microdosing developed in Niger to address problems such as the impact of drought on food supplies. Each microdose consists of a 6-gram mix of phosphorus and nitrogen fertilizer, which just fills the cap of a soda bottle—an item that is easy to obtain. The cap of fertilizer is then poured into each hole before the seed is planted; it equates to using 4 kilograms (kg) of phosphorus per hectare, three to six times less than used in Europe and North America, but still very effective.[7] Decreasing rates of fertilizer application by over one sixth leads to an average 55 to 70 percent increase in millet yield and at lower cost.[8] The International Crops Research Institute for the Semi-Arid Tropics (ICRISAT) has also shown that by reducing the amounts of nitrogen (N) and phosphate (P) fertilizer used by farmers, crops are more able to absorb water.[9] This is a good example of where an appropriate technology is a win–win proposition.

The same principle can be applied to herbicide use. Far too often, herbicides are sprayed relatively indiscriminately, killing not only weeds but other wild plants and sometimes damaging the crops themselves. Yet Africa suffers from serious weed problems: one is *Striga* (or witchweed), a devastating parasitic weed that sucks nutrients from the roots of maize, sorghum, and other crops (Box 7.2).

In principle and indeed in practice, there are a number of ways to make conventional technologies and their products more appropriate for developing countries, either by limiting or integrating their use so as to reduce their cost and their environmental footprint. Nevertheless, many conventional technologies remain inappropriate for a range of reasons. In some cases it makes more sense to make use of traditional technologies, which are adapted to the local environment from the outset.

Box 7.2   Control of *Striga* with Minimal Herbicide Use

Striga infests as many as 40 million hectares of farmland in Sub-Saharan Africa and causes yield losses ranging from 20 percent to 100 percent. The livelihoods of more than 100 million people are affected, causing $1 billion in annual crop losses.[10]

Striga is readily controlled by a herbicide, imazapyr, but this tends to damage or kill the crop. Recently, a mutant gene in maize has been discovered through tissue culture that confers resistance to the herbicide and is being bred into local maize varieties. The maize seed is dipped into the herbicide before being planted. When the parasitic weed seeds germinate, they attach to the maize roots to suck out the nutrients and in the process take up the systemic herbicide from the maize. The *Striga* is killed, while allowing the maize to grow with little or no impact from the herbicide. This technology has allowed maize yields to increase by 40 to 80 percent in comparison with traditional maize varieties.[11]

The Relevance of Traditional Technologies

Traditional technologies are approaches to problems that have been used by people for hundreds, if not thousands, of years. They can be thought of as having "stood the test of time." Examples range from scarecrows to sophisticated home garden systems. Some have clearly worked and still work today; for others, there is no scientific evidence that they are effective.

One ubiquitous practice is the cultivation of traditional crop varieties or of "wild" plants. Their origins may lie in the early days of agriculture. They may be domesticated to an extent and improved by farmer selection but usually retain their essential characteristics. Examples include the local rices of northern Thailand (Box 7.3).

The Development of Intermediate Technologies

Intermediate technologies have been defined in a number of ways, but fundamentally they are traditional technologies improved in appropriate respects by their integration with modern conventional technologies. The most effective are those combining the best of conventional and traditional technology.

Box 7.3   Local Rices in Northern Thailand

Although rice is a self-pollinating crop, it experiences 0 to 1 percent out-crossing.[12] As a result, many areas of Asia maintain a rich genetic diversity of rice that can be utilized by farmers, for example to counter pest or disease outbreaks, providing outcomes that do not involve a trade-off between conservation and yield or quality or profit.[13] Most of the rice lands of East and Southeast Asia are now under modern varieties, but local varieties remain important in rural areas isolated from national or international breeding programs.[14]

In northern Thailand, the local varieties and landraces are largely sticky (or glutinous) rices preferred by consumers in the region. The genetic diversity present in the landraces allows farmers considerable potential for selection, in particular, against the rice gall midge (*Orseolia oryzae*), a serious pest that deforms the growing point on the rice plant, making it unable to develop a panicle and thus produce grain. There is a growing incidence of gall midge infestations at elevations higher than their normal range (400m to 800m) which may be connected to climate change.[15] The best local varieties can yield twice as much as modern varieties that are not resistant to gall midge.[16]

Such landraces can exhibit a high degree of plasticity that permits them to adapt to new challenges.[17] Farmers cultivating these varieties aim to increase production and the returns from farming while maintaining high in situ genetic diversity.[18]

Such technologies also comprise a wide range of applications, extending from the various ecologically based systems of intercropping (described in Chapter 6) to mechanical devices; some are simple, others more sophisticated.

A good example of a relatively simple intermediate technology is the development of an affordable and reliable treadle pump. For many years, engineers have been developing pumps that allow farmers to replace the arduous task of lifting irrigation water from shallow wells by bucket. Oil-driven pumps are expensive to purchase and to run. The modern treadle pump is ideal in many respects; it is efficient and easy for farmers to use and maintain and is virtually fool-proof. It relies on human rather than oil or electric power and is also relatively cheap due to a combination of public subsidies with private manufacture and servicing, and community involvement (Box 7.4).

Box 7.4 The Treadle Pump[19]

First innovated by local people in Bangladesh in the early 1980s, the treadle pump is a simple machine that a farmer can stand on and pump with his or her feet to irrigate a field. The pump produces the suction or pressure needed to raise water from a natural source or dug well.

The technology has been improved on by a number of enterprising engineers, producing a variety of effective and easy-to-maintain designs that can be used to irrigate up to 2 acres of land without any motor or fuel. International Development Enterprises (IDE) began in 1984 by refining the pump design they encountered in Bangladesh. After successfully creating a market and supply-chain for the product there, they have expanded the approach to countries across Africa and Asia.[20]

Another design, known as the "Super MoneyMaker" pump, was designed in 1998 by the NGO KickStart for farmers in Kenya who needed a device that could pump water uphill, useful when water sources are at the bottom of a hilly plot or for filling overhead tanks. As of 2011 the pump has been sold to over 189,000 farmers in Kenya, Tanzania, and Mali.[21]

Pumps like these help farmers by extending the traditional growing season and expanding the number of crops that they can cultivate. For example, Nazrul Islam, a farmer in Bangladesh, applied for a microloan of about $20 to cover the costs of digging a well and buying a treadle pump. He was able to repay the loan and buy additional land and livestock within the first year of use.[22]

Advanced Technologies for the Future

Finally, new scientific "platforms" for innovation have recently been developed, derived from fundamental discoveries in the physical, chemical, and biological sciences. These have the potential to be developed simultaneously for the needs of the industrialized and the developing world, and in the right circumstances can be appropriate for the needs of developing country farmers. The new platform technologies, discussed below, include the following:

- Information and communication technology
- Nanotechnology
- Biotechnology

While these platforms have their origins in advanced research laboratories, developing country scientists and especially those in the emerging countries are rapidly exploiting their properties in a wide range of appropriate applications.[23]

Information and Communication Technology

Traditionally, information and communication technology (ICT) in developing countries has been based on indigenous forms of storytelling, song, and theater, the print media, and radio. Such technology has often been used instrumentally in extension programs to inform farmers about new varieties or other technologies. Radio soap operas about rural life based on the BBC's long-running *The Archers* are popular in Rwanda (*Urunana* [Hand in Hand]) and Afghanistan (*Naway Kor, Naway Jwand* [New Home, New Life].[24] The latter is so popular, with 35 million listeners (over 70 percent of the population), it was thought to have been partly responsible for the reluctance of the Taliban to ban radio, despite their bans on television, music, and dancing. The thrice-weekly aired soap is set in the village of Bar Killi and tackles the dramas of blood feuds, forced marriages, landmines, and opium addiction as well as gender and humanitarian issues.[25]

Most modern ICT, however, is based on electronic communications and the Internet. The widespread use of mobile phones has been the most dramatic outcome of the ICT revolution. It is a good example of an imported technology that has been widely and successfully adopted in developing countries. Mobile phone use has exploded in recent years, across Asia and Latin America, and especially in Africa. In 1995, there were 650,000 mobile phone subscribers in Africa; by 2008 there were 350 million subscribers.[26] Although not specifically designed for, or with the involvement of, poor people, they have rapidly seized on the potential of mobile phones. Interestingly, in many cases, the primary advantage to poor people is unrelated to the key purpose for which it was designed; namely, mobility.

New users are employing mobile phones in an extraordinary variety of situations. Mobiles are being made available to virtually everyone, including those who are poor, remote, and/or illiterate. They can be shared between individuals and households

Box 7.5   Mobile Phones for Farming[27]

> Expanding mobile phone use has begun to remove many long-standing obstacles for farmers in developing countries. Mobiles can be used to find out the location and prices of inputs, such as seeds and fertilizers, and of crops being purchased by market dealers (see Chapter 8). As a consequence, farmers have been able to get much better prices for their products, in some cases in anticipation of the harvest.[28]
>
> Mobiles are also ideal sources of technical agricultural information. New services such as *AppLab*, run by the Grameen Foundation in partnership with Google and the provider MTN Uganda, are allowing farmers to get tailored, speedy answers to their questions. The initiative includes platforms such as *Farmer's Friend*, a searchable database of agricultural information; *Google SMS*, a question-and-answer texting service; and *Google Trader*, a SMS-based "marketplace" application that helps buyers and sellers find each other.[29]

and made public at booths as a relatively cheap pay-service. Their use has not only improved communication and the flow of information in developing countries, but has spurred a variety of novel ways of pursuing agricultural objectives (Box 7.5).

Alongside mobile phones, the Internet is a powerful communication tool for connecting people and groups and for accessing up-to-date information from around the world. In the town of Veerampattinam in the south of India, the M.S. Swaminathan Research Foundation has put up loudspeakers to broadcast information derived from the Internet, such as weather and ocean-wave forecasts, agricultural and fishing techniques, market prices, government programs, and local bus schedules. This allows citizens to access accurate information without even touching a computer or phone.[30]

Internet access in the developing countries is growing fast—with usage increasing between 2000 and 2010 by over 2,300 percent in Africa, over 1,000 percent in Latin America, and over 600 percent in Asia—and the use of computers for education, communication, and information processing is steadily expanding.[31] This will be further helped by improvements in infrastructure, such as the new 17,000-km-long underwater fiber-optic cable installed by the African-owned company Seacom along the eastern coast of Africa. The cable, which went live in July 2009, creates a much-needed digital link between Eastern Africa, South Asia, and Europe, and will bring higher-speed, lower-cost broadband to millions of users.[32] The key challenge now is to improve ICT access by the rural poor.

The Potential of Nanotechnology

Nanotechnology involves the manipulation of matter at atomic and molecular levels (i.e., at a scale of a billionth of a meter) to produce a great variety of materials and devices. At this scale, materials often have unique characteristics.

While the United States is currently leading the way in nanoscience research, many emerging countries have also established strong programs. Applications may include energy storage, production, and conversion; disease diagnosis; drug delivery systems; air and water pollution detection and remediation; and food processing and storage. Most of these applications are being developed for wealthy countries, but many of them have potential applications in developing countries.

One of the most promising applications is the use of nanomembranes, nanosensors, and magnetic nanoparticles for *water purification*, especially of irrigation water—allowing for desalination, detoxification, remediation, and detection of contaminants and pathogens. Research is in the early stages and many of the products are still too expensive for developing country applications, but the ability to immediately treat even very contaminated water to a high quality, at the source, without the use of electricity, heavy chemical dosages, or high pressure, together with the scale of predicted water shortages, makes it an area worth pursuing.[33]

Biotechnology

Traditionally *biotechnology* has been associated with the centuries' old practice of fermentation used in the making of bread, beer, and spirits. A modern, more inclusive, definition encompasses any technological application that uses biological systems, living organisms, or their derivatives to make useful products or processes.[34] Driving modern biotechnology is the revolution in cellular and molecular biology that occurred in the second half of the twentieth century.[35] In particular, it uses our knowledge of DNA and RNA to identify the genetic basis of useful traits in animals and plants.

Modern plant breeding—derived from the discovery by Gregor Mendel of the particulate nature of inheritance and developed over much of the last century—has transformed agricultural production through the selection and crossing of crop varieties (see Chapter 9). However, it is often an uncertain and lengthy process: Discovery of mutations with desirable properties is serendipitous, and incorporating them into new varieties involves many crop generations and hence years of careful breeding, sometimes with limited success.

An example of successful, conventional breeding is the development of "quality protein maize" (see Chapter 9). A discovery in the 1960s of maize mutants with high levels of desirable amino acids started a breeding program that, after several decades, has successfully incorporated these desirable traits into better maize varieties for developing countries.[36] By contrast, biotechnology makes this process speedier and more effective.

The most prominent form of crop biotechnology is the production of genetically modified (GM) crops. The process is described in detail in Chapter 9. In brief, a gene of potential usefulness is isolated and then inserted into the cell of a crop plant by a process of transformation, either by using a naturally occurring bacterium (*Agrobacterium tumefaciens*) that infects plants or by coating gold particles with the gene and shooting the particles into crop plant cells with a gene gun. The transformed

Box 7.6 *Bt* Cotton

Bt cotton contains a gene from a common soil bacterium, *Bacillus thuringiensis,* that produces an insecticidal protein in the plant that kills major cotton pests such as the cotton bollworm (*Helicoverpa armigera*) and pink bollworm (*Pechnophora gossypiella*). These are serious worldwide pests of cotton, maize, vegetables, and other crops; their long-distance migrations and the variety of their hosts can lead to large outbreaks. The cotton bollworm is estimated to cause annual crop damage of as much as $2 billion across the globe.[37]

Nearly thirteen million "small and resource poor" farmers are now growing *Bt* cotton.[38] Burkina Faso, the largest cotton producer in Africa, adopted *Bt* cotton on a commercial scale in 2008. A year later it was being grown on over 100,000 ha, with yields up to 50 percent higher than conventional cotton and with the number of pesticide sprays reduced from an average of eight to at most two.[39] A World Bank review of the benefits of *Bt* cotton in Argentina, China, Mexico, India, and South Africa showed increases in yields of between 11 percent and 65 percent, and increased profits as high as 340 percent with significantly reduced use of pesticides and pest management costs (Table 7.1).[40]

Table 7.1 Economic and environmental benefits of growing *Bt* cotton in developing countries

Benefit	Argentina	China	India	Mexico	South Africa
Added yield (%)	33	19	26	11	65
Added profit (%)	31	340	47	12	198
Reduced chemical sprays (number)	2.4	—	2.7	2.2	—
Reduced pest management costs (%)	47	67	73	77	58

Source: World Bank[41]

cells are then cultured and grown into whole plants that are tested in the green-house to ensure that the transferred gene, the transgene, functions properly. Not all transgenic plants will express the trait or gene product well. But once the trait is stable, it can be bred using conventional plant breeding methods into cultivars with adaptation to the environmental conditions where the crop is produced.

One example is a crop, such as cotton, engineered to express a bacterial gene that controls certain insect pests, so reducing the need for harmful synthetic pesticides (Box 7.6).

However, despite these well-documented benefits, recombinant DNA or GM technology is controversial and has attracted considerable opposition, especially in Europe. I discuss this more fully in Chapter 9.

The Returns to Research

All of these technologies are the products of "research" whether by farmers themselves, by a research institute or university, or by the big life science companies. We know from the history of the Green Revolution (Chapter 3) just how important agricultural research can be for agricultural development. It was not the whole cause of the revolution's success, but without it few, if any, of the achievements would have been realized. Andrew Dorward, of the School of Oriental and African Studies (SOAS), and colleagues found that in areas where the green revolution was considered successful in transforming agriculture and reducing poverty, local research and extension almost always took place.[42]

Calculating rates of return from agricultural research is notoriously difficult, but a 2000 meta-analysis suggests median rates of return for different regions of the world to lie between 44 and 50 percent.[43] Rates of return are much higher for research alone, but lower for research and extension and lower still for extension alone. There is no evidence that these rates of return are declining over time.[44]

The benefits have been especially high in Eastern and Southern Africa, with economic rates of return of 117 percent for sorghum, 30 to 80 percent for rice, 51 percent for wheat, and 29 percent for livestock. Calculations by Joachim von Braun and his colleagues at IFPRI suggest that a doubling of investment in public agricultural research in Sub-Saharan Africa would increase growth in agricultural output from 0.5 to 1.1 percent and reduce poverty by 282 million people. The effect on East and Southeast Asia would be increased growth from 0.5 to 1.55 percent and poverty reduced by 204 million.[45] Yet public spending on agricultural research in Africa is 0.7 percent of agricultural GDP; by contrast developed countries spend an average of 2.5 percent of agricultural GDP.[46] Despite high expectations that the private sector would fill the gap, this has not occurred: Funding from the private sector currently contributes only 2 percent of African agricultural research.[47]

Research Systems

Traditionally, research systems in developing countries have been based on those established under colonial rule, whose main aim was the commercial expansion of export crops. Some of the research institutes were of a very high quality, staffed by first-class researchers and delivering results of practical use. A good example was the Rubber Research Institute of Malaya whose research helped to keep natural rubber competitive with synthetic rubber. I began my career in the Agricultural Research Service of the Colony of North Borneo, later the state of Sabah in Malaysia. I was the entomologist and with my colleagues in soil science, agronomy, breeding, and plant pathology worked on such crops as cocoa, oil palm, and manila hemp for the nascent estate crop industry, although we also carried out research on some of the problems facing the predominant smallholder farming: identifying appropriate soils, finding solutions for pests and diseases, and providing good-quality planting material.[48]

In the newly independent nations, research relied heavily on donor funding but continued with the aim of developing large-scale commercial farming as a means of promoting economic growth. The food crisis of the 1960s and the impact of the Green Revolution helped to shift the focus of research to food staples and to the needs and problems of smallholder producers. However, since the 1980s research expenditure has slowed, and the number of researchers has increased, resulting in smaller budgets per researcher (the average expenditure per researcher in 2000 was 50 percent lower than that of 1970) and increasing inefficiency.[49]

Global Innovation

At the same time there has been a transformation in the way global scientific innovation is carried out and how it bears on the problems of development. Jeff Waage, Sarah Delaney of the London International Development Centre, and I recently wrote a book describing these changes.[50] We contrasted the old linear model of development research with a more sophisticated global model. Under the linear model, new knowledge is gained in a research laboratory; for example, knowledge of the genetics of disease resistance gained from basic research on a laboratory animal. This may lead to translational research on livestock to determine whether similar genes exist that convey useful resistance. If this research is successful, industry may use it to develop products, in this case using livestock breeding methods to incorporate genes conferring resistance into specific commercial breeds for sale to farmers.

However, there is growing recognition that scientific innovation is not always a linear process; it often involves interplay among basic, translational, and applied research stages. Basic research may not be the driver (Figure 7.2).

Modern research typically involves a diverse system of players and institutions that make up what is called a "science innovation system." The players may come from companies, universities, government, and civil society. In addition to scientists, other stakeholders include policy makers, banks, and investors. Moreover, science no longer functions in isolation at a national level. Scientists from around the world collaborate with each other to access the best expertise, resources, and partnerships, and funding and institutions have adapted accordingly.[52] This is as true of agricultural research as it is in other fields. It can be a very powerful system of innovation, but it poses a challenge for the developing countries: How do they access the results of global innovation systems? Inevitably it means building up their own capacity, primarily through the development of local national innovation systems (Box 7.7).

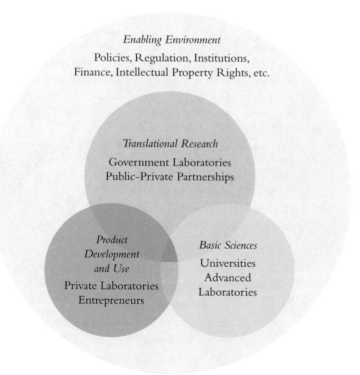

Figure 7.2 A modern science innovation system.[51]

Box 7.7 National Innovation Systems[53]

At the heart of national innovation systems are "networks of institutions in the public and private sectors whose activities and interactions initiate, import, modify and diffuse new technologies."[54] They involve universities and research institutes; small and large companies; financial and legal institutions; industrial associations; government ministries; NGOs of various kinds; and individuals as both stakeholders and innovators in their own right. These systems generate seven key functions:[55]

- Competence building
- Financial support
- Provision of regulatory frameworks and measures
- Facilitation of information exchange
- Stimulation of demand and creation of markets
- Reduction of uncertainties and resolution of conflicts

Successful innovation systems also depend on the creation of clusters of competing and complementary companies and research institutions. A key component is the existence of a vibrant small- and medium-sized enterprise (SME) community with access to venture capital, either indigenous or foreign. Banks and other financial institutions are critical, as are supportive economic incentives. Governments are important, providing funding for technology incubators, export processing zones, and production networks, as well as helping with skills training.

Allied to SMEs is the need for entrepreneurs. These may be scientists or engineers with strong experience in the practicalities of applying inventions to real-life problems. But frequently, they are individuals with a business background, an appreciation of the market opportunities, and a capacity to understand the essentials of the technologies on offer. In other words, they have a flair for spotting winners and seeing the process through from the embryonic stage to production and sale. These entrepreneurs can range from small farmers to senior scientists or engineers in companies or universities.

Consultative Group on International Agricultural Research

The major global player in agricultural research for development is the Consultative Group on International Agricultural Research (CGIAR). It was created in 1971 as an informal association of donors to provide financial support to four international agricultural research centers that the Ford and Rockefeller foundations had established over the previous decade. In the 1980s and 1990s, new centers joined the CGIAR, enjoying the benefits of core research funding provided by the donor group. These extended the research activity of the CGIAR beyond food crops to include livestock; living aquatic resources; forestry and agroforestry; water resources; agriculture capacity building and policy.

Since their establishment in the 1960s up until 2001, the centers, supported by the CGIAR, have spent about $7 billion (in 1990 U.S. dollars). Cost-benefit studies indicate that, by conservative criteria, including a very limited set of directly attributable impacts, the overall economic benefits attributable to CGIAR research have been at least double the research costs. Another cost-benefit meta-analysis using broader criteria estimated that by 2011, the benefits would be as much as 17 times greater.[56]

Sixty-five percent of the global area under improved varieties of the ten most important food crops is planted with varieties derived from the CGIAR-funded research. Without the CGIAR contribution, it is estimated world food production today would be 4 to 5 percent lower and in developing countries 7 to 8 percent lower.[57]

Fundamental to the success of the CGIAR centers has been the development of partnerships with national scientific centers that form the core of the national innovation systems. Box 7.8 describes the way in which the Centro Internacional de Mejoramiento de Maíz y Trigo (International Maize and Wheat Improvement Center; CIMMYT) has partnered with the Kenya Agricultural Research Institute (KARI).

Box 7.8   The Partnership between CIMMYT and KARI

The International Maize and Wheat Improvement Center (Centro Internacional de Mejoramiento de Maíz y Trigo; CIMMYT) is in effect a large-scale international research and training consortium with offices in Asia, Africa, and Latin America and over 600 staff. It has built up considerable expertise on maize crops; maize varieties, developed by CIMMYT and its partners, are now planted on nearly half of the area sown to improved varieties in nontemperate areas of the developing world. CIMMYT's gene bank holds 25,000 unique collections of native maize races.

To effectively bring this high-level knowledge to the national and local levels in Kenya, CIMMYT has been working with the Kenya Agricultural Research Institute (KARI). Established in 1979 by the Kenyan government, KARI brings together national research and dissemination efforts in food crops, livestock management, and land and water use.

In one project CIMMYT researchers found maize plants resistant to the maize stem borer in the center's gene bank, in maize seed originally from the Caribbean. Working with Kenyan farmers, KARI scientists used conventional plant breeding techniques to cross the introduced varieties with maize varieties already adapted to the conditions found in eastern Africa, selecting for traits that the farmers valued. KARI then facilitated the testing of the new varieties through the Kenya Plant Health Inspectorate Services

(continued)

Box 7.8 (*continued*)

and helped to create a dissemination process involving local seed distributors and extension agents. One of the new varieties, named *Pamuka 1*, is now being distributed to farmers. It is not only high yielding but resistant to stem borers and drought tolerant.[58]

It is a good example of a global innovation system in operation, driven by the applied problem defined by KARI with CIMMYT acting as a translational research partner. Nevertheless in Sub-Saharan Africa, as of 2000, only around 10 percent of cropland was planted with improved varieties, and technology adoption is only in the tens of thousands of hectares.[59]

The CGIAR has not been without its challenges. The critical core funding from donors has declined over time and centers have operated independently, missing opportunities for synergy. More important, investment in the CGIAR centers has not been matched by investment in national research capacity in many regions, particularly Africa. This has led in some cases to research resources flowing disproportionately to CGIAR centers, as well as to a brain drain from national programs into better-supported center jobs, undermining the partnerships between centers and national scientific research systems. All of these problems are the target of an ambitious plan, launched in 2008, to reform the CGIAR system, bringing centers and partner institutions together around a single strategic results framework with more sustainable funding.[60]

What Makes Technologies Appropriate

As I argue in this chapter, all kinds of technology can be appropriate. Each needs to be considered in its own right, irrespective of source, but with sufficient attention to the outcomes in terms of productivity, stability, resilience, and equitability. There are no magic bullets. One technology is better than another depending on the environmental and socioeconomic circumstances. The question of scale is also important. A technology may be successful in terms of the chosen criteria on a small scale—in an experimental plot or on a farm or cluster of farms—but the real test is whether it can be scaled up to benefit hundreds, if not thousands, of farmers and still retain its desirable characteristics. Much depends on the efficiency and indeed, appropriateness, of the "enabling environment," which includes the provision of research and extension services. In Chapter 8 I focus on other aspects of the enabling environment for technology, in particular the provision of markets, both for outputs and inputs. Without these, technology will not deliver.

8

Creating Markets

I did not realize how much I was being cheated until I took part in this sale where every single bag weighs exactly 50 kg. . . . It's good to know I can get more money for my maize.

Lydia Myantekyiwaah, a farmer from Ghana's Ashanti region,
participating in the Purchase for Progress program[1]

The success of a Doubly Green Revolution does not just depend on appropriate technologies but also, as the original Green Revolution demonstrated, on the larger policy and institutional context. New technologies will be produced and farmers will take them up only if there is an enabling environment that ensures investors and farmers get a reasonable return. More broadly, the poor will escape hunger and poverty, and the virtuous cycle will take off and accelerate, only if the conditions are right.

In this chapter I address these questions:

- What is the nature of an enabling environment and why is it important?
- Why are markets so crucial to agricultural development?
- How can small farmers take advantage of markets?
- How can small farmers be supplied with microinsurance and microcredit?
- How can connectivity to markets be improved?

The concept of an *enabling environment* has been defined in many ways, but here I take it to be the sum total of the macroeconomic policies that favor markets and trade, the provision of inputs and related physical infrastructure (e.g., roads, irrigation), and social infrastructure (e.g., education, research), together with the accompanying institutions and regulations.[2]

In Chapter 4, I described the evolution of the political economy of food security and agricultural development. Today economic liberalism is the predominant consensus. There is a strong belief that free market systems are best placed to deliver development and to reduce hunger and poverty. Within this consensus, the key international and national policies include:

- Economic policies that do not discriminate against agriculture
- Liberalized private markets for farm inputs and outputs
- Efficient rural financial institutions, including adequate access by farmers to credit, inputs, and marketing services
- Land tenure reform or redistribution as appropriate
- Adequate rural infrastructure
- Investments in rural education, clean water, health, nutrition, and family planning
- Attention to the needs of women and of ethnic and other minority groups, and securing their legal rights[3]

Although listed separately these policies are intimately interconnected, feeding off each other in the virtuous circle of agricultural and rural development. Thus while liberalized markets are necessary prerequisites for significant agricultural growth, they are not sufficient to create the other goals of the Doubly Green Revolution; namely, increased stability and resilience and greater equitability of access to the benefits. This is well illustrated by the example of Madagascar (see Chapter 4).

Accelerated growth in agricultural output cannot be maintained without adequate investments in rural infrastructure and in agricultural research and extension. Equally important, liberalization will not contribute significantly to poverty alleviation and the reduction in inequity, at least in the short term, unless the poor are deliberately targeted. Essential steps include the creation of employment for the land poor and landless, increased production on small- and medium-sized as well as large farms, and provision of nearby input and output markets.[4] Important, too, is recognizing where the rural poor are located and attending to regions of lower agro-climatic and resource potential, not just the best.[5]

Without such investment the results of liberalization policies will fall short of expectations; as a result they may justifiably set governments and people against market-oriented approaches. Some of the needed investments are public goods and provided by governments, but others are the province of the private sector. Which investments are needed and which are most cost-effective can be determined by calculating total factor productivity (TFP), as for the Green Revolution in Chapter 3 (see Box 3.4).

The Role of Markets

Given the singular nature of the core technologies, it is perhaps not surprising that the calculation of TFP for the Green Revolution showed that research and extension made the most important contributions. Today, it is likely to be a different set of factors. Multiple and varied technologies are needed and, as the example of liberalization in Madagascar demonstrates (Box 4.5), putting them into effect appears to be as much about the nature and efficiency of markets, both the markets where farmers sell their products and those where they buy the inputs they need to produce them.

Markets are also key to reducing poverty; only through access to markets can poor farmers increase the income from their labor and lift themselves and their families out of poverty. Yet, most poor farmers are not involved significantly in agricultural markets.[6] Smallholders, in particular, often have little contact with the market and hence a poor understanding of, and ability to react to, market forces.

In the colonial era, markets were well developed, but targeted on the key high-value commodities demanded by the colonial powers: tea, coffee, cocoa, sugar, rubber, jute, indigo. Much of the production and trade was in the hands of large private traders, although there was significant government involvement. After independence, the private sector was usually excluded from marketing. Instead, parastatal marketing boards were created that offered farmers guaranteed prices for their products. In some countries, the system incentivized farmers to increase production, while in others its aim was to guarantee cheap food for urban populations and act as a type of taxation on small farmers.[7]

Government intervention in food markets is frequently inefficient; the marketing costs of private traders in India are 30 percent lower than the costs incurred by the parastatal, the Food Corporation of India. Government intervention can also exacerbate food price distortions and fluctuations: In Pakistan, the real prices of commodities without government intervention remained more stable than the prices of commodities with some degree of government intervention.[8] For a variety of reasons, government intervention tends to create uncertainty.[9] Officials often lack relevant and accurate information; as a consequence, policies can be reactive and short-term, focused on the immediate situation and lacking coherence with other policies.[10]

Market Reforms

By the early 1980s, it was evident that the parastatal model was costly (and in some instances corrupt), had contributed to real declines in producer prices, and had failed to promote agricultural growth. The World Bank embarked on a program of encouraging reform. Marketing of traditional export crops such as cocoa, coffee, and cotton was liberalized and international stocks and price management mechanisms were dismantled. But this led to increased short-term variability in commodity prices.[11] Markets were also opened to the private sector for major food staples, and barriers on imports of foodstuffs were reduced. Many smallholder farmers benefited from these changes; many others—especially those in the more remote, poorly connected areas—lost reliable markets and incomes. Agriculture became riskier.[12]

At the same time, markets were changing under the influence of a myriad of factors—urbanization, population growth, increasing per capita incomes, changes in consumer preferences, the modernization of food processing and retailing, and improvements in transport and communications infrastructure.[13] Adding to this mix, developing country farmers were becoming sources of commodities for large, multinational agrofood companies. Global trade in agricultural goods has grown rapidly; by 2005, Asian agricultural imports, dominated by India and China, made up a quarter of global trade, mostly for cereals, oil crops, meat, and horticultural products.[14] One consequence has been the growing importance of quality standards for agricultural produce.

As a result of these changes agricultural produce markets are now highly differentiated, ranging from village markets selling locally produced, locally consumed products to global markets selling packaged, off-season vegetables. This spectrum of markets offers new opportunities for smallholder farmers, but at the same time poses heightened risks and new and difficult barriers to surmount.[15]

Improving the Benefits from Value Chains

In this changing environment with both greater participation and increased concentration of market power, it is important to try and shift more of the value of the produce to the hands of the farmers who produce it.

A *value chain* in practical terms is a sequence of production, marketing, and consumption where at each stage of the process there is an opportunity for value to be added to the commodity and this is passed to, or captured by, those involved in the process at that stage. In the last chapter I described a value chain, extending from the cell in the crop plant through to the cell in the human being consuming the crop product (see Fig. 7.1). The part of this long chain of special importance to small farmers is the so-called supply chain typically referred to as "farm to fork"—the chain of processing, transporting, buying and selling that extends from the grain or other commodity after it is harvested to when it is on a consumer's plate or in his or her bowl. For the average farmer in a developing country, this is a process where the rewards are low and the risks are high. Often the situation is made worse by the high numbers of middle-men in market chains who transfer the risks to those with the least power, smallholder farmers. Power asymmetries in emerging markets are particularly common; in Brazil, four firms control 75 percent of the Brazilian hybrid maize markets, and in El Salvador two mills dominate 97 percent of the market for wheat.[16]

Supply chains have become increasingly complicated and sophisticated in developing countries, with growing urbanization, the creation of urban supermarkets and the links to international trade. The spread of supermarkets from developed countries began in Latin America, where they now typically account for 60 percent or more of retail food sales, but has only recently reached Africa. The dominance of supermarkets tends to be very uneven geographically, as is their size, from family businesses to global retail chains.

Traditional markets, despite the rise in supermarkets, remain important outlets for both producers and consumers, particularly for poor households in developing countries. The differences between traditional and modern markets, particularly in terms of quality of produce, can mean that if the two coexist, farmers are able to sell higher-quality goods to supermarkets and lower-quality to traditional markets (Table 8.1).[17]

Table 8.1 Main features of traditional versus modern supply chains for agricultural and food products

Traditional	Modern
Low elasticity of demand	High elasticity of demand
Traders or processors dominate	Retailers dominate
Low value to volume ratio	High value to volume ratio
Quality defined by basic grades	Quality defined by private standards
Limited need for quality and safety assurance	Quality and safety assurance critical
Many products have low perishability	High perishability
Low levels of product processing and transformation prior to export	High levels of processing
High risk and transaction costs	Low risk and transaction costs within a short supply chain
Numerous specialist small businesses	Limited numbers of specialist businesses
Little or no traceability/identity preservation through supply chain	Enhanced need for traceability/identity preservation through supply chain
Need for basic logistical capacity	Advanced logistical capacity

Source: IFAD, Adapted from Henson 2006[18]

Commodity Exchanges

Typically smallholders have small quantities of low-quality goods that they need to sell immediately for cash. One answer lies in creating village-level "grain banks," owned and run by an association of farmers, where they can deposit their grain. The store is usually fumigated against pests, some grain is kept in case the owner needs it later in the year, and the rest is sold when prices seem right.

In such a system in Kenya, the marketing depends on having a countrywide network of small and large markets. The network is supported by the Kenya Agricultural Commodity Exchange (KACE), a private-sector firm that provides farmers with prices and other market intelligence accessible to smallholders using a mobile phone short message service (SMS) system.[19]

A more elaborate exchange has been created in Ethiopia, where market trading has previously been in the hands of the government (Box 8.1).

While international markets are opening up to smallholders, domestic markets at present offer more opportunities for smallholder producers. The whole urban market of Africa in 2002 had a value of nearly $17 billion for smallholder producers, while the export market was valued at only $4 billion.[22] Domestic markets foster new opportunities through the growth in demand for high-value, labor-intensive products and through increasing the scale of trade across regions within the continent.

These are important trends with significant consequences. Farmers with sufficient land, the necessary skills, and access to both output and input markets are moving away from an exclusive production of staples and into a more commercialized

Box 8.1   The Ethiopia Commodity Exchange (ECX)

The ECX was established in April 2008 as a marketplace through which all actors in the value chain—farmers, traders, processors, and retailers—can come together to trade domestic and major export earning commodities (coffee, sesame, haricot beans, teff, wheat, and maize). The first of its kind in Africa, its inception was led by former World Bank economist Eleni Gabre-Madhin.

The ECX provides market data, clearly defined rules of trading and dispute settlement procedures, a central trading system with remote electronic trading centers, storage and warehouse delivery centers, product grade certification and quality assurance, and clearing banks. In theory, these lower the costs and risks associated with traditional trading.[20] As of February 2011 its successes have included the trading of over $1 billion of commodity value and the grading, handling, storing, trading, and delivery of 4.6 million bags of produce.[21]

production of higher-value crops. Farm commercialization in the developing countries is most advanced in East and Southeast Asia, followed by Latin America, while much of South Asia and Sub-Saharan Africa is in the earlier stages of the transformation. Along with commercialization comes a reduction in on-farm labor, a shift from human or animal labor to mechanization, and the use of synthetic chemical inputs instead of natural sources such as farmyard manures.[23] An analysis by Joachim von Braun of IFPRI also found that in the majority of cases, commercialization led to increased employment opportunities and an improvement in nutritional status for children resulting from increased incomes.[24] But governments need to carefully monitor these developments, acting where necessary to protect the poor and reduce the gap between smallholders and larger commercial producers.[25]

Food Safety

A key element in the supply chain is the quality of the commodity, in particular the safety of food products. Over the past three decades the proportion of processed food exports as a share of total world food exports increased from about 4 percent in 1980 to over 60 percent in 2006. This increase has come largely from developing countries (where processed food exports as a proportion of total world exports tripled between 1980 and 2006) exporting to the developed world. However, the main contributors have been predominantly in the upper-middle and middle income countries.[26]

One of the barriers preventing low-income countries from increasing their share is the imposition of increasingly stringent food safety standards, set by developed countries.[27] The standards result from a growing consumer awareness of food-borne diseases, increased globalization and industrialization of agriculture, and better technological ability to detect food contamination.[28] Processed foods, especially, tend to face higher food safety standards than basic staples because they contain multiple components, are often highly perishable, and are often classified as "ready to eat." They are also subject to protocols that limit pesticide usage and residues, which define handling and storage operations and processes of traceability.[29] Inevitably these require large investments on the part of the farmer—for chemical stores, spraying equipment, and grading sheds as well as training in "standards, practices, controls and traceability requirements."[30]

Smallholders face formidable obstacles in complying with these requirements: They do not have sufficient scale to generate the large and fixed costs of compliance, they lack the necessary information to comply, and, unlike large established farms, they are unable to offer consumers a reputation for providing high-quality and safe products.[31] Sometimes the standards can force producers out of the market. In Kenya, a reorganization of the fresh vegetable export market in the 1990s to meet the standards of European supermarkets led to high costs ($20,000 for a group of forty-five farmers to certify their products) and to the rapid decline in the number of participating smallholder producers. In the district of Machakos in Kenya, the number of growers supplying a key exporter of green beans declined from 1,200 in 1991 to less than 400 in 2004.[32]

Farmer Associations

One possible way for smallholders to access the benefits from value chains is to be part of a producer association.[33] These farmer associations are characterized by the common pursuit of higher incomes and reduced risk through economies of scale, and they can take many forms.[34] Cooperatives are usually formal and registered; other collective groups are more informal. Examples of farmer associations include food, fodder, seed banks, or other storage facilities managed by local communities; savings and credit groups; and even extended families.

Faso Jigi, a farmers' cooperative association in Mali, demonstrates the benefits of such organizations (Box 8.2).

Box 8.2   Faso Jigi in Mali

Faso Jigi was set up in 1995 with the help of the Canadian International Development Agency (CIDA) and the Quebecois agri-agency L'Union des Producteurs agricoles, as well as Développement International (UPA-DI) in Mali. Its aim was to assist smallholder producers of cereals and shallots in marketing their products. It did this by:

- Reducing transaction costs through economies of scale in storage and transportation
- Disseminating market information to smallholders
- Enabling access to technical advice
- Making collective purchases of inputs
- Advancing credit to smallholders against a commitment to deliver (Faso Jigi borrowed the collective amount of total loans from a financial institution using its marketing fund as guarantee)
- Establishing an insurance fund

Since its establishment over 5,000 farmers in 134 cooperatives have become involved, and over 7,000 tons of cereal valued at $3.5 million in 2009 is sold per year. Wholesalers prefer sourcing from Faso Jigi and are willing to pay higher prices because the association offers centralization of stocks, better quality in storage facilities, and accessibility.[35]

There are, however, shortcomings associated with community groups; they can be exclusive, leaving out marginal groups such as widows, AIDS-affected households, or ethnic minorities. They may also have to depend heavily on state support or on development agencies for their implementation and regulation.

In Chile, where, through government efforts in the 1990s and early 2000s, 780 rural producer organizations were formed, only 20 percent of them were considered practicable, with 45 percent having costs higher than their revenues and a third being reliant on subsidies and grants for over 60 percent of their total income.[36] "Those that were successful shared three common attributes: they served as vehicles for members to innovate and change their farming practices; they networked, linking their members to ideas, resources, incentives, and new opportunities; and they sought to transmit undistorted market signals—costs and benefits—to their members who could then respond."[37]

Contract Farming

Another, sometimes complementary approach to accessing the benefits of value chains is through contract farming, which involves formal contracts between producers and farmers; the former may agree to purchase products at a certain price and may provide inputs to the farmer who agrees to provide a certain amount of the produce. The earliest examples were in Asia in the nineteenth century, but it has subsequently spread to other continents. In Africa, contracting, particularly in fruits and vegetables, began to grow in the 1930s, and by 1985 sixty schemes were operating in sixteen African countries. Contract farms are not widespread, however; only 7 percent of farmers in a selection of countries were under contract.[38]

Contract farming usually comprises a central processing or exporting agency that purchases the produce of a number of independent farmers, although the structure of the systems, ways in which they operate, and overall objectives (e.g., social, economic, or political) can vary. The terms of the partnership are detailed in a contract, often with the agribusiness providing inputs, equipment, and technical advice to the farmer in exchange for a consistent supply of a set standard of commodity produced.[39] Such arrangements exploit the economies of scale in both purchases and sales. They address the failure of traditional spot markets (i.e., cash markets in which commodities are traded for immediate delivery) to transfer information to producers on quality, future demand, and consumer preferences, and they minimize the risks associated with fluctuating prices and market shocks.[40] Although evidence is scarce, contract farming can increase net revenue; for example, by reducing transport and transaction costs. Dairy and spinach producers in India can increase their net revenue by up to100 percent compared to independent producers.[41]

However, unsuccessful schemes outnumber the successful. One criticism is that contract farming may further marginalize those farmers and communities who do not (or cannot) participate. Farmers who do participate can lose their autonomy and become overly dependent on their contracts. In the state of Punjab in India, contract farming, which has been expanding since the 1980s, tends to operate to the benefit of larger farms. Fewer than 15 percent of the farmers under contract farming had less than two hectares of land.[42] Large farms are more able to (1) produce a greater consistency of quality and supply, (2) adopt technology, and (3) access credit—and they require less monitoring.

One route to involving smallholders is to make contracts with producer associations. In Uganda, external support to the Nyabyumba United Farmers group enabled the farmers to become a supplier of potatoes to Nando's fast-food restaurants in Kampala. Members of the group, 60 percent of whom are women, have gone from being essentially subsistence farmers gaining income from activities off-farm to being commercial producers and net sellers.[43]

Fair Trade

The examples of farmer associations and contract farming have had mixed success. This experience suggests that getting better deals for farmers in the supply chain may require more aggressive measures, through trade unions, state intervention, or fair trade agreements. Organic or fair trade markets can offer smallholders price premiums and long-term contracts. However these are niche markets; they form a very small proportion of agricultural trade (global organic markets, the larger of the two, composed only 1 to 2 percent of total food sales in 2008), and they require certification at a large cost to the smallholder. Fair trade, often adopted as part of a global business's corporate social responsibility agenda, may offer a greater chance of engaging and supporting smallholders (Box 8.3).[44]

Box 8.3   Fair Trade

The fair trade movement aims to rebalance the uneven global trade relationship in favor of the disadvantaged, through developing trading partnerships based on transparency, equity, and respect, and a focus on improving prices, working conditions, and sustainability for producers.[45] Typically, fair trade companies pay sustainable prices, higher than the market price.[46]

In 2008, despite the global recession, worldwide sales in fair trade grew by 22 percent with much of this increase coming from growing markets for tea, cotton, coffee,

(continued)

and bananas. The *fair trade premium*—income additional to the production price and used for community development projects—contributed over $17 million to coffee farmers and almost $16 million to banana farmers in 2008.[47]

In the poorest region of Belize, the Toledo Cacao Growers Association, a cooperative formed in 1986, has helped its members to increase the quality of cocoa grown and the prices farmers receive. They have adopted organic and environmentally sustainable farming practices and improved their living conditions. A halving of the price of cocoa in the early 1990s, however, meant income no longer covered the cost of production. Then the chocolate company Green and Black's offered the association a long-term contract in 1994 to purchase all the organic cocoa it could produce at an above-market price, a supply used to create the Fair Trade Maya Gold Chocolate.[48] This contract has allowed individual farmers to not only return to farming but to send their children to school and afford medicines.

Smallholders and the Value Chain

In summary, if value chains are to offer pro-poor opportunities for growth, then those markets in which smallholders can have a "comparative advantage" need to be identified and the producers have to be actively helped to capture a fair share of the improved value and equity. Although marketing is essentially a function of the private sector, government support is necessary to (1) establish smallholder-friendly regulatory and political environments, (2) encourage private-sector investment in rural producers, (3) help producers organize themselves, and (4) support rural infrastructure development. These are all key elements in the enabling environment for smallholder advancement.

Agrodealers

Having efficient, fair, and transparent output markets is crucial if smallholders are to significantly participate in agricultural development, but by itself this is not enough. Farmers also need efficient, fair, and transparent *input markets* where they can purchase the seed, fertilizers, machinery, and other inputs they need in the quantities that are appropriate for them, and at the same time obtain relevant advice on new products and opportunities.

One approach pioneered initially by the Rockefeller Foundation and developed by the Alliance for a Green Revolution in Africa (AGRA) is to facilitate the creation of village-level agrodealers (Box 8.4).

Box 8.4   Portrait of an Agrodealer[49]

> Flora Kahumbe . . . owns two agro-dealer shops near Monkey Bay, Malawi, at the south
> end of Lake Malawi. She was trained by the Rural Agricultural Market Development
> Trust (RUMARK), a local NGO that gets support from the Bill & Melinda Gates Foun-
> dation. RUMARK makes sure that agro-dealers like Flora know about proper storage for
> seeds and chemicals, safe application of crop-protection chemicals like pesticides and the
> appropriate ways of applying the right types of fertilizer for maximum effect.
>
> Yet, Flora is more than just a shop-owner; she's really almost an extension agent that
> provides valuable knowledge to farmers on how to get the most out of their seed. With
> three employees in each store, Flora is creating stable employment in her community and
> ensuring that the seed she sells does its best to feed Malawi's growing mouths.
>
> To date, the Alliance for a Green Revolution in Africa (AGRA) has trained and sup-
> ported over 5,000 agrodealers in eastern and western Africa. Although the stores are
> small (what Americans call "mom-and-pop stores") they are collectively having a major
> impact, providing $45 million worth of improved seeds, fertilizers, and other inputs in
> 2008. They sell key inputs to farmers in small, affordable quantities, and, most signifi-
> cantly, they reduce the distances farmers have to go to get inputs—in one area of Kenya
> from 17 km in 2004 to 4 km in 2007.[50]
>
> RUMARK has so far educated around 1,100 agrodealers on seed and chemical stor-
> age and on pesticide and fertilizer safety and application.[51] Agrodealers are also trained
> in business management, product knowledge, and crop husbandry and share this knowl-
> edge with farmers through 1,300 demonstration plots.

The Kenya Agrodealer Strengthening Program (KASP), funded by AGRA, has built a network of agrodealers that covers eighty-five districts in Kenya's agricultural areas, accessed by 1.4 million farmers. KASP has been instrumental in improving agrodealers' access to finance through local microfinance institutions and in providing guarantees and matching investment facilities. It also advances agricultural policy advocacy through a Ministry of Agriculture think tank and by helping to create associations that advocate on behalf of small business agrodealers.[52]

Experiments in Crop Insurance

Functioning input markets inevitably require farmers to take the risk of purchasing inputs. This risk is lessened if they have some form of insurance should the crop fail.

Most small farmers are preoccupied with risk—whether arising from the weather, pests, and disease or the market—and, consequently, they have evolved many ways of reducing its potential effect.[53] In a survey of 250 households in Bangladesh, India, and South Africa, an average of ten financial instruments were used within a single year. These were predominantly informal tools, commonly borrowing from friends and family, while other tools, less commonly used, were formal instruments such as microfinance, savings, loans, credit in shops, and insurance.[54]

The paucity of formal instruments is due to a lack of access and to the reluctance of formal financial services to serve small farmers. An estimated 2.2 billion people in developing countries do not have access to formal financial services. Part of the problem is also the nature of agricultural risks, which do not conform generally to the criteria for *insurable risk*:

- The risk is associated with an easily quantifiable probability of occurrence.
- The resulting damage is easy to connect to the risk event and value.
- The event shouldn't happen with a frequency to make insurance unaffordable.
- The occurrence of the event or the ensuing damage should not be affected by the farmer's behavior. [55]

It is axiomatic that for an insurance scheme to be profitable, the price farmers pay for premiums has to exceed the payments made to farmers. Public crop insurance programs rarely conform to this. Membership is low and many are heavily subsidized, resulting in excessive risk taking by farmers; for example, growing unsuitable crops in high-risk regions. Sometimes membership is made compulsory.[56] By contrast, private crop insurance schemes have an estimated total premium income around $1 billion per year. However, they predominantly sell to larger, commercial farms and do not address the social issues of small farmers.[57]

Microinsurance

The most promising route to reducing risk for small farmers is through the growth of microinsurance designed to provide low premium insurance cover for poor people, which today reaches an estimated fifteen million people with low incomes. There is a great diversity of such schemes depending on the agricultural product and what is being insured e.g. the yield or the loss of income.[58]

Index insurance is a popular approach; it has the advantage of not needing farm visits, relying instead on monitoring through weather stations. In India approximately 2.1 million index-insurance products have been sold since 2003.[59] A recent example in Africa is an index-insurance scheme for livestock launched in 2010 in the Marsabit area of northern Kenya (Box 8.5).

These various microinsurance experiments are promising, but there are limitations. They have high administrative costs because they deal with the claims of many

Box 8.5   Livestock Insurance[60]

Index-based livestock insurance developed by the International Livestock Research Institute (ILRI) helps small- or medium-scale livestock farmers or pastoralists insure against the risk of investing in areas of regular drought. The novelty of this scheme is that it lowers transaction costs by allowing the insurance company to pay out based on predicted livestock mortality. NASA satellite images, released every ten days at an 8x8km resolution, allow measurement of the level of photosynthetic activity in the vegetation in a given location. These measurements are used to predict livestock mortality using household-level livestock mortality data that has been collected monthly since 2000 in the region.[61] Livestock owners have the value of their herds assessed based on tropical livestock units (TLUs) where one cow is equal to 1 TLU and one goat is equal to 0.1 TLU (1 TLU is currently valued at 15,000 Kenyan shillings [Kshs]).

Premiums are set at 5.5 percent (of consumer price) for the northern, higher-drought risk, part of the district and 3.25 percent for the southern part. The insurance brokers pay out twice a year (after each of the two dry seasons) if the predicted mortality is over a certain threshold, called a "strike level" (currently set at 15%).

One benefit of the scheme is that it discourages herders from "self-insuring (i.e., having more cattle than they really need). The scheme also encourages better animal husbandry and helps the Kenyan government, which normally must make big pay-outs to farmers after droughts to keep them out of extreme poverty. It is particularly helpful for pastoralists who have managed to accrue greater than 15 TLU, the level deemed enough to make a sustainable livelihood. The insurance prevents them from falling back below this line.

smallholders, designing the contracts is costly, and there is a lack of reliable data or weather stations for determining the indices. Some form of state or nongovernmental organization (NGO) involvement is usually required initially to set up insurance markets that will entice private investment.[62]

Microcredit

Farmers often do not have the cash available to buy inputs. They hope the return at harvest will more than cover the input costs but need to use their own savings or a loan to tide them over until the harvest is in—ideally linked to an insurance scheme should the crop fail.

The government credit schemes set up in the early years of the Green Revolution proved effective in the high-potential lands where high returns to investments could be achieved. More problematic has been the provision of credit in the lower-potential areas, and particularly for the poorest households, where the returns are low and the risks high. Typically, such loans are very small, yet require careful attention and hence are relatively costly to service. They are not attractive to commercial banks and are difficult for more bureaucratic government agencies to handle. The traditional alternative is the money lender, often charging very high rates of interest that create long-term, increasing indebtedness. While governments have attempted to provide small loans, the majority of these schemes have proved unsustainable with costs far outweighing the benefits.[63]

However, there is now sufficient experience, going back over three or more decades, to demonstrate the effectiveness of local, self-managed credit groups run by NGOs or private banks. The key to the formation of these groups is often a coming-together of households on a small development activity that satisfies a collective need.[64] In southern India, the NGO Myrada organizes savings and credit groups around small projects such as desilting a tank or creating a drainage system.[65] In one case, a group started from digging a trap to catch a marauding elephant. The collective physical activity and the experience of cooperative planning and management in the project lays the basis for trust and self-confidence. Although there are guidelines, each scheme tends to evolve along different lines, reflecting its origin and local circumstances. Members determine the rules, set interest rates, decide on the types of loans, and vet and approve the loan applications. Most of the loans are less than $3 and used to pay for a marriage or funeral, to buy food ahead of harvest, or to purchase one or two sheep.

Perhaps the best known of all microcredit schemes is the Grameen Bank, established in Bangladesh in the 1970s and encompassing over eight million members.[66] In his book, *Jorimon of Beltoil Village and Others,* Muhammad Yunus, the founder of the Grameen Bank, presents real-life stories of poor women in rural Bangladesh (Box 8.6).[67]

Box 8.6   The Story of Koituri[68]

Koituri was married at thirteen to Joynal, a twenty-year-old who worked as a laborer. Joynal turned out to be a cruel husband; he beat his wife, constantly demanding a dowry although this had not been part of the marriage negotiations. Although she provided two sons, the beatings continued and eventually she could stand it no longer. She moved to her father's compound and worked in the houses of better-off neighbors, husking paddy, cleaning the cowsheds, and doing domestic chores. For this she received two meals a day and a kilogram of rice.

(continued)

Box 8.6 (*continued*)

Koituri, unlike the Pals whose story is told in Box 2.6 in Chapter 2, did not migrate to the city. Her life began to improve when the Grameen Bank opened a nearby branch. Some women formed groups in her village. At first she was nervous, but she joined a group of nine women and received training from the Bank. She took out a small loan to buy a goat and a kid. She tethered the goat where she worked and cut grass to feed it. But it did not prosper, and she sold it for a loss.

However, with the money she leased a tiny plot and planted lima beans on bamboo scaffolding. These she successfully sold; she took a new loan, rented some more land, bought a plow, paid for irrigation and labor, and began to grow an IRRI rice variety. Eventually she paid off all her loans, acquired a new loan with which she bought a calf, and, when Mohammed Yunus was writing about her, she was contemplating a loan to replace her hut with a small house. In Yunus's words: "The Bank loan has not only changed her financial status, it has also brought about a change of her outlook. She used to feel frightened all the time. . . . Now she does not feel insecure any more. She has her friends from the group to look after her. Previously, she worked as a maid servant to others and was treated like one. Now she has money in her hand. People come to her in their trouble and she can help them with small loans."

In the global scale of affairs, Koituri's achievement is small and insignificant, but for her it has meant the difference between life and death and brought dignity and self-confidence.

Besides belonging to a savings and credit group, smallholders can also access credit from other sources: from a trader to whom the producer will sell their goods, from input suppliers who are repaid at harvest, from a marketing or agro-processing company, or from a contracting agribusiness firm. Both credit and re-payment can be in goods and services as opposed to cash. Farmers can also secure loans with the help of a third party, such as rural producer organizations, to add to the guarantee or by using their physical assets as collateral for accessing credit.[69] Warrantage and warehouse receipt systems are two common sources of credit (Box 8.7).

Microfinance institutions (MFIs), which specialize in providing financial ser-vices for the poor, predominantly offer credit; only about a quarter offered savings services.[72] That is now changing, partly due to the global financial crisis and partly because transaction costs can be reduced by using mobile banking (detailed below under ICT infrastructure) and the services of local shopkeepers and agrodealers. In India the MFI Cashpor is working with the Grameen Foundation and a savings bank to offer services to those who are traditionally marginalized.[73]

Box 8.7   Sources of Smallholder Credit

Warrantage systems

Under warrantage systems, farmers pool their grain for storage after harvest, against which cash loans are issued, allowing the collective to buy inputs. When grain prices rise, farmers sell their stored stock to pay back loans. The FAO have established 50 warrantage schemes in the Sahel, including 250 input stores and hundreds of grain stores benefiting 13,000 farm households over the past fifteen years.[70]

Warehouse receipt systems

Warehouse receipt systems involve the storing of a predefined quality of produce in a warehouse operated by a third party, allowing farmers to use the receipt of this storage as collateral. These systems help smallholders access financial institutions that are more willing to make loans because their transaction costs (e.g., ensuring product quality) are reduced. A warehouse receipt system in Tanzania was first piloted in 2001 and is supported by two IFAD-funded programs: the Agricultural Marketing Systems Development Programme, which established and managed the storage facilities, and the Rural Financial Services of Savings and Credit Cooperative Societies (SACCOs) formed by local communities. Once a farmer has stored his/her produce, he/she can obtain a loan from a SACCO of up to 70 percent of the value of the stored goods. The Tanzanian government passed the Warehouse Receipt System Act 2005 to expand the system across the whole country.[71]

Connectivity

The final element in creating efficient, fair and transparent markets is connecting them to farmers, either *physically* (e.g., through all-weather roads) or *virtually* through some form of information and communication technology (ICT). These can be thought of as, respectively, the hard and soft components of connectivity.

Many, and in some locations most, smallholders lead isolated lives, connected with their neighbors in their local community but with few links to the outside world. Some in the industrialized world argue that this is an ideal state. For a hundred years or more, groups of people in the developed countries have tried to set up small self-sufficient communities, often in remote rural locations. Some have been successful, for example, the Amish and the Shakers in the United States, because they have retained key links with the larger world; others, however, have ended in disaster.

For the poor of the developing world, this is not an option. Remaining remote, even if they are self-sufficient, will continue to trap them and, more important, the next generation in poverty. The young of the developing countries deserve to choose a better future. To achieve this, rural communities need connectivity, especially with urban resources and with opportunities.

Physical Infrastructure

Evidence suggests that physical infrastructure helps households move out of poverty.[74] In India, every additional million rupees spent on rural roads during the 1990s was found to lift 881 people out of poverty. Villages in Bangladesh with better road access had higher levels of input use and agricultural production, greater incomes, and greater wage-earning opportunities.[75]

The benefits of roads and other forms of transportation are obvious. They:

- reduce the cost and time involved in transporting agricultural products to markets and in receiving inputs to farms;
- facilitate labor migration and expand opportunities for off-farm employment;
- allow greater market integration;
- reduce price volatility.

Transportation needs to be seen as a key element in the value chain. Transportation costs can be a large proportion of total marketing costs (e.g., in Benin, Madagascar, and Malawi between 50 and 60 percent) and can be prohibitive for smallholders, particularly because they are often fixed and tend to be a higher proportion of the end value of the product for small producers. Even in India with relatively well-developed transport systems, the costs of transporting and marketing milk and vegetables can be up to 15 percent of a farmer's gross revenue.[76]

The construction of transportation systems can itself bring significant benefits, particularly if the ownership and maintenance is local. Between 1998 and 2002, Tanzania improved up to 530 miles of rural roads and constructed 107 bridges under a USAID program. This led to an average 40 percent reduction in transportation costs and played an important role in building capacity through training local communities to select their private contractors and supervise both the initial works and ongoing maintenance.[77]

Despite the obvious benefits to building physical infrastructure, a major challenge is raising the necessary capital, most notably in Sub-Saharan Africa where the transportation coverage is significantly less than in Asia (Figure 8.1).[78]

In Sub-Saharan Africa a major goal is to create regional trade networks in agricultural products, linking small farmers to the supermarkets and exporters throughout their local region. Historically, road links in Africa both between countries and within the continent as a whole have been weak and intermittent, preventing

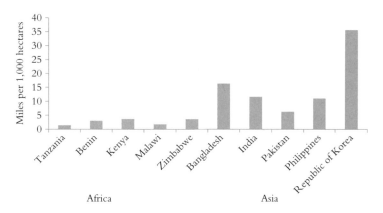

Figure 8.1 Miles of road and rail per 1,000 hectares of agricultural area.[79]

effective trade and cooperation. The Trans-African corridor program, or highway network, developed by the United Nations Economic Commission for Africa (UNECA), the Africa Development Bank (AfDB), and the African Union (AU), comprises nine transcontinental roads equaling 56,683 km that link, or closely pass, most continental African states. It is estimated to generate $250 billion over fifteen years in overland intra-African trade.[80]

Despite construction of the network being impaired by civil conflict and the sheer logistical scale of such an endeavor, around half of the planned roads have been paved so far. Crucial missing links exist, however, especially across central Africa where the construction and maintenance of highways has low national priority.

The corridor program provides a unique opportunity to stimulate and strengthen internal trade within the continent and to attract investment. In addition to significantly reducing transport costs, the corridors enable medical and educational services to reach previously inaccessible areas, so demonstrating the multiplier effect of infrastructural provision. The corridor program provides a major opportunity to stimulate agricultural growth and trade (Box 8.8).

Box 8.8 Beira Agricultural Growth Corridor (BAGC)[81]

The BAGC aims to connect the countries of Southeast Africa—Zambia, Malawi, Zimbabwe, and Mozambique—to the port of Beira through both roads and railway lines. The project was initiated by Yara (a global chemical company) in 2008 in partnership with a wide variety of stakeholders and investors such as the Mozambique government,

(continued)

Box 8.8 (*continued*)

AGRA, the Norwegian government, TransFarm Africa, and AgDevCo to name a few, with the investment blueprint presented in 2010.

Initial studies have found the area around the corridor to have enormous farming potential; for example, of the total 10 million hectares of arable land, only 1.5 million hectares is currently farmed and only 0.3 percent of this land is farmed commercially. The corridor is predicted to bring substantial benefits to the farming sector of the area and specifically to smallholder farmers: with an investment of $250 million over five years, 200,000 smallholder farmers in Mozambique will directly benefit from improved yields and increasing incomes, 350,000 new jobs will be created, and up to one million people will be moved out of extreme poverty.

ICT Infrastructure

In recent years ICT systems have taken over some of the same benefits of physical infrastructure; mobile and postal networks, for example, help share information regarding prices, markets, and the sourcing of inputs and labor. Systems are being developed by the private sector, with government providing subsidies for the most remote areas. Although roads in the past were critical to disseminating information to farmers, the rapid growth in the use of mobile phones as well as their increasing affordability enable farmers access to information related to prices, markets, disease outbreaks, job markets, weather forecasts, and technical advice as well as to services such as banking and contacting buyers (see Chapter 7). For example, Maasai herders can obtain information about the locations of grazing lands and watering holes, contact doctors and vets, and use their phones to trade and find the best deals when selling their cattle. Due to their remote location they use solar panels or diesel generators to charge mobile phones.[82]

Mobile banking systems allow users to convert cash in and out of "stored value" accounts linked to their mobile phone. Farmers can use the stored value to receive payment for crops or to pay for goods, transferring stored value between their own and other people's accounts. This simple, secure service reduces transaction costs, dispensing with the expense and time of traveling to and waiting in a bank and avoiding the risk of theft.

A highly innovative system for using mobile phones for microbanking is the M-PESA scheme (*M* for mobile and *Pesa* meaning money in Swahili), which is having considerable success in Kenya by opening up banking services to previously marginalized groups.[83] It was originally a pilot funded jointly by Vodafone and UK DFID.[84] The service was launched by Safaricom in 2007 in Kenya, where

Safaricom controls 80 percent of the mobile market.[85] M-PESA is now used by over 50 percent of the adult population of Kenya, and has seven million local users in Tanzania. Future development of M-PESA includes further growth in Tanzania, Afghanistan, and South Africa and the provision of an index–insurance product to protect farmers from drought.[86]

Mobile phones are also having a positive impact on gender equity by reducing the need for female growers to depend on men for marketing. They also give women control of their own money and a greater sense of safety when traveling alone. One estimate puts mobile phones as worth the equivalent of two to four extra years of women's education in terms of reducing gender inequalities.[87] Literacy requirements may present a barrier to accessing ICT services for many poor rural people, but this may be shortly overcome by newer technologies that can translate text to voice and back again.[88]

The Enabling Environment

This chapter has discussed one of the key components of an enabling environment: the creation of efficient, fair and transparent markets and the connectivity that makes them work for smallholders. Other key components of an enabling environment include the provision of other forms of physical infrastructure such as irrigation and power (see Chapters 14 and 16), extension (see Chapter 11), and the establishment of research and development (partly described in Chapter 7)

In the next chapter I discuss one of the critical outputs of research and development; namely, the production of new crop varieties and livestock breeds through advanced breeding technologies.

Part III

9

Designer Crops

In a continent that is hungry, the GM debate should be very different. The technology provides one of the best ways to substantially increase agricultural productivity and thus ensure food security to the people.

Blaise Compaore, president of Burkina Faso, National Peasants Day, 2010[1]

Plant and animal breeding is an art, nearly as old as agriculture itself. The early farmers selected seed from vigorous, high-yielding plants and used them for sowing in the following season. By degrees, wild grasses were transformed into domestic cereals—wheat, barley, maize, rice—the process of selection creating distinctive varieties adapted to local conditions and needs. Farmers soon came to look for promising mutants and natural hybrids. Bread wheat, a natural cross between emmer wheat and a wild goat grass that arose about 7,000 years ago somewhere to the southwest of the Caspian Sea, was recognized and cultivated by the early farmers, becoming a cornerstone of European and, eventually, world agriculture.

For thousands of years plant and animal breeding was a family affair, conducted in and around the farm dwelling by men and women, using simple ground rules and relying on intuition, personal experience, and wisdom passed on from generation to generation. The first systematic studies of plant hybridization were conducted in the eighteenth century, but the science of plant breeding owes its origins to Charles Darwin who, in his discussions of pigeon and dog breeds, explained the basis of selection and to Gregor Mendel, using the garden sweet pea, who identified the particulate nature of the basis of heredity.[2] One consequence of their scientific discoveries has been the emergence of professionals, breeders in institutes and research stations, who identify and explain the underlying mechanisms and hence make breeding more predictable and efficient. Nevertheless, although plant and animal breeding is now considerably more sophisticated than a hundred years ago, in its essentials, it has little changed. Mendel, if alive today, would recognize the processes of hybridization and selection that led to the creation of the Green Revolution wheats and rices.

In this chapter I address the following questions:

- How did crop breeding underlie the success of the Green Revolution?
- What has been the importance of hybridization and its application in maize and rice and to livestock?
- What are the limitations of conventional breeding?
- How can biotechnology remedy these limitations?
- What are some of the hazards of biotechnology?

Breeding for the Green Revolution

The dwarfing genes for wheat originated in Japan.[3] In 1935, Japanese breeders had crossed a Japanese semi-dwarf variety containing a dwarfing gene with Mediterranean and Russian varieties to produce a new cultivar, *Norin 10* (Chapter 3). This in turn was crossed with other U.S. varieties, and one of these crosses was sent to Norman Borlaug for crossing with the Mexican tropical and subtropical varieties. The key genes have subsequently been identified as alleles (forms) of the RHt (*reduced height*) gene, which acts by interfering with the gibberellin growth hormone that affects stem elongation, so producing the dwarfing effect.[4]

The rice dwarfing genes, originally from the Chinese variety *Dee-geo-woo-gen*, are alleles of sd-1 (*semidwarf-1*), which acts on the same growth hormone, reducing stem elongation. These were first used in the 1950s by Taiwanese breeders to develop a semidwarf variety known as *Taichung Native* 1 (TN1) and later at IRRI in crosses with the *indica* rice variety *Peta* to produce the "miracle rice" IR8.

The hybridization process was relatively straightforward. The semi-dwarf genes are now present in short-strawed, fertilizer-responsive wheat and rice varieties throughout the world. It has been a simple and powerful process, reliant for its success on the existence of naturally occurring genes capable of being easily transferred from one plant to another by the traditional methods of plant breeding. Most of the subsequent successes have been of a similar nature. Breeders have progressively improved a relatively small set of "mainstream" varieties by crossing them with uncommon, local varieties, or in many instances wild relatives, identified as having desirable characteristics—resistance to pests or diseases, tolerance of drought, improved milling quality or flavor. More fundamentally, plant, and in particular animal, breeding has been directed at improving the overall structure and physiology of crops and livestock. The aim, in simple terms, has been to increase the *harvest index*—the proportion of the plant's or animal's energy and materials, in particular carbohydrate and protein, that goes into the final, harvested product. Beef cattle and sheep have been bred for a high "dressing-out" percentage, the proportion of beef or lamb carcass after slaughter in the total weight; dairy cows for a high milk production; and chickens for the number of eggs per day. The harvest index in cereals is the ratio of grain to straw, achieved by introducing the dwarfing genes that had the double effect of allowing a higher uptake of nutrients and ensuring they went primarily to the grain.

Hybrid Maize

Some of the most exciting developments in conventional plant breeding have been the production of hybrid maize and rice.

Maize (referred to as *corn* in the United States) is a cross-pollinating crop, so that the seed collected by the farmer from his or her crop at the end of the season for planting is hybrid seed and is usually highly variable with poor outcomes. As early as the nineteenth century, U.S. plant breeders and farmers had been experimenting with deliberately hybridizing maize.[5] But significant progress was not made until the beginning of the twentieth century with the invention of *detasseling*—the removal of the tassels, comprising their male flowers and their pollen from the maize plant. This prevents the plants from fertilizing themselves, making controlled and predictable crosses possible. In 1909, George Shull at the Carnegie Institute's Cold Spring Harbor Research station showed that, by inbreeding using self-pollination for several generations to produce pure lines and then crossing the lines, new hybrids could be achieved combining the best features of the lines plus significantly increased yields—an expression of hybrid vigor or heterosis.[6] Some of the first commercial hybrids were bred by Henry Wallace, a key figure in the Green Revolution (Chapter 3). So successful were the early hybrids that the proportion of hybrid maize in Iowa grew from less than 10 percent in 1935 to well over 90 percent only four years later.

The process of hybridization has become increasingly sophisticated, and it is costly to produce hybrids for particular environments.[7] Today developing country farmers have a choice between *open pollinated varieties* (OPVs) and *hybrids*, each with distinct advantages and disadvantages.[8] Farmers produce OPVs by letting their maize cross in the field and retaining the seed. The next generation of plants is usually genetically diverse and not very uniform. But plants are related and over time will become adapted to the local environment. Their cost is very low, although the typical yields are 10–25 percent less than the hybrids. Hybrids, in contrast, are very uniform, but their seed needs to be purchased each year and the costs may be unaffordable. Modern certified OPVs are produced by breeders for specific environments and offer a compromise.

Modern maize varieties have had significant impacts in the developing countries. Mexican production, for example, stands at 22 million tons in 2011 compared with 6 million tons in the 1960s.[9] Nevertheless, the uptake overall has been poor. Less than 50 percent of the tropical maize area is sown to hybrids or OPVs (the rest is low-yielding local varieties), and the yield gap is often very high.[10]

Hybrid Rice

By contrast, hybrid rices have been developed more recently in China and have had a dramatic effect on rice production both there and elsewhere (Box 9.1).[11]

The CMS system described in Box 9.1 is complex, but was given high-priority support by the Chinese government to speed up development. In 1972, a series of male sterile lines and corresponding maintainer lines were developed and several restorer lines were identified. In 1974, the first rice hybrid, *Nan-You 2*, was bred, and in the winter of 1975, "the largest group of hybrid rice researchers and technicians in China's agricultural history went to Hainan to produce hybrid rice seeds in more than 4,000 ha of land."[15] This enabled China to produce enough hybrid seeds for large-scale commercial production in 1976.

For some time there has been disappointing progress in improving the yields of tropical rices. However, the new hybrids have a high-yield potential when grown in tropical lowland environments (Figure 9.2). They account for about 50 percent of the rice area in China with average yields of 7 tons per hectare.

Box 9.1   The Creation of Hybrid Rices[12]

Hybridization in rice is difficult because it is self-pollinated and because the tiny florets—with male and female organs in the same floret—flower for only a short time. There is some degree of cross-pollination, but it is small and carried out, at least in northern Thailand, by tiny bees.

The solution, as with maize, is to develop a rice with sterile male flowers. Under the leadership of Yuan Longping—dubbed the "father of hybrid rice" and currently the director general of China's National Hybrid Rice Research & Development Center—the Chinese developed a cytoplasmic male sterility (CMS), or three-line system, using wild abortive germplasm found on Hainan Island by one of Yuan's team. The system consists of crossing three lines:

1. *Seed parent* line: female (male sterile) and used as the source of the hybrid seed.
2. *Maintainer* line: as a pollinator to maintain male sterility (CMS) in the seed parent. It has viable pollen grains and sets normal seed.
3. *Pollen parent or restorer* line: restores fertility in the first offspring (F1) when it is crossed to the seed parent line.

(continued)

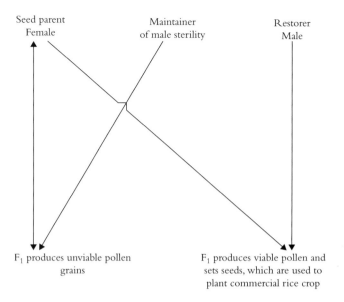

Figure 9.1   The different breeding lines for creating hybrid rices.[13]

The maintainer lines are repeatedly crossed with the female (male sterile) parent line until a stable sterile plant is achieved. This is the called the *CMS plant*, which is then crossed with the restorer line so that fertile seed are produced for farmers to plant.[14]

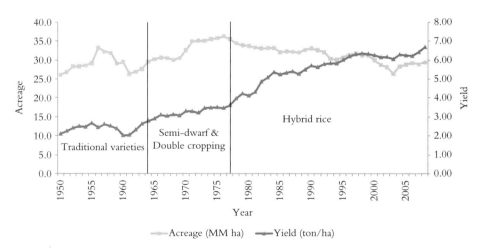

Figure 9.2   Growth of Chinese rice yields.[16]

Several other crop species, for example sorghum, sunflowers, onions, tomatoes, and other horticultural crops, have been successfully subject to hybrid breeding using techniques similar to those employed in rice or maize. Hybrid wheat has proved difficult to attain.

Livestock Breeding

In some respects livestock breeding is a more ancient science. Farmers have been selecting for useful traits in their herds for thousands of years, since the first domestication of animals. Cattle have been selected for such obvious characteristics as size, color and the shape of their horns; sheep for their wool; chickens for the amount of eggs they produce; and livestock of all kinds for the taste and quality of their meat. Animals are also selected for resistance to disease. This process, like conventional breeding in plants, can be very effective but tends to be slower than for plants because the time to reproductive maturity in animals is longer.

One way to speed up the process is through *artificial insemination* (AI)—the mechanical insertion of fresh or stored semen into the female reproductive tract. In principle it is relatively simple and has been used for more than fifty years to economically and efficiently improve herds, especially dairy cattle. In the developed world, the technology is highly sophisticated, involving efficient semen collection and dilution as well as cryopreservation techniques. A single bull can now be used simultaneously in several countries for tens of thousands of inseminations a year.[17] In this way a very small number of top bulls can be used to serve a large cattle population resulting in a rapid increase in the rate of genetic improvement in dairy cattle.[18]

So far AI has been used less in developing countries. The technology, although apparently simple, requires relatively high levels of skill and experience; for example, success depends on accurate heat detection and timely insemination. However some 34 million inseminations were carried out in India in 2007.[19] AI has been particularly useful in cross-breeding.

Cross-Breeding Livestock

Crossing very different parents has been as important in animal as in plant breeding. The main focus in the developing countries has been on improving the milk production of dairy cattle. There are two dominant species of cattle in the world—the taurine cattle (*Bos taurus*) of the temperate climates of Europe, North Asia, and West Africa and the humped zebu cattle (*Bos indicus*) of the hot arid and semiarid regions of Africa and Asia. Although these two species can naturally cross, they are today very different because of the selection pressure for milk production in taurine cows,

172

which has led to the dominance of such very high producers as the Holstein-Friesian types.[20] In contrast, natural selection has produced zebu cattle with a high degree of heat tolerance, resistance to many tropical diseases, and the ability to survive long periods of feed and water shortage. Their dairy potential is poor; they have low milk yield, are late maturing, and usually do not let down milk unless stimulated by the sucking of a calf.[21]

A number of sophisticated cross-breeding schemes have now been introduced with considerable success.[22] India, for example, is now aiming to produce 20 million cross-bred cattle. In the Kilifi plantation near Mombasa in Kenya, a rotational crossing scheme alternating with bulls from the two breeds has resulted in Sahiwal x British Ayrshire crosses producing 3,000 kg of milk per year, mainly from pasture.[23]

Limitations of Conventional Breeding

Conventional plant and animal breeding techniques have clearly contributed a great deal to food security over the past 150 years. They still have much to offer and will remain a mainstay of breeding for the foreseeable future. They are powerful techniques in the hands of plant breeders in national research institutes, in the centers of the Consultative Group on International Agricultural Research (CGIAR), and in the hands of small and multinational seed companies. While companies such as Monsanto are recognized for their work on genetically modified (GM) crops, much of their activity lies in conventional plant breeding, albeit conducted on a large, semi-industrial scale. Crops such as soybean where yields have risen year after year owe their success as much to conventional breeding as to GM technologies.

Nevertheless, as has been long recognized, conventional techniques have practical limitations. Conventional breeding in the hands of the multinational seed companies tends to focus on key commercial crops to the detriment of crops grown and consumed by poorer individuals and to the stability of agricultural systems overall.[24] It is also not as efficient or speedy a process as it might be. The crossing of two parent plants, each with desirable characteristics is essentially a random process. While some desirable characteristics may emerge, others may be lost. Potential yield may increase, but often at the expense of pest or disease resistance or some other characteristic that is already present, such as superior grain quality. There are also natural limitations to conventional plant breeding. Traits that a breeder wishes to incorporate in a plant or an animal may not be present in any variety, breed, or species with which a cross can be made, although they may occur in quite unrelated species. Conventional plant breeding is also a relatively slow process. For some of the major cash crops of the developing countries—rubber, oil palms, cocoa, coffee, and coconuts— it takes decades to achieve yield advances that in rice would be realized in at most two to three years. Plant breeders are accustomed to this pace of advance, but for agronomists and farmers it can be frustrating.

The Power of Biotechnology

The revolutionary achievements of cellular and molecular biology are crucial in the context of the limitations of conventional breeding.[25] Modern cellular and molecular techniques open up a world in which breeders can deliberately design and engineer new plant and animal types, speedily and with much less reliance on random processes. Under the general heading of *biotechnology*, these techniques are already having a significant impact on both medicine and plant and animal breeding, benefiting poor people as well as those who are better off. The application of this knowledge to crop breeding is pursued through three practical techniques:

- *Marker-assisted selection*—based on the ability to detect the presence of particular DNA sequences at specific locations in an organism and link these to the presence of genes responsible for particular traits
- *Cell and Tissue culture*—permits the growth of whole plants from a single cell or clump of cells in an artificial medium
- *Recombinant DNA* (also known as genetic engineering [GE] or genetic modification [GM])—enables the direct transfer of genes from one organism to another

As described below each of these is beginning to bring significant benefits to developing country farmers, both large and small.

Marker-Assisted Selection

The idea of selecting for valuable traits through identification of genetic markers began to gain traction only with the development of new techniques for identifying sequences of DNA. For the breeder the trick is to find a region on the plant genome that can be readily identified and is closely inherited along with a gene or genes that code for a trait of interest or value. Typically these markers do not behave like normal genes, and they usually have no biological effect. They are identifiable sequences of DNA found at specific locations and transmitted by the standard laws of inheritance from one generation to the next. There are many different kinds, each with their distinct advantages and disadvantages.[26]

Once a number of markers are identified, they can be assembled into marker maps of a genome; the genes are then located by statistical association with the markers.

Sometimes only one or two genes determine a trait, but most important agronomic traits are governed by *quantitative trait loci* (QTL); that is, they are controlled by many genes each tending to have a relatively small effect. Yield is a typical quantitative trait in plants, and milk production in livestock is another.

A major benefit of marker-assisted selection (MAS) is the possibility of determining the presence of a trait at the seedling or even the seed stage. A small piece of the seed or seedling is cut out and subject to DNA analysis. Monsanto has invented a machine called a chipper that can cut a small chip from a seed so that many thousands of seeds can be analyzed—the seeds germinate as usual. In the past, breeding for, say, insect resistance meant the seeds from a cross with a resistant parent would have to be sown in a nursery and, when the plant was mature, infested with insects to see which crosses had the resistance gene. Using MAS greatly speeds up the breeding process and makes it cheaper—there is no need for the construction of large insect-rearing facilities. New varieties can be brought to near commercialization in four to six generations instead of ten.

Most of the applications of MAS of relevance to developing countries have been for maize, partly because production of hybrids permits the capturing of the intellectual property rights and hence a return on the costs of applying MAS.[27] An example is in breeding for maize streak virus (MSV), the most serious disease of maize in Africa, affecting 60 percent of the planted area and losses of over five million tons/year.[28] Genetic resistance to MSV has been known for over twenty years, but it has not been widely deployed in local maize varieties because few national breeding programs can afford to maintain the insect colonies and other infrastructure necessary to measure for resistance against insect-vectored viral diseases. Now, using DNA markers that flank the gene, it is possible to identify the precise location of the resistance gene and hence to backcross it into local varieties.

Deepwater and Submergence-Tolerant Rice

One of the most striking applications of MAS has been the development of deepwater and submergence-tolerant rices.[29] Although much of rice in Asia is grown in standing water, rice plants cannot withstand being completely submerged for more than a few days. Such deep flooding affects more than 25 percent of global rice-producing land and is likely to increase with global warming. In some areas, water levels rise progressively by several meters during the growing season; in others, flash flooding, particularly during the seedling stage, fully submerges the rice plants for up to several weeks (Box 9.2).

These examples illustrate the significant progress in applying MAS to developing country problems. The costs of marker discovery and analysis are coming down fast,

Box 9.2 MAS Applied to Breeding for Deepwater and
Submergence-Tolerant Rices

So-called deepwater rice is known for its ability to elongate its internodes. These have hollow structures and function as snorkels to allow gas exchange with the atmosphere, and thus prevent drowning. In 2009, a pair of genes responsible was identified by a team led by Motoyuki Ashikari at Nagoya University in Japan.[30] They were named *SNOR-KEL1* and *SNORKEL2*. Under deepwater conditions, ethylene—a plant hormone—accumulates in the plant and induces expression of the two genes. Their products then trigger remarkable internode elongation via gibberellin growth hormones, so causing the rice plant to grow by up to eight meters in the presence of rising water levels.

A second discovery was made in 2006 by a team led by David Mackill at the International Rice Research Institute (IRRI) in the Philippines.[31] The team identified a quantitative trait locus (QTL) in rice called *Submergence 1 (Sub1)* that allowed submerged plants to survive for more than two weeks. The resulting rice was named Scuba rice. In *Sub1* there are three genes that are responsive to ethylene; the allele of one of them, *Sub1A,* responds by limiting the elongation of the internodes that is activated by ethylene. This conserves carbohydrates so permitting regrowth when the flood recedes.[32] The rice becomes dormant during the flooding then continues growing once floodwaters recede.

In effect these discoveries have provided two sets of genes, both responding to ethylene in deep water, but in different ways: one resulting in internode lengthening, and the other limiting the elongation. The potential is to utilize both sets of genes so that high-yielding rice varieties can withstand both flooding that is deep and quick, where *Submergence* genes are appropriate, and floodwaters that climb in a progressive and prolonged fashion, for which *Snorkel* genes are suited.[33]

MAS–based breeding is already underway. Markers for the *Sub1* locus have now been used to introgress *Sub1A* from an IRRI deepwater rice into a widely grown Indian variety, Swarna. The resulting crosses when grown in the field in the Philippines exhibited the submergence tolerance, but the yields, plant height, harvest index, and grain quality remained the same. New submergence-tolerant varieties are now being produced in this way in Laos, Bangladesh, and India, and in Thailand where a submergence-tolerant jasmine rice is being bred.[34] In one farmer's fields during IRRI's Indian field trials 95 to 98 per cent of the scuba rice plants recovered while only 10 to 12 per cent of the traditional variety survived. Within one year of its release it has been adopted by over 100,000 Indian farmers.[35]

and an appropriate combination of MAS and traditional phenotypic selection can reduce land and labor costs and bring about quicker results.[36] Researchers at the International Maize and Wheat Improvement Center (CIMMYT) have summarized their experiences and listed circumstances in which MAS is likely to be advantageous to developing country breeding programs.[37] Immediate benefits are likely to come from breeding for pest and disease resistance.

Tissue Culture

Because all the somatic cells of a plant contain the full complement of genes, it is often possible to grow individual cells and pieces of tissue into full healthy plants. The requirement is to grow them *in vitro;* that is, under sterile conditions in a suitable medium.[38] This may be a liquid or solid, containing a range of macro- and micronutrients together with growth regulators that induce rooting and shoot formation. Tissue culturing is a technique that has long been practiced, under the title of micropropagation, to produce large numbers of clones of a plant. Because they are isolated and raised in contained and sterile media, this is a highly effective way of producing pest- and disease-free planting material.

Although commonly used for high-priced ornamental and horticultural crops, the technique has been very successful in producing staple bananas in East Africa free from diseases (Box 9.3). Bananas are a major source of food and income throughout the tropics, and especially in East Africa. Ugandans, for example, are the largest consumers of bananas in the world, eating on average nearly 1 kg/person/day.

Box 9.3   Breeding Healthy Bananas through Tissue Culture[39]

Banana plants are susceptible to disease because new plants are grown directly from cuttings from a "mother plant," thus transferring any disease present, even if it is not visible. The Black Sigatoka fungus, a leaf spot disease, has been particularly devastating to banana crops worldwide since its first outbreak in Fiji in 1963. It arrived in East Africa in the 1970s, decreasing productivity by as much as 40 percent. The fungus can be controlled with fungicides, but it has developed increasing resistance over the years, making this option both expensive and damaging to the environment.

After the end of apartheid in South Africa, Kenyan agricultural scientist Florence Wambugu visited there to observe the previously closely guarded work on tissue culture bananas. She applied the technique in Kenya and found she could quickly generate healthy new plants that could be planted in the field. She persuaded the Kenyan Agricultural Research Institute (KARI) to undertake trials on local varieties in the mid-1990s, and a training program was initiated.

During the last decade, over six million tissue cultured banana stems have been planted in Kenya, producing an additional income of 5.5 billion Kenya shillings ($64 million) to banana farmers. A business model known as Wangigi and piloted by Africa Harvest has greatly increased access to tissue culture banana outlets for 3,500 farmers, with some farmers being trained to train others in the technology. A farmer-owned marketing company, Tee Cee Banana Enterprises Limited (TCBEL), has been established to handle everything from postharvest handling and storage to setting industry standards.[40]

Embryo Rescue

More recently, cell and tissue culture has become a powerful tool in modern plant breeding, especially in the production of wide crosses. Often the embryo produced by a wide cross may fail to develop, but will do so if cut out and grown in a culture medium. An example is the successful crossing of the Asian rice *Oryza sativa* with the African rice *Oryza glaberrima* to produce a wide range of new high-performing rices for Africa—the NERICAs (Box 9.4).

Rice consumption is growing dramatically in West Africa, fueled by population growth, rising incomes, and a shift in consumer preferences, especially in urban areas. Local production, while increasing, is falling ever further behind demand, and the region is now importing over half its requirements, some 6 million tons annually at a cost of over $1 billion.[41]

NERICAs have the potential to help meet the huge and growing demand for rice in Africa, by increasing yields rather than through expansion onto ever more

Box 9.4   Production of the New Rices for Africa (NERICAs)[42]

The traditional African species of rice (*Oryza glaberrima*) has very low yields of about 1 ton per ha, compared with 5 tons or more for the Asian species (*Oryza sativa*). But, using tissue culture technologies, including embryo rescue and anther culture, Monty Jones, a Sierra Leone scientist working at the Africa Rice Center (WARDA), was able to cross the two species, producing hundreds of new varieties. Initially the embryo rescue technique did not work well, but collaboration with Chinese scientists provided a new tissue culture method involving the use of coconut oil that proved highly successful.

The rice varieties produced in this manner, known as the New Rices for Africa (NERICAs), share many of the characteristics of their African ancestors. They grow well in drought-prone, upland conditions. Their early vigorous growth crowds out weeds. They are resistant to local pests and disease and tolerant of poor nutrient conditions and mineral toxicity. Yet as they mature, they take on some of the characteristics of their Asian ancestors, producing more erect leaves and full panicles of grain. And they are ready for harvesting in 90 to 100 days, which is 30 to 50 days earlier than current varieties. Under low inputs, they yield up to 4 tons per ha.

The success of NERICAs includes their planting on almost 200,000 ha in Nigeria and on 35,000 ha in Uganda in 2007. As a result, Uganda was able to reduce its rice imports by half between 2002 and 2007 and is now exporting rice. A shift from maize to NERICA production in Uganda with proper crop rotation was found to increase income by $250 per ha.[43] In Benin, where NERICAs were introduced in 1998, the impact on rural household income has been positive and significant.[44]

178

marginalized lands. Evidence suggests that the need for less weed control and the shorter growing season with NERICAs is reducing the demand for child labor and improving school attendance in the areas it is being grown.

Advanced Cell Culture

Highly sophisticated forms of cell and tissue culture are now being developed to tackle intransigent breeding problems. In contrast to the production of the NERICAs by culturing embryos derived from sexual crosses, it is now also possible to raise plants from fusion of nonsexual (i.e., somatic) cells, which can greatly increase the diversity of the population of plants that eventually result.[45] Tissue culture also allows for a speeding up of the process of plant breeding by culturing the anthers—the male reproductive organs—of the cross.[46] The cells in the anthers are haploid, containing only one member of each chromosome pair. They can be placed in a culture where treatment, for example using colchicines, causes the chromosomes to double so producing a young plant that will now produce identical offspring containing the new desirable characteristics. In this way breeders can produce true breeding lines in two seasons—as opposed to about seven seasons by conventional means.[47]

Recombinant DNA

Recombinant DNA technology relies on a variety of naturally occurring enzymes, such as restriction enzymes, which act as molecular scalpels, and DNA lignase, which acts as a molecular suture. In effect this is a very sophisticated process of plant microsurgery that enables genes to be taken from one DNA helix and spliced into another and hence transferred between chromosomes. Although an artificial act, it is, in essence, the same process as occurs when the plant or animal breeder crosses one plant with another or one animal with another. During the crossing process, chromosomes transfer pieces from one to another, but the new combinations are usually randomly determined. The great advantage of recombinant DNA technology is that the new combinations are determined beforehand and, with skill and care, are precisely achieved. As a result, the plant breeder is no longer restricted to the genetic variation that arises in traditional breeding programs.

The process of transferring a gene to a new plant or animal can follow one of several routes (Plate 4).[48] One of the earliest and most successful techniques employs *Agrobacterium tumefaciens,* a bacterium that naturally invades potatoes, tomatoes, and alfalfa among other plants, as a vector to deliver new genes to plants. It is a surprisingly easy process: first recombinant DNA techniques are used to replace the genes

A. tumefaciens normally integrates into plant chromosomes with genes for desired new traits (e.g. insect resistance), then pieces of leaves are dipped into a suspension of bacteria, and cultured to produce whole plants containing the new gene. This technique now works effectively on cereals as well as broad leaf plants. Alternatively the new genes can be injected more forcefully. One approach is to apply a coat of DNA to gold particles, which are then fired into the plant cell with a microparticle gene gun. It is customary for millions of cells to be treated, and those in which the new gene have lodged are identified by means of a marker and cultured.

Genetic engineering can create new plant varieties and animal breeds that not only deliver higher yields but contain internal solutions to biotic and abiotic challenges, reducing the need for chemical inputs such as fungicides and pesticides and increasing tolerance to drought, salinity, chemical toxicity, and other adverse circumstances. Some of the most valuable applications lie in the conferment of resistance to bacterial and virus diseases and to insect pests. Resistance to viruses can be achieved by transferring to plants certain viral genes, such as those that encode for the virus coat protein. This interferes with normal replication of the virus and inhibits the spread of the infection. The technique has been used against several important viruses—rice stripe virus, alfalfa mosaic virus, and potato leafroll virus. The application of genetic engineering to insect pest control is described in Chapter 12.

In the developed countries, some of the most successful applications have resulted from engineering herbicide tolerance in crop plants, making it safe for herbicides to be applied to growing crops so that the weeds are killed but the crops are left unharmed. This has been achieved for crops such as maize and soybeans treated with the herbicide glyphosate. It is less relevant for developing countries, in part because hand weeding provides a cheaper form of control. So far, it has been Brazil and other South American countries that have employed such crops on a large scale, although maize double stacked with insect resistance and herbicide tolerance has begun to be grown in the Philippines.

Biofortification

The application of genetic engineering can be as valuable for the poor as for the better off. A good example lies in *biofortifcation*—the production of crops with enhanced nutritional value. Perhaps surprisingly, most cereals and other staples are deficient in a number of proteins and other micronutrients. For instance, maize is deficient in the amino acids lysine and tryptophan, which are essential for building proteins in the body. However, the capacity to produce these amino acids exists in the maize genome, requiring only the right genetic background for expression. In this case, a suitable mutant (known as *opaque 2*) has been found and used in a long, meticulous program of conventional breeding carried out by Evangelina Villegas and Surinder Vasal at CIMMYT to produce new high-yielding maize with these amino acids present.[49] This quality protein maize (QPM) is now being crossed with a wide range of locally adapted maize varieties using MAS. [50]

HarvestPlus, a challenge program of the CGIAR, is developing seven crops with enhanced levels of three critical micronutrients: zinc, iron, and vitamin A. In most cases, conventional breeding, assisted with MAS, has been used. Vitamin A–enriched sweet potato is now available in Uganda and Mozambique.[51] Over the next couple of years, HarvestPlus planned releases include vitamin A cassava and maize, iron bean and pearl millet, and zinc wheat and rice.[52]

Sometimes, however, the relevant genes are not present in the crop or in the critical components of the crop; for example, in rice where beta-carotene the precursor of vitamin A is not in the grain endosperm (it is present in the leaves and stalks, and minute amounts are contained in brown unmilled rice). In Asia, and many parts of Sub-Saharan Africa, poor families consume rice as the basic staple of their diet, and babies are often weaned on rice gruel. In this case, the approach has been to find suitable genes elsewhere and to use recombinant DNA to insert them into a new variety of rice, named "Golden Rice" after its distinctive color (Box 9.5).

The second generation of Golden Rice has already been developed into locally appropriate varieties in the Philippines and India. It is expected to be available in other countries in the next two to four years.[54] The Bangladesh Rice Research Institute expects to have conducted all necessary experiments on golden rice by 2012.[55] In an assessment of the economic impact golden rice could have on Asia, the benefits to Asia's GDP were estimated at $18 billion annually.[56]

Box 9.5 Producing Golden Rice[53]

In an effort to incorporate beta-carotene into grains of rice, Ingo Potrykus of the Swiss Federal Institute of Technology and his colleague Peter Beyer of the University of Freiburg first transferred two daffodil and one bacterial gene into rice. The biochemical pathway leading to beta-carotene is largely present in the rice grain but lacks two crucial enzymes: phytoene synthase (*psy*)—provided by a daffodil gene—and carotene desaturase (*crtI*)—provided by a bacterium gene. In the greenhouse, this transfer gave beta-carotene levels of about 1.6 μg/g, significant but not large. Subsequently, scientists at Syngenta have found new versions of the *psy* gene in maize; which, when introduced to rice, increased the beta-carotene levels to 31 μg/g. Given a conversion ratio of beta-carotene to vitamin A of 4:1, the new golden rice (Golden Rice 2) will be able to provide the necessary boost to daily diets, even after six months of storage.

Economic Benefits

Every year the International Service for the Acquisition of Agri-biotech Applications (ISAAA), under the leadership of Clive James, an agricultural scientist previously the deputy director general at CIMMYT, produces an independent estimate of global progress on the adoption of genetically engineered (GM) crops (referred to in their reports as "biotech crops").[57] In 2010, ISAAA reported that the accumulated area of biotech crops since 1996 had reached one billion ha in a total of twenty-nine countries (Figure 9.3).

Nineteen of the countries growing biotech crops were developing countries, and over 90 percent of the growers (14.4 million) were resource-poor farmers, including over six million each in India and China. Five developing countries—India, China, Brazil, Argentina, and South Africa—grow 43 percent of the global total.

There is no doubt that GM crops have brought significant benefits, despite the actual and perceived downsides. Area, number of different crops and traits deployed, and the number of farmers growing them continues to increase. Biotech crops are the "fastest adopted crop technology in the history of modern agriculture."[59] While it is relatively easy to calculate economic gains for a specified crop in a particular place, obtaining a global figure for the benefits of growing biotech crops is difficult. Clive James, however, has come up with a figure of economic gains at the farm level of about $65 billion since 1996. Of this amount just less than half, 44 percent,

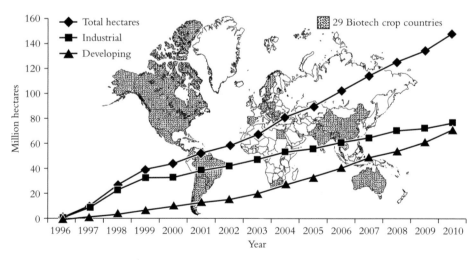

Figure 9.3 Global area of biotech (GM or genetically engineered) crops.[58]

resulted from reduced production costs (less plowing, fewer pesticide sprays, and less labor) and just over half, 56 percent, from substantial yield gains of 229 million tons. The latter needs some qualification. To date there has been little focus on inserting genes for yield increase; most of the yield growth has come from better control of pests and weeds and from incorporating yield-enhancing genes by conventional breeding in the lines containing the GM genes.

There are significant benefits from the reduction in pesticide use, especially where hazardous pesticides are being phased out. The accumulative reduction in pesticides for the period 1996 to 2009 has been estimated at 393 million kilograms (kg) of active ingredient (a.i.)—a saving of 8.8 percent in pesticide use.[60] There is also evidence of significant positive externalities. For example, some ten million Chinese farmers growing crops such as maize and soybean are benefiting from the control of bollworms on *Bacillus thuringiensis* cotton, where the same bollworm species attacks all three crops (see Chapter 12).[61]

The Hazards

The potentials for genetic engineering are almost endless. But there are hazards, some apparent, others more perceived.[62] In the developed countries, especially in Europe, there has grown a vociferous and fairly widespread opposition to the cultivation of GM crops. The reasons vary, from perceptions of serious health hazards, to environmental concerns, to opposition to the power of multinational seed companies. So far the benefits of growing GM crops are not perceived by European farmers to be sufficient for them to vigorously lobby for the restrictions on GM to be lifted. Supermarkets are also unwilling to break their promises not to sell GM foods in the face of often highly emotive opposition. In 1992 Paul Lewis wrote to the *New York Times* opposing GM tomatoes, referring to them as "Frankenstein foods," a description that continues to be used by the tabloid press.[63]

People have been consuming GM foods for fifteen years—in the United States often on a daily basis—without any obvious signs of ill health, but concerns over health hazards persist. Many studies purporting to show deleterious health risks in laboratory animals or humans are quoted, but few survive the peer review process and appear in the scientific literature (see an extensive review by Peggy Lemaux of the Department of Plant and Microbial Biology in the University of California, Berkeley).[64] Genetically modified foods and the products made from them are under regulatory control of three federal agencies in the United States: the Food and Drug Administration (FDA), the Environmental Protection Agency (EPA), and the U.S. Department of Agriculture (USDA).[65] The foods and their products undergo safety testing by the companies or institutions that develop them, and the data is reviewed by the federal agencies. Frequently they are also tested independently with reports in peer-reviewed journals. This process is comparable to safety assessments done for pharmaceutical drugs.

The daily human intake of DNA in food is estimated at 0.1–1.0 g.[66] Assuming 50 percent of the diet is from GM foods, the intake of transgenes is 0.5–5 μg/day.[67] DNA is chemically identical regardless of its source and is mostly degraded during industrial processing and in the digestive tract. Some small fragments can be detected in certain body tissues, but there is no evidence of consequent harm. A large number of studies, including multigenerational studies, of rats and mice fed glyphosate resistant soybeans have revealed no adverse effects on litter size, histological appearance of tissues, or numbers of deaths of progeny.[68] In the case of *Bt* proteins, whose use in GM I describe in Chapter 12, they occur naturally in the soil and have been used safely for forty years; for example, as bacterial sprays by organic farmers and others. The production of *Bt* proteins in GM crops has also passed the requisite safety tests.[69]

Penny Lemaux's conclusion from her exhaustive review of the potential health hazards of genetically engineered foods is that

> development of GE crops to date seems to have been responsible and regulatory agencies have, in general, proceeded with caution in releasing GE varieties. Although no human activity can be guaranteed 100 per cent safe, the commercial GE crops and products available today are at least as safe in terms of food safety as those produced by conventional methods. This does not mean we should relax our vigilance in investigating products resulting from this new technology as well as the time-honoured methods. But, we should not hold the new GE products to standards not required for food and feed products produced by other technologies and methods.[70]

Perhaps of greater concern is the possibility of genes moving from GM crop plants to other crops, such as organic varieties or to wild relatives, creating unforeseen consequences.[71] Genes can indeed move in this way. Many crops have wild relatives, and hybrids may naturally occur, which would permit the new genes to pass from the crop. In Canada hybridization between transgenic canola (oilseed rape, or *Brassica napus*) and neighboring weedy *Brassica rapa* has occurred.[72] The hybrids declined over time, but the herbicide-tolerant transgene persisted in the *B. rapa* population without herbicide applications from 2003 to 2008. The *B. rapa* populations can be controlled by other herbicides, but this means that weed control options are reduced.

Rice plants are self-pollinating, but not exclusively so. A degree of cross-pollination does occur under natural conditions, both among cultivated rice and between cultivated and wild rices. There is thus a likelihood in Asia of *Bt* transferring to the wild relatives, particularly *Oryza nivara* and *O. rufipogon*.[73] These could then become serious weeds, although the evidence suggests they are not presently limited by stem borers or other caterpillars.

The developed countries are better equipped to assess these hazards than developing countries. They can call on a wide range of expertise, and most have now set up regulatory bodies and are insisting on closely monitored trials to try and identify the likely risks before GM crops and livestock are released to the environment. The developing countries are now rapidly following suit, although they are under pressure from some European NGOs and indeed from some European governments to legislate in such a way that growing of GM crops becomes virtually impossible; for example, by inserting strict liability terms in the regulations.[74]

Finally there is concern that the technology has been captured by large, multinational corporations and is consequently exploitive. Monsanto is usually regarded as the arch villain, and it is true that in the early years of the exploitation of GM technology it gained considerable bad publicity for its prosecution of farmers who ignored their contracts. In 1999, I addressed the board of Monsanto in Washington, DC, and recommended a number of actions to improve their standing, including an agreement not to commercialize the so-called terminator gene that would produce sterile seeds from GM crops.[75] To their credit they agreed to most of these recommendations and in subsequent years have done much to improve their image; for example, by donating drought-tolerant genes of maize for use by African plant breeders, free of charge (see Chapter 15). Part of the answer lies in greater investment in such technologies by the public sector and ensuring that the technologies and their products are readily available to developing countries and to poor populations.

One example is the current effort to produce GM bananas resistant to the new, devastating scourge banana wilt (BXW) disease. Scientists at the Ugandan National Banana Research Program are currently conducting field trials of three banana varieties resistant to BXW with results expected at the end of 2011. Scientists obtained two different sweet pepper genes one encoding the hypersensitivity response and one a plant ferredoxin-like protein, which were inserted into bananas in the government laboratory. These transgenes were donated by Academia Sinica of Taiwan, free of charge for use in Sub-Saharan Africa.[76]

Fundamental Challenges

If we are to significantly increase yields and produce new crop varieties and animal breeds that will perform at close to the agroclimatic frontier, we are going to need to produce varieties that make much more efficient use of sunlight, nitrogen and other nutrients, and water.[77] The aim is not only to increase productivity but to do so in a way that is both stable and resilient—and is also accessible to smallholder farmers. I return to these challenges in the chapters ahead.

10

The Livestock Revolution

A revolution is taking place in global agriculture that has profound implications
for our health, livelihoods, and environment.

Christopher Delgado, Mark Rosegrant, Henning Steinfeld,
Simeon Ehui, and Claude Courbois[1]

The Green Revolution was not the only agricultural transformation of the twentieth century. Less well understood and celebrated was a revolutionary growth in livestock production, distinguished by the great diversity of products it embraced: meat,
milk, cheese, and eggs from, variously, cattle, sheep, goats, pigs, poultry, and other
more exotic creatures.

There are nearly 4,000 species of domesticated animals, but only twelve dominate
global livestock production.[2] In Africa, livestock owners rely on cattle, sheep, goats,
donkeys, and camels; in central Asia, they keep horses, cattle, goats, sheep, donkeys,
and, in some areas, Bactrian camels. Yaks dominate production in the highlands of
Asia. Llama and alpaca are common in Andean systems of South America. Water
buffalo are important in India and elsewhere in Asia.[3]

Most of these are ruminants: They possess a rumen in front of the stomach that
provides a slow pregastric fermentation of plant fiber by bacteria, protozoa, and
fungi. As a result, ruminants can survive on fibrous, low-protein feeds that are poorly
digested by nonruminant animals. Another advantage is that the microbes in the
rumen can detoxify plant toxins.[4]

In this chapter I address the following questions:

- What has been the nature of the Livestock Revolution?
- What have been its benefits?
- What are the evolutionary relationships among grazing, mixed systems, and
  industrial production?
- What are the main challenges ahead?

The Nature of the Revolution

Christopher Delgado and his colleagues at the International Food Policy Research Institute (IFPRI) and at FAO documented the dramatic growth in livestock husbandry that occurred in the second half of the twentieth century and coined the phrase "Livestock Revolution" to highlight its importance. In some respects the Green and Livestock Revolutions were similar: They were truly transformative in their technologies and in their social and economic consequences. They both had their origins in developed countries but profoundly affected the developing countries. The significant difference was that the Green Revolution was a supply-led response to hunger and the shortage of calories, whereas, at least initially, the Livestock Revolution was a demand-led response to the rapid increase in the consumption of livestock products in the developed and emerging countries. Unlike the Green Revolution the technologies were already mostly in place.

In simple terms, the Green Revolution provided farmers with calories, while the Livestock Revolution provided rapidly increasing incomes (Table 10.1). In the roughly twenty-five years from 1971 to 1995, the additional meat, milk, and fish consumed in developing countries was greater than in developed countries, although it was only two-thirds as important, in terms of quantity, as the increase in wheat, rice, and maize. The Green Revolution provided many more calories than the increase in meat, but the latter was worth, in monetary terms, almost three times as much.

Table 10.1 Increase in food consumption of meat, milk, fish, and cereals from 1971–1995

Commodity	Consumption increase (in million metric tons)		Value of consumption increase[a] (in billion 1990 US$)		Calorie value of consumption increase (in million kilocalories)	
	Developed	Developing	Developed	Developing	Developed	Developing
Meat[b]	26	70	37	124	38	172
Milk	50	105	14	29	22	64
Fish[c]	5	34	27	68	4	20
Major cereals[d]	25	335	3	65	82	1,064

Source: Delgado et al.[5]

[a] Calculated using 1990 world prices expressed in constant, average 1990–1992 US$

[b] Beef, sheep, and goat meat, pork and poultry

[c] Marine and freshwater finfish, cephalopods, crustaceans, molluscs, and other marine fish

[d] Wheat, rice, and maize used directly as human food

Since 1995 consumption per capita of meat, milk, and eggs has continued to increase in the developing countries, and, not surprisingly, it is the emerging nations, especially China and Brazil, who have become the big producers. Between 1980 and 2007 China increased its production of meat more than sixfold, accounting for nearly 50 percent of meat production in developing countries and just over 30 percent of world production. Brazil expanded meat production fourfold, now contributing 11 percent of developing country meat production and 7 percent of global production.

For developing countries other than China and Brazil, the figures are much lower. In 2006, there were twenty-six developing countries consuming less than 10 kg per capita per annum of meat and twenty-three countries where the consumption levels were less than those ten years earlier.[6] Although India has doubled meat production since 1980, meat production levels remain low in a global context. In contrast, after more than tripling milk production, India now produces some 15 percent of the world's milk. Production of meat, milk, and eggs also increased in Sub-Saharan Africa but more slowly than in other regions.

The Benefits of Livestock

The popularity of livestock production and consumption is not difficult to comprehend. A list of the myriad of benefits makes the point (Box 10.1).

Box 10.1   The Benefits of Livestock Production and Consumption[7]

- Contribute 40 percent of the global value of agricultural output
- Support the livelihoods and food security of almost a billion people
- Provide food and incomes
- Contribute 15 percent total food energy and 25 percent dietary protein, globally
- Provide essential micronutrients (e.g., iron, calcium) that are more readily available in meat, milk, and eggs than in plant-based foods
- Are a valuable asset, serving as a store of wealth, collateral for credit, and an essential safety net during times of crisis
- Are central to mixed farming systems: consume agricultural waste products, help control insects and weeds, produce manure and waste for cooking, and provide draft power for plowing and transport
- Provide employment, in some cases especially for women
- Have a cultural significance, as the basis for religious ceremonies
- Confer status

The Importance of Animal Source Food

Probably the most significant, yet still controversial, benefit of livestock is their contribution to the human diet.[8] Animal source foods (ASFs) have several unique characteristics.[9] They are energy-dense and are good sources of protein and of many key micronutrients. They can measurably enhance nutritional quality in diets, especially for nutritionally vulnerable groups such as young children and pregnant and lactating women. In many cases, nutrients in ASF (e.g., iron and zinc) exhibit greater bioavailability than those from plant sources. Moreover, meat and fish are sources of heme-iron which is better absorbed than the non-heme found in fortified cereals and plants such as lentils and beans.[10]

As I described in Chapter 2, the focus of nutrition interventions has evolved over the past century from control of protein deficiency, to concern about protein-energy deficiency, and now to the prevention and treatment of micronutrient deficiencies.[11] Three broad categories for improvement in microdeficiencies in the diet are supplementation, fortification, and dietary improvement:

- Supplementation often fails to supply all necessary nutrients. Individuals in nontargeted groups are usually neglected and compliance is poor especially when the supplements need to be taken very frequently and for extended periods of time.
- Food fortification programs can lead to sustainable improvements, but so far they have yet to reach the poorer and undernourished segments of the population. There have also been some technical barriers to fortification, although these have been largely overcome.
- This has led Lindsay Allen, a nutritionist at the University of California, Davis, and others to argue that increasing micronutrient intake by improving dietary quality is the ideal approach and specifically to improve the uptake of ASF (see Chapter 2).[12]

In the early 1980s, an observational study under USAID's Nutrition Collaborative Research Support Program was conducted in parallel in Egypt, Kenya, and Mexico, relating outcomes to measurements of dietary intakes for different categories of individuals: pregnant women and their infants, preschoolers, school children, adult men, and nonpregnant/nonlactating women. Although protein quantity and quality was measured as adequate, it became evident there were serious microdeficiency intakes in each locality. As a result the intake of animal source protein, although a small contributor to energy intake, became the best predictor of growth, lactation outcome, and cognitive function.[13]

The next step was to undertake intervention trials. The results of one of these, carried out on Kenyan schoolchildren, are described in Box 10.2

Box 10.2   Effects of Animal Source Food on Kenyan Children[14]

The experiment

Twelve rural Kenyan schools in Embu District were selected. Five hundred and fifty-four children were randomized to four feeding interventions using a local vegetable stew as the vehicle. The groups were designated as Meat (daily supplement of two ounces of meat), Milk (a cup of milk each day), Energy (oil supplement of equivalent energy), and Control (no school intervention feedings). Feeding was carried out on school days for seven terms over twenty-one months, and the measurements of each child were repeated at intervals over two years.

The baseline

Baseline data revealed stunting and underweight in 30 percent of children and widespread inadequate intakes of micronutrients, particularly of iron, zinc, vitamins A and B-12, riboflavin, and calcium. Little or no ASF was eaten, and fat intake was low. Malaria was present in 31 percent of children, and hookworm, amebiasis, and giardia were widely prevalent.

The outcomes

Growth—all the supplements increased weight gain, and the addition of meat increased their lean body mass. "The meat group had 80 per cent more increase in muscle mass over the two years of the study, and the milk and energy group had 40 per cent more increase."

Cognitive development—all children with meat supplementation significantly outperformed on a cognitive test that measures on-the-spot reasoning and problem-solving ability. The children "were more active in the playground, more talkative and playful, and showed more leadership skills." There were no group differences on tests of verbal comprehension.

Conducting trials of this kind is difficult due to many compounding factors—such as the health status of the children, the quality of the school, and the influence of the parents. For example, more educated mothers provide more ASF to their children.[15] Nevertheless, the results proved clear enough for the authors to conclude that "In developing countries where food security is low and children are mild to moderately undernourished with multiple micronutrient deficiencies, agricultural interventions to assist families in raising small animals for family consumption are likely to have significant positive impacts on children."[16]

Needless to say, those campaigning for vegan and vegetarian diets believe this to be unnecessary.[17] They maintain that any micronutrient deficiencies in such diets can be rectified by fortification. This may be true for developed countries, but providing ASF may be easier in situations where nutrition in general is very poor. There is also a growing argument that a meat-eating diet is unsustainable and will add to the pressure on land and other resources. A meat diet is likely to be more demanding of land, water, and energy than a vegetarian (plus milk and eggs) diet, but by how much is not clear.[18] Some European countries may shift significantly to a lower intake of meat, but this seems unlikely for countries such as China and Brazil where meat eating is ubiquitous. As we saw in Figure 1.5 there appears to be an inexorable trend to greater meat and other livestock consumption as per capita incomes grow.

Livestock Livelihoods

The number of poor people who depend on livestock is not well established, but the commonly cited figure, which is somewhat out of date, is about a billion or some 70 percent of the world's "extreme poor."[19] FAO's Rural Income Generating Activities (RIGA) database of eighteen countries in Africa, Asia, Latin America, and Eastern Europe reveals that 60 percent of rural households keep livestock, and this is true of even the poorest households.[20]

Small, poor farmers engage with the livestock system in a variety of ways, and the degree of livelihood dependence on livestock and their associated products covers a wide spectrum. In some respects, livestock are the ideal core of a smallholder livelihood. They are multifunctional—providing income, quality food, fuel, draft power, building material, fertilizer, and, crucially, a safety net for when times are hard. For some farmers, livestock are one element of a basic subsistence; others may choose to sell high-value products such as milk and eggs and buy lower-cost staple foods; and for others, raising livestock may be solely a commercial enterprise.

Given this range of beneficial functions, the question is: Who benefits from the Livestock Revolution? The rise in meat consumption as a function of higher per capita incomes also holds true within countries; it is the better off in the developing countries—especially in the emerging countries, such as China and Brazil—that have been benefiting from the greater availability of livestock products. Larger producers have tended to benefit most, but the issue is whether small, poor farmers have also been beneficiaries. In a review of over 800 livestock projects and programs by the UK Department for International Development (DFID) in 1998, the lack of significant impact on poverty for the majority of projects was concluded to be due to inadequate transmission of, often inappropriate, technologies to the poor. The benefits were mostly captured by wealthier producers.[21] Nevertheless, keeping a few goats or chickens or a dairy cow is practiced by people with few land assets, typically poor in rural contexts. Ther are few other comparable income-earning opportunities of this significance open to poor people in rural areas. In the view of FAO, growth in the livestock sector can foster wider economic growth and can benefit development if policies support smallholders, particularly those that rely on livestock as a safety net.[22] A good example is the dairy goat project described below in Box 10.6.

Livestock and the Environment

Alongside the benefits, some of the consequences of the Livestock Revolution have been negative or threaten to be so. For the most part, these reflect the impact of livestock on the environment. Free-range animals, especially cattle and goats, can have serious effects on ecosystems—on vegetation, soils, and water and on biodiversity—if not managed well. In addition to suffering from climate change, livestock are also direct and indirect producers of greenhouse gases. Although they contribute less than 2 percent of global gross domestic product (GDP) they produce 18 percent of global greenhouse gas emissions (see Chapter 16).[23] Finally, livestock often harbor zoonotic diseases that may be transferable to humans. Recent examples include avian and swine influenzas.

Most of these negative consequences are relatively small in their impact and can be contained when livestock numbers are low, but many have become considerably more serious with the rapid increase in livestock numbers. Some of the solutions are technical (see Chapter 16), but structural solutions may also be necessary as Jacques Diouf, director general of the FAO, in his introduction to the 2009 FAO report, points out, "The rapid transition of the livestock sector has been taking place in an institutional void. The speed of change has often significantly outpaced the capacity of governments and societies to provide the necessary policy and regulatory framework to ensure an appropriate balance between the provision of private and public goods."[24]

I will refer to these problems and concerns in various parts of this and later chapters. But I will begin by discussing the diversity of livestock systems that exist in developing countries.

The Diversity of Livestock Systems

Livestock systems vary enormously in terms of scale and intensity of production. At one end of the spectrum are traditional pastoral systems involving nomadic herds of animals; at the other, highly industrialized systems of production in dedicated built structures. It is difficult to capture this range in a simple classification, but the FAO categories are a useful approximation:

- *Grazing systems*—utilizing pastures or rangeland (extensive or intensive)
- *Mixed systems*—combining livestock and crops (rain-fed or irrigated)
- *Industrial systems*—livestock in buildings or pens

Grazing Systems

Grazing systems occupy some 26 percent of the earth's ice-free land surface.[25] The most extensive forms exist in the dry areas of the world that are marginal for crop production. They typically support ruminants—cattle, sheep, goats, and camels— grazing mainly grasses and other herbaceous plants, often on communal or open-access rangelands. Both animals and humans may move considerable distances across the grazing lands, and the owners may sometimes be nomadic.

In the developing countries, grazing systems are commonly characterized by low or erratic precipitation, poor drainage, extreme temperatures, rough topography, and other physical limitations that render them unsuitable for cultivation. Variation in available moisture and nutrients produces a mosaic of systems, some with rich natural grass swards, others covered with rough, sparse vegetation. Most have varying extents of shrub and tree cover and grade into woodlots and forests. Collectively they support most of the developing world's population of cattle, sheep, and goats; they are also an important source of fuelwood and a variety of other natural resources.[26]

The Problem of Overgrazing

Much of the grazed range land is, depending on the purpose to which it is put, degraded. For many years the conventional wisdom has attributed this to overgrazing, but recent ecological research has suggested the concepts of overgrazing and degradation have been seriously oversimplified. At the heart of the problem has been the misapplication of management theories developed to maximize beef production on temperate grasslands in the developed countries (Box 10.3).

As the box makes clear, "overgrazing" is relative to the aims of the farmer, as is "degradation." There is no objective measure and the state of the vegetation will vary considerably. The critical question is how resilient is the livestock system; can the soil and vegetation conditions be reversed simply by reducing the stocking rate? Because the rainfall is highly variable and unreliable, there is no permanent target point on Caughley's curve to aim at. The curves move with the rainfall; the state of the vegetation changes dramatically from year to year, often more as a result of the

Box 10.3   Carrying Capacities of Grazing Systems

In 1979, Graeme Caughley, then at the Division of Wildlife Research of CSIRO in Australia, proposed a model of the relationship between animal numbers and vegetation that distinguishes ecological and economic carrying capacity (Figure 10.1).

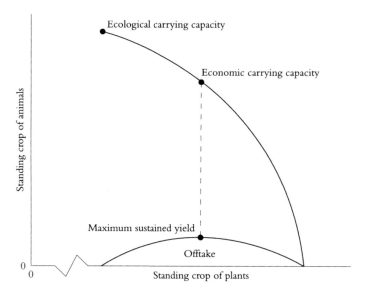

Figure 10.1   The relationship between animals and plants in a grazing system.[27]

In the upper curve, the ecological carrying capacity occurs when the production of forage equals its consumption and the animal population ceases to grow. At this point there will be many animals, but they may not be in good condition and the vegetation will be much less abundant.

The lower, offtake curve indicates the various combinations of animals and vegetation that will produce a sustainable meat yield; its maximum is the economic carrying capacity. For beef production in temperate pastures, this is an appropriate goal, but in developing countries other targets may be more appropriate. For pastoralists, where the aim is to maximize a range of livestock products—milk, blood, traction power, and transport—other than meat, it is profitable to maintain a large stocking rate since the off-take does not usually require slaughter.

rainfall than the animal stocking density. Stable conditions are rare, and carrying capacities difficult, if not impossible, to estimate. In addition to climatic variability, outbreaks of diseases, political upheavals, and policy changes add to the disturbances.

Pastoralism

Pastoral livestock populations thus tend to go through periodic cycles of "boom and bust." They are also under threat due to increasingly stringent sanitary standards, loss of access to grazing areas through competition with arable cultivation or through conservation measures, and restrictive border controls.[28] Poor pastoralists also lack sufficient labor and have limited access to information, technologies, and markets.

For far too long, African and other developing country rangelands have been subject to policies based on temperate, beef-producing pasture systems. These have led to the encouragement of individual land ownership, reinforced by construction of fencing and other measures to restrict free-range grazing. But such an approach is rarely appropriate. The large numbers of people involved in pastoralism, the diversity of products being sought, and the great unpredictability of climatic conditions dictate an approach that is flexible and opportunistic. Pastoralism, with systems

Box 10.4   Masai Pastoralism in Kenya

The Masai are one of the best known nomadic herders, but in recent years many have become sedentary pastoralists, partly due to the lack of protection the government has afforded their lifestyle.[30] This has also caused declines in natural resources and wildlife.[31] For example, the fencing of habitat for agriculture prohibits the migration of wild animals such as wildebeest and increases illegal poaching and retaliatory killings in defense of agricultural holdings and livestock.[32]

In a study of the Kitengela plains a team from ILRI found coexistence between wild animals and livestock to be possible and even beneficial for the wildlife, given their preferences for feeding on nutritious and short grass found in and near pastoral lands.[33] Maps produced with the aid of Geographic Information Systems (GIS) helped community monitoring of animal distributions and were used to determine payments to farmers for leaving land unfenced so protecting corridors for animal migration.[34] ILRI then went on to develop a conservancy model that enables Masai landowners working with tourism companies to manage settlements and livestock and to receive payment for ecosystem services.[35] Three conservancies now manage around 350 square km of land generating an income of $3 million per year for around 1,000 local households.[36]

of management evolved for these circumstances, needs improvement and support rather than replacement (Box 10.4).[29]

In their review of rangeland management in Africa, Roy Behnke and Ian Scoones of IDS argue for a change in attitudes. Explicit recognition of the inherent instability of pastoral systems should encourage governments to support supplementary feeding during severe droughts, through the provision of feed and the establishment of feeding pens, and to facilitate rapid destocking and restocking at the beginning and end of droughts through livestock markets designed for this purpose.[37]

Mixed Systems

The other major livestock system in the developing countries is the *mixed system* where cropping and livestock rearing are carried out on the same holding with some degree of integration. FAO defines these systems as either where more than 10 percent of the dry matter fed to animals comes from crop by-products or stubble or where more than 10 percent of the total value of production comes from nonlivestock farming activities. There is a great diversity of such systems, and in practice there are no hard-and-fast distinctions between mixed systems and grazing or industrial systems.[38]

However, a fundamental principle of mixed systems is the mutual utilization of waste—crop waste to feed animals, and livestock waste to provide nutrients for crops—but there are many other cross-benefits. They may vary from providing a continuity of labor demand to providing grazing vegetation under trees. Grazing cattle and goats under oil palm in Malaysia produces better oil palm harvests and saves on the costs of weed control, as well as increasing sources of income.[39]

In recent years there has been a progressive evolution from grazing toward mixed farming systems. It is driven partly by population pressure and partly by changing markets. Arable farmers need to increase production; the presence of animals on a farm can provide wastes for fertilization and a useful form of traction, allowing for cultivation of more land. Crop residues are increasingly collected from the arable fields and transported to the homestead to feed animals that are kept in kraals or stables. Eventually, farms adopt zero grazing, with animals fed on residues or grain, grown on the farm or purchased. At the same time, pastoralists left with less and less land for grazing, often of poor quality, migrate to the cities or adopt arable farming.[40]

Dairy Cattle in Mixed Farms

In many developing countries, the most integrated and commercialized mixed farms are those centered on the keeping of dairy cows. Milk consumption in the developing countries increased by 105 million tons (mmt) of liquid milk equivalents (LMEs) between 1970 and the mid-1990s, two times the increase that occurred in the developed world.[41] The Tetra Pak Dairy Index predicts a 30 percent growth in the consumption of liquid dairy products from 270 billion liters in 2010 to 350 billion liters in 2020, driven largely by economic growth in Asia.[42]

India is now the world's largest milk producer, following a fourfold increase in milk production from cattle and buffalo between 1963 and 2003. Milk is thus India's largest "crop." Although over 80 percent of farmers in India, or about 93 million farms, crop less than two hectares, the number of farmers engaged in dairy production has increased by around 40 percent between 1991 and 2000. At the same time, average herd size has declined; 60 percent of dairy farms now support only one to three head of cattle or buffalo.[43] There has been a rapid increase in numbers of buffalo and crossbred cattle at the expense of local cattle breeds; over a third of the increase in milk production is due to the higher productivity of the crossbred animals.

The growth of smallholder dairy production in India owes much to the active support of government-sponsored programs, such as Operation Flood, and a major effort to market milk in urban areas (Box 10.5).[44]

Box 10.5 Operation Flood[45]

The story of the extraordinary growth of the dairy sector in India in the second half of the twentieth century ranks alongside the Green Revolution and, indeed, owes much to it.[46]

Its origins, however, go back further to the freedom movement of the 1940s when "two village cooperatives producing 246 litres of milk" were established with the express aim of ending the exploitation of small milk producers by middlemen and the agents of the large dairies.[47] Amul (the Sanskrit word meaning "priceless"), as the cooperative movement came to be called, grew rapidly, attracting the attention of political leaders and influenced by the visionary leadership of Dr. Verghese Kurien, chairman and founder of the National Dairy Development Board in India.

For India as a whole, however, per capita milk production remained stagnant and sometimes declined. The government of India responded by embarking on a plan,

(continued)

---

named Operation Flood, in the 1960s to transform the whole dairy sector based on the Amul experience. Funds from the sale of European surplus milk powder and of butter oil provided through the World Food Program were used to create primary village cooperative societies linked to district and state cooperative unions for milk collection and processing. The program also improved animal health and breeding services, including a rapid rise in artificial inseminations that resulted in an increase in the percentage of crossbreeds from nearly 5 percent in 1982 to over 13 percent in 2003.

Initially the focus was India's major milk producing areas, the "milksheds," which were then linked to the four main cities of Mumbai, Kolkata, New Delhi, and Chennai. Average milk procurement through cooperatives increased from less than 2.5 million liters per day to nearly 11 liters. However, this was only about 6 to 7 percent of India's total milk production. The cooperatives of Gujarat and Maharashtra remained successful and productive, and at the forefront of new technologies, including the spread of crossbred cattle. But in other states the success of Operation Flood cooperative development was mixed, partly due to political interference in cooperative governance and competition from informal, traditional, and private-sector players. Significantly, milk production grew even faster in Pakistan where demand was high and there were no public cooperatives.

After Operation Flood ended in 1996, the government enacted a new set of policies that favored increasing privatization of the dairy industry, opening up the industry to competition and ending the protection of the milkshed-based cooperatives. However, Amul has continued to flourish, today comprising 13,000 village societies, 2.5 million milk-producing households, and, at the peak, procuring over eight million liters of milk a day.

The Value of Dairy Goats

Cattle and buffalo are not the only producers of milk. Goat milk is highly nutritious, with a composition similar to human milk and thus easily digested by infants.[48] It is an excellent source of calcium, phosphorus, and vitamin A. Goats are also efficient producers of milk in conditions where cattle do not thrive. They are "often raised at the margins of communities. This is true in the physical sense that goats . . . are well adapted to the harsh conditions and poor quality feed found at the interface between deserts, mountains and cultivable land, and on 'waste' land within cropped areas where poor and landless people are also found."[49] The downside is their capacity to inflict severe environmental degradation through overgrazing, and it is partly for this reason that attempts are being made to bring them into mixed farming systems with low- or zero-grazing. Since this requires relatively little investment, it should be within the reach of poor farmers (Box 10.6).

Box 10.6   Dairy Goats in Ethiopia[50]

Ethiopia suffers from widespread hunger: It is not self-sufficient in animal products and is a net importer of food. Only 7 percent of food energy is derived from animal source foods and daily per capita protein and fat consumption is 7 and 8 g, respectively, far below recommended levels. Annual per capita milk consumption fell from about 20 liters in 1993–1994 to 16 liters in 2003, while annual meat consumption is as low as 11 to 12 kg per person.

The Dairy Goat Development Project was initiated in Ethiopia with the help of an international NGO, FARM-Africa, to increase both incomes and milk consumption. The focus was on women because they are the traditional keepers of goats. They were encouraged to grow fodder, trained in goat management and the poorest women identified by the community, each given two female goats on credit (repaid through returning a weaned kid to the project in order to lend to another woman). As the project progressed, the women organized themselves into groups to handle a credit disbursement and repayment scheme. They also established joint saving schemes, which enabled them to gain access to a source of finance. Over time, and with further extension support and animal health training, group members were able to use exotic goats to improve their existing stocks.

At the start, in one of the projects in the Eastern Hararghe in the southern province of Oromia, 20 percent of households had no access to milk and nearly 60 percent of households consumed no meat. Many of the households gave infants and young children water mixed with sugar and fenugreek juice as a substitute for milk. These children exhibited signs of ill health; for example, chronic diarrhea, itching, and skin infections.

Three years later there were clear improvements, the benefits reaching over 5,000 households. The women proved to be adept at managing crossbred goats, and milk yields increased from 200 ml per day to over 2 liters, while lactation length increased from 2 to 3 months to over 12 months in some cases. Each participating household was milking its lactating goats twice a day and per capita milk consumption averaged 15 liters a year. Goats could be sold during periods of drought, reducing communities' dependence on aid.

"Each of the households has managed to generate steady annual income from goat sales. Through this, they have since acquired diversified assets such as cows, oxen and donkeys, have been able to start up small businesses of their own, or have invested in improved agricultural technologies, such as inorganic fertilizers, improved poultry stocks and select grain seed. Such initiatives have raised crop and animal production and enabled the households to send their children to schools as well as improve their welfare."[51]

This is a clear example of the virtuous circle at work.

Industrial Livestock Systems

Just as there has been an evolutionary trend from grazing systems to mixed systems, there has been a similar trend, and for similar reasons, from mixed to industrial livestock systems. Operations have become progressively larger and more intensive, benefiting from technical advances and economies of scale, particularly in processing and marketing. There has been a strong tendency toward vertical integration where all parts of the supply chain are united in a common enterprise. Such integration is particularly pronounced in poultry production. Although typical of industrial countries, such systems are also increasingly abundant in developing countries, especially in the emerging economies.

FAO defines *industrial livestock systems* as those that purchase at least 90 percent of their feed from other enterprises—usually grain or industrial by-products. They are mostly intensive and are often found near large urban centers. As demand increases, large-scale operators emerge co-locating the livestock facilities, slaughter houses, and processing plants, often near urban areas because of the perishable nature of the products. One of the downsides is the greater risk of disease and of environmental pollution.

In emerging economies the increasing demand for livestock products is being met by large-scale, intensive, and technologically sophisticated producers. The trend is for fast turnover, large quantities, and low price coupled with strict food safety standards. A good example is the rapid increase in production of chickens in China for meat and eggs,[52] where the driving factors include the following:

- Increased feed conversion ratios for broilers in large-scale enterprises to levels comparable to those in the industrial countries
- Improvements in transport infrastructure, with railways important for feed distribution and roads for transport of products
- Creation of large-scale (usually private) operations (with annual output of more than 10,000 birds) controlling one-half of production
- Domination by large, integrated companies that control the entire production and marketing chain
- Rise of contract rearing as the norm, with the integrator supplying feed and chicks, together with various services and advice, and buying back finished birds

One large, integrated operation in Fujian Province produces 50 million broilers a year and employs 4,000 employees—one job for every 12,500 birds produced annually.[53]

Animal Foodstuffs

With the transition in the developing countries from pastoralism to mixed farms has come a greater emphasis on improved fodder crops; at the same time, the transition to intensive industrialized farms has increased reliance on purchased feed concentrates.

Arguably the biggest constraint on mixed livestock farming is the availability of fodder crops. This is particularly true of rain-fed systems in regions with a dry season when fodder crops on the farm are in short supply or absent. Inevitably, the harvested fodder has to be bought in, and transport costs can be prohibitive. One solution is to convert fodder into a semi-concentrated form. The plains of the northern Indian state of Uttarakhand produce abundant fodder, but above them in the steep Himalayan foothills there are considerable shortages. R. J. Sharma of the College of Veterinary Science at the Agricultural University of Pantnagar has devised a simple mechanical compress that produces 4-kg "bricks" of fodder made up of 50 percent chopped straw or cane tops blended with 39 percent concentrate and 11 percent molasses, which increases palatability and acts as a binding material. The bricks are cheaper to transport by truck than bulk fodder and last the six months of the dry season. One brick will feed a small cow for a day.[54]

Increasing attention is also being paid to developing better fodder crop varieties or finding new exotic fodder crops, such as leguminous trees. The cultivation of multi-purpose crops—such as the cowpea in West Africa—is one solution being explored. Until recently, different varieties of cowpea have been grown for different purposes: early-maturing varieties cultivated for the peas during the relatively short growing season, with later-maturing varieties grown for fodder. Now there are dual-purpose varieties, developed by scientists at IITA and ILRI for the semi-arid northern state of Kano in Nigeria. These are drought tolerant and mature much later, producing both peas and fodder in the dry season.[55]

Another possibility is to grow sweet sorghum—for human food, biofuel, and animal feed (Box 10.7).

The use of feed concentrate, usually high in energy and protein, instead of fodder makes handling easier. It also removes the need to source locally because feed concentrates are internationally traded. To date, the use of feed concentrates as a percentage of all feeds is only about 12 percent in developing countries, compared with 40 percent in developed.[57] But total use there is rising rapidly, having more than doubled between 1980 and 2005 from 240 million tons to just over 600 million tons (in developed countries it has remained stable at around 650 million tons).[58] The ingredients used are similar, although cereals predominate in developed country feed, and roots, tubers, and oil cake are more common in developing countries that favor monogastrics such as pigs and poultry.[59]

Box 10.7   The Multiple Benefits of Sweet Sorghum[56]

Sweet sorghum is the common name given to sorghum varieties with high sugar content. They are drought tolerant and highly water use efficient—needing far less water than sugar cane. Ordinary sorghum is an important grain and fodder. Recently attention has turned to sweet sorghum, for a third purpose, as a source of ethanol biofuel. Once the juice for ethanol production has been extracted from the stalks, the stalk residue (bagasse) can be combined with leaves stripped from the stems and returned to the farmers, providing a nutritious animal feed, readily palatable and digestible, rich in micronutrients and minerals.

Preliminary estimates suggest farmers could make an extra $40 per hectare compared with ordinary sorghum. However, transport costs are likely to be a limiting factor. A brick similar to that discussed above is being developed. The alternative may be to establish decentralized juice extraction and syrup-making units serving farmers in and around a cluster of villages. The sweet sorghum stover for biofuel production could remain on-site, making it easier for farmers to benefit from the by-products, while the syrup would be transported to a central distillery to be converted to ethanol.

Breeding Better Forage

Finally, there is an urgent need to develop better-quality fodder and feed and improved livestock breeds if the growing demand for livestock products is to be met without increased pressure on land and environmental resources.

There has also been a long history of breeding aimed at improving the productivity and quality of fodder crops. Most of this has taken place in the temperate climates of the developed countries, but there have been significant efforts at breeding tropical forages in Australia and in several parts of the developing world.[60] Some of the most advanced work has been in Latin America, conducted by CIAT (Centro Internacional de Agricultura Tropical) in conjunction with national research institutions. In the humid and semi-humid tropics of Latin America and elsewhere, most pastures consist of native species of inferior nutritional quality, which limits the intensification of livestock production. However in recent decades a number of tropical grasses and legumes have been identified that are highly productive, of superior nutritional quality, and are well suited to marginal lands characterized by low soil fertility and drought.

For the last thirty years CIAT has been developing improved forages for three Latin American agroecosystems—the savannas, the forest margins, and the hillsides. In each case they are seeking forages that not only improve livestock production but also reduce erosion, help to control weeds, and reverse land degradation.[61]

At the forest margins and on the steep hillsides, free grazing can rapidly degrade the land, and, consequently more land is cleared, creating a vicious cycle of clearance and degradation. The goal is to find improved grass and legume forage that will support stable, sustainable livestock smallholdings without further land clearance.

Over the years CIAT has built a collection of over 22,000 forages. Often it is legume–grass mixes that perform particularly well; for example, a mixture of the legume *Arachis pintoi* when combined with grasses increases milk yields by 10 to 15 percent.[62] Among the grasses, recent breeding has resulted in improved varieties of *Brachiaria* (Box 10.8).

Box 10.8   Improving *Brachiaria*[63]

Brachiaria is a predominantly African grass genus with about a hundred species; some perennial and well suited as forage grasses. Several species were introduced into tropical America some thirty years ago and are now grown extensively—over 70 million hectares in the region—composing more than a third of the total savannas. They have a great variety of common names in English, Spanish, and Portuguese, including signal grass, Brachiaria de Abisinia, and Brizantão. Providing phosphorus, lime, and various trace elements are added to the soil, *Brachiaria* can be established and will be very productive (live weight gains of 100 to 300 kg/ha/year for several years).

But it faces a number of problems, some due to the speed of introduction. First, the high carbon to nitrogen ratio in the abundant litter produced by the grass results in nitrogen deficiency; the soil bacteria compete with the grasses, resulting in degradation of the pasture. The solution appears to be growing the *Brachiaria* with an N-fixing legume such as *Arachis pintoi*.

A second problem is that several of the cultivars have serious defects. Thus, some cultivars are prone to attack by a spittle bug and others lack tolerance to the acid soils common in the savannas.

Brazilian researchers have recently released a cultivar (*Marandu*) of *B. brizantha* that is resistant to the spittle bug but not tolerant of acid soils. In contrast, CIAT workers have found another cultivar (*Basilisk*) of *B. decumbens* that is tolerant of the soils and results in high weight gains. The challenge has been to combine the two. Unfortunately *B. brizantha* and *B. decumbens* normally reproduce asexually, by *apomixis*; that is, an embryo is produced without fusion of male and female gametes, so the offspring are true clones. The apomixis is associated with tetraploidy (four sets of chromosomes), but at least one species, *B. ruziziensis*, is diploid and reproduces sexually. A solution developed by scientists at the University of Louvain in Belgium is to treat this species with a natural chemical, colchicine, to produce a sexual tetraploid that can then be crossed with two apomictic tetraploids, so providing a bridge between them.

The many crosses produced are now being evaluated in the field to help generate spittle bug-, acid-soil tolerant clones for commercial release. Further improvements are being made through marker-assisted selection.

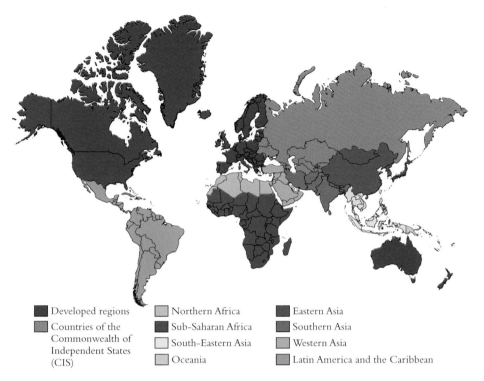

Plate 1   UN regional country groupings.

Source: UN. 2009. *The Millennium Development Goals report 2009.* New York: UN.

Legend:
- Developed regions
- Countries of the Commonwealth of Independent States (CIS)
- Northern Africa
- Sub-Saharan Africa
- South-Eastern Asia
- Oceania
- Eastern Asia
- Southern Asia
- Western Asia
- Latin America and the Caribbean

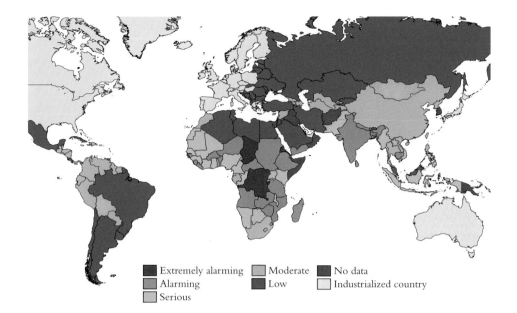

	Extremely alarming		Moderate		No data
	Alarming		Low		Industrialized country
	Serious				

Plate 2 2009 Global Hunger Index (GHI) for individual developing countries.

Source: von Grebmer, K., Ruel, M., Menon, P., Nestorova, B., Olonfinbiyi, T., Fritschel, H., Yohannes, Y., von Oppeln, C., Towey, O., Golden, K., and Thompson, J. 2010. *Global hunger index. The challenge of hunger: focus on financial crisis and gender inequality.* Bonn: International Food Policy and Research Institute, Concern Worldwide and Welthungerhilfe, October 2010. Reproduced with permission from the International Food Policy Research Institute www.ifpri.org. The GHI 2010 can be found at this web address: http://www.ifpri.org/publication/2010-global -hunger-index.

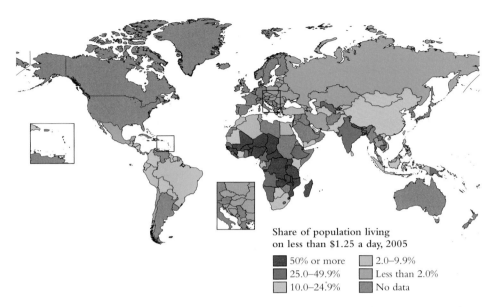

Plate 3   Proportion of people living in extreme poverty (under US$1.25 a day) in 2005.

Source: World Bank. 2008. *World development indicators. Poverty data: a supplement to the world development indicators 2008.* Washington, DC: World Bank. http://siteresources.worldbank.org/DATA STATISTICS/Resources/WDI08supplement1216.pdf.

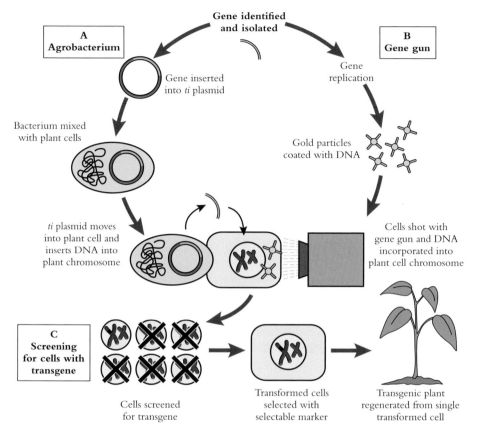

Gene identified and isolated

A Agrobacterium

B Gene gun

Gene inserted into *ti* plasmid

Gene replication

Bacterium mixed with plant cells

Gold particles coated with DNA

ti plasmid moves into plant cell and inserts DNA into plant chromosome

Cells shot with gene gun and DNA incorporated into plant cell chromosome

C Screening for cells with transgene

Cells screened for transgene

Transformed cells selected with selectable marker

Transgenic plant regenerated from single transformed cell

Plate 4   Two processes for crop transformation.

Source: Peel, M. 2001. *A basic primer on biotechnology.* NDSU Extension Service. http://www.ag .ndsu.edu/pubs/plantsci/crops/a1219w.htm.

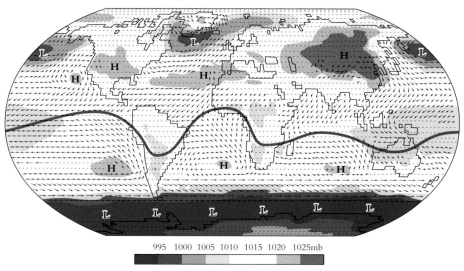

Sea-level pressure and surface winds, January

995 1000 1005 1010 1015 1020 1025mb

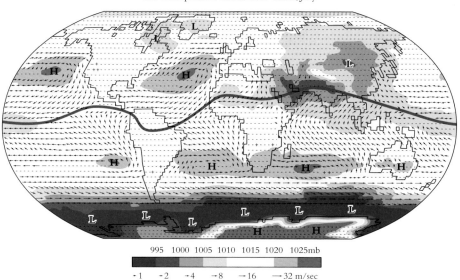

Sea-level pressure and surface winds, July

995 1000 1005 1010 1015 1020 1025mb

•1   •2   •4   •8   —16   —32 m/sec

Plate 5   The position of the ITCZ (Intertropical Convergence Zone) in January and July (red lines).

Source: Rohli, R. V., and Vega, A. J. 2008. *Climatology*. Sudbury, MA: Jones & Bartlett. Data from: NCEP/NCAR Reanalysis project 1959–1997 climatologies.

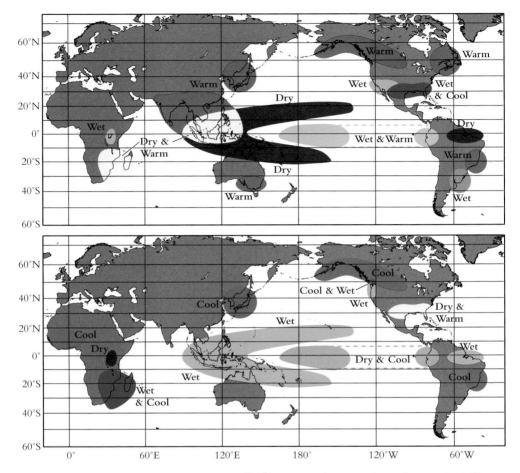

Plate 6 The El Niño–La Niña Oscillation. *Top*: El Niño phase; *bottom*: La Niña phase.

Source: NOAA, National Weather Service. 2005. *Warm (El Niño/Southern Oscillation – ENSO) episodes in the tropical Pacific.* http://www.cpc.ncep.noaa.gov/products/analysis_monitoring/impacts/warm.gif.

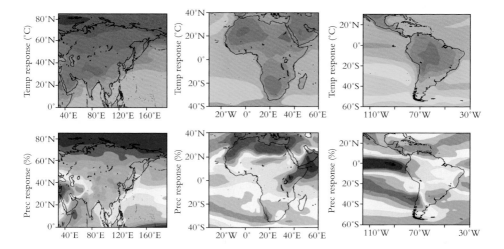

Plate 7 Annual temperature and precipitation changes over Asia (left), Africa (center), and Latin America (right). *Top row*: Annual mean temperature change between 1980 to 1999 and 2080 to 2099, averaged over twenty-one models. *Bottom row*: Same as top, but for fractional change in precipitation.

Sources: Climate change 2007: The physical science basis. Contribution of Working Group I to the Fourth Assessment Report of the Intergovernmental Panel on Climate Change. Cambridge, UK: Cambridge University Press.

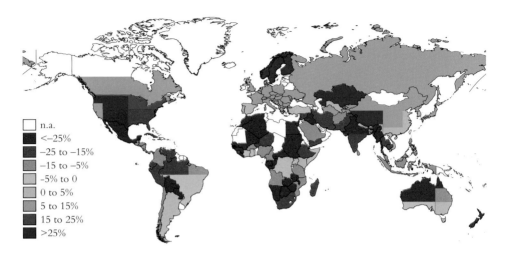

Plate 8 Projected impact of climate change on 2080 agricultural production, assuming a 15 percent carbon fertilization benefit.

Source: Cline, W. 2007. *Global warming and agriculture: impact estimates by country.* Washington, DC: Center for Global Development/Peterson Institute for International Economics

Increasing Lysine and Reducing Lignin

The quality of tropical forages can be improved either by increasing desirable compounds, such as lysine, or by reducing the concentration of undesirable compounds, such as lignin.[64] These improvements are likely to result in greater intake and digestibility of fodder.

Pigs and chickens need lysine in their diets but the content in cereals is poor compared with that in legumes, such as peas and lupins. On the other hand, legumes are poor in the sulfur amino acids, methionine and cysteine, necessary for cattle and sheep if meat, milk, and wool productivity is to be increased. High levels of these amino acids are present in sunflower seeds and the chicken egg protein, ovalbumin. The encoding genes have now been cloned and inserted into peas. The next step is to insert the same or similar genes in forage legumes, such as alfalfa and clover, with suitable promoters so that they are expressed in the foliage.

In tropical forages the carbohydrates are often protected from the rumen fungi and bacteria by the presence of lignin and so are not available to the ruminant animal eating them. There has been some success in reducing lignification using conventional breeding and selection, but attention is now turning to biotechnology. So-called brown midrib mutants block the pathway that produces lignin and sometimes occur in maize, sorghum, and millet; when this occurs, 50 percent less lignin is produced and the digestibility is 10 to 30 percent higher. Research is now aimed at cloning these genes and inserting them into forage legumes.[65]

Controlling Diseases

Livestock diseases have a serious impact on the livelihoods of farmers in developing countries.[66] Many of these diseases are of global significance; for example, foot and mouth disease of cattle and Newcastle disease of poultry. Outbreaks can be eradicated in wealthier countries, but the same diseases are often endemic in poorer countries, causing substantial losses and limiting their capacity to trade.

Livestock diseases can be managed by modifying the environment of the pathogen or by targeting the pathogen directly. Successful vaccination campaigns against animal diseases include the eradication of rinderpest ("cattle plague"), one of the most long-standing of livestock diseases.[67]

Considerable progress has already been made in the production of genetically engineered vaccines. Hitherto vaccines have consisted of dead or alive, but attenuated, viruses or other pathogens. The pathogens in this form are harmless, but they provide the antigens that stimulate production of antibodies in the vaccinated animal and hence give it protection if invaded by the live, virulent pathogen. However this traditional method of vaccination can be hazardous. Live vaccines, in particular, carry material other than the antigens that may cause undesirable side-effects. The advantage of genetically engineered vaccines—produced by cultures of bacteria that contain antigen-encoding genes—is that they consist solely of the required antigens. A practical advantage is that it should be possible to develop vaccines that do not require refrigeration—a major benefit for rural smallholders.

Finally, in disease control, attention is increasingly turning toward improved surveillance, particularly of zoonotic diseases affecting humans and the need for integrated disease control packages. These include multiple strategies, such as environmental control, community education, vaccination, and treatment medications applied together to control diseases with more complex transmission patterns. Such integrated systemsl intimately involve farmers themselves, even if the technologies come from advanced laboratories. In Chapter 11 I examine more closely the role of farmers in the adaptation and innovation of appropriate technologies.

11

Farmers as Innovators

> The overarching challenge is through participation to enable those who are
> most marginalized, powerless and poor to achieve a better life for themselves.
>
> Robert Chambers[1]

Much of the content of previous chapters has represented a "top down" approach to agricultural development. Farmers are seen as recipients of inputs and knowledge, yet as anyone who has worked with farmers in developing countries is aware, they are skilled and knowledgeable and often highly innovative. Where development has not worked, it is often because their needs are not appreciated and their knowledge is ignored.

In this chapter I address the following questions:

- What is Agroecosystem Analysis (AEA) and how did it arise?
- How can AEA and other participatory techniques help farmers to become better analysts?
- How can farmers be encouraged to be better experimenters and innovators?

Traditional Extension

A top-down approach has been characteristic of the process of *agricultural extension*, a key component of the enabling environment for agricultural development that I discussed earlier. During the Green Revolution, extension was a relatively straight-forward task. There were a limited number of messages that had to be communicated to farmers. Extension workers were briefed on the messages and then sent out to communicate them to individual farmers or farmer groups. Such an approach came to be known as "training and visit."[2]

Today, the task is considerably more complicated. The lands where most of the rural poor households are located tend to be less well endowed, more heterogeneous and complex, and also inherently riskier than the typical Green Revolution lands

Box 11.1   Local Extension in Bangladesh[3]

Bangladesh is a net importer of wheat, but uptake of new varieties has been poor, in part because the Department of Agricultural Extension has hosted farmer demonstrations and training on land belonging to richer farmers; those farmers less able to take the risks in adopting new practices are marginalized in the process.

In a pilot study, 45 mainly marginal farming families from two different villages were required to rent or own a plot of arable land of 0.08 ha and then trained as wheat seed producers and traders. Training was provided by a local NGO, Dipshika, with support from scientists of the Wheat Research Centre, part of the Bangladesh Agricultural Research Institute. The farmers were taught wheat management practices and how to select and store high-quality wheat seed for periods of five to six months in order to maximize profits. They were also given a loan of 2 kg of a new variety (*Shatbadi*) that was to be replaced by seeds from subsequent harvests and the opportunity to access credit to cover the recommended inputs. Although one family made a loss (due to saving most of their seed for their own use), all other families made a profit (of between $12 and $80) with yields of over 3,000 kg/ha equivalent.

The poorest households were able to use the proceeds of the grain sales to pay off debts and to invest in such things as school fees, medicine, livestock, land, tree saplings, bicycles, tin roofs, and water pumps.[4] The pilot scheme was then scaled up during the 2006–2007 season to 545 families; wheat yields ranged from 1,000 to 5,025 kg/ha and the average immediate income was around $30. The new wheat variety was sold to over 1,500 farmers who would previously not have had access to it.[5]

described in Chapter 3. There are few, if any, simple messages and, because of the heterogeneity, advice for one farmer often differs from that given to another, even a close neighbor. Ideally each farmer requires their own extension worker. That is clearly impossible.

One approach to this problem has been to contract out extension to nongovernment organizations (NGOs), local community organizations, and the private sector at the local level (Box 11.1).

This local approach involves a more fine-grained, interactive extension system, but it is still predominantly top down in style. The alternative is to encourage farmers to be analysts and innovators, and for some to be formal or informal extension agents for their communities.

Agroecosystem Analysis

What would this begin to look like? There has long been a tradition, aptly named by Robert Chambers, of the Institute of Development Studies (IDS), as "rural development tourism," in which periodic visits to the field by experts of various kinds are seen as providing sufficient "feel" for conditions.[6] But typically such visits are confined to the accessible roadside areas, to meetings with headmen and other local officials and usually conducted at those times of the year when travel is easiest. They inevitably produce biased impressions that can be seriously misleading. Much rural poverty remains unperceived. Such visits are no substitute for intimate and systematic analyses of the circumstances and livelihoods of rural households.

To address this need, I and several colleagues at the University of Chiang Mai in northern Thailand developed the technique of Agroecosystem Analysis (AEA) based on the agroecosystems concepts described in Chapter 6.[7] A Ford Foundation grant had been given to the university in 1968 to create a Multiple Cropping Project (MCP) aimed at designing advanced, triple-crop, rotational systems with which farmers could capitalize on the government irrigation schemes recently installed in the Chiang Mai valley. The Foundation also gave many of the young staff scholarships to go abroad for further graduate training. They returned in the late 1970s, eager to use their new skills and experience to help the farmers of the valley. But, as they soon became aware, much of the work of the MCP in the intervening years had not proved particularly relevant. Although the MCP had developed some half dozen, apparently superior and productive, crop systems there were few cases of adoption by the farmers; on the other hand, the farmers themselves had developed a large number of triple-crop systems in response to the new opportunities provided by the irrigation systems.

This raised questions in their minds as to the role they, as university researchers, could most effectively play. In terms of helping the farmers of the valley, where did their comparative advantage lie? Should they continue to design new systems; if not, what kind of research should they undertake? These questions, they realized, could not be answered until they had a better idea of the existing farming systems in the valley and the particular problems the farmers were facing.

AEA begins with a series of extended field trips, employing direct observation and interviews with farmers to produce a set of maps and other simple diagrams that summarize the patterns of events and activities in the local agroecosystems. The diagrams are then used, in intensive, multidisciplinary workshops, to identify and discuss the key agroecosystem issues that I described in Chapter 6: productivity, stability, resilience, and equitability.

Out of the workshops come a set of key questions and hypotheses requiring further investigation and research. For example, by overlaying the various maps produced as part of the AEA of the Chiang Mai valley it became clear that the intensive triple cropping developed by the farmers was largely confined to those areas under traditional or mixed (traditional plus government) irrigation systems, rather than

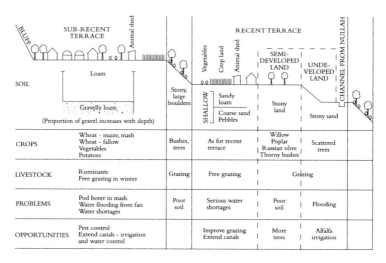

Figure 11.1   A transect of a village in Hunza, Northern Pakistan.[9]

the areas supplied by the government (Royal Irrigation Department) systems alone. Farmers found they could not rely on the government irrigation supplies. A key research question for the group was how to improve the control and reliability of irrigation in the valley. In the thirty years since AEA was developed by the Chiang Mai group, key questions such as this have provided the basis for a successful program of research targeted at the needs of the farmers in the valley.

Although originally a technique designed for university and research station workers, subsequent applications in both developing and developed countries have been designed to meet the needs of government agencies, extension workers, and NGOs.[8] As the technique has spread, it has evolved; the original repertoire of diagrams has been added to, as individuals have developed new ways of representing their observations and findings. Transects (Figure 11.1), map overlays, seasonal calendars, impact flow diagrams, Venn diagrams, preference rankings, and decision trees, to name a few, now constitute a rich array of tools for analysis. Experience has increasingly shown the power of simple diagrams to generate productive discussion among researchers from different disciplinary backgrounds and, most significantly, to stimulate genuine interchange between researchers and farmers.

Farmers as Analysts

These exercises were designed primarily for experts—researchers, extension workers, and aid officials. The analyses, although deliberately aimed at understanding the ecological and socioeconomic nature of local agroecosystems, were still designed and driven in a top-down fashion. However, the increasing involvement of farmers—not just as sources of information but as participants in analysis—began to suggest to a number of us that a more revolutionary approach was possible.

In 1987, I, along with Robert Chambers of IDS, Jenny McCracken of the International Institute for Environment and Development (IIED), and Constantinos Berhe of the Ethiopian Red Cross, took part in an analysis of two villages in the Wollo province of Ethiopia, site of a severe drought and related mortality only four years previously. The teams of analysts included government officials, agricultural and forest experts, Red Cross activists, and local leaders. Although it began as a standard AEA exercise, Robert Chambers brought to the process experience from Rapid Rural Appraisal (RRA), an approach developed in the 1970s of which he was one of the pioneers.

Although RRA, like AEA, is a method of extracting information from rural people, it is more informal in style, relying on semi-structured interviewing, participant observation, gaming, and extended discussion.[10] It was developed, in part, as a reaction to the standard questionnaire approach of much of rural development analysis, in which fixed sets of questions are asked of rural people, the interviewers often being hired for the job and confining themselves to obtaining answers, whether accurate or not. By contrast, semi-structured interviewing takes place in the normal surroundings of the interviewee, does not involve a written questionnaire, and, while some of the questions are predetermined, new questions or new lines of questioning arise as the interview proceeds. The intention is to create genuine dialogue in which experience and knowledge are compared and exchanged. Sometimes individuals are interviewed, on other occasions groups of people are brought together—village leaders, key informants, rich farmers, poor farmers, women, and the elderly, many of whom are excluded or marginalized from more structured approaches.

In Wollo, we began to combine semi-structured interviewing and diagram making, and in the process we became even more aware of the richness of analytical understanding that is common, if not normal, in rural people in the developing countries. One exercise that I carried out with Ethiopian farmers produced a ranking of tree species for cultivation. A set of six species was assessed by presenting them to the farmers in every pair-wise combination. The farmers were asked to indicate which they would prefer if they could grow only one of the pair and to give the reasons. Their accumulated ranking revealed the great range of usage for which different species were suitable.[11] When the process was repeated with a group of foresters, the ranking was very different. The foresters emphasized the ease and reliability of species in the nursery, which was their main responsibility, while the farmers placed a higher premium on versatility.

Table 11.1 Innovation assessment in Wollo province, Ethiopia

Option	Productivity	Stability	Sustainability	Equity	Cost	Time to benefit	Technical feasibility	Social feasibility
Reforestation	+	++	++	○	☐	☐	■	◪
Agroforestry	+	++	++	++	◪	◪	■	◪
Home gardens	++	++	++	○	■	◪	■	■
Small-scale irrigation	+	+	+	+	◪	◪	◪	■
Short-cycle varieties	++	++	++	−	◪	◪	◪	◪
Credit	○	++	++	++	◪	◪	☐	■

		Cost	Time	Feasibility
−	Negative impact			
○	No impact	☐ High	Long	Low
+	Positive impact	◪ Medium	Medium	Medium
++	Very positive impact	■ Low	Short	High

Source: Ethiopian Red Cross[12]

These rankings revealed not only knowledge but a capacity to make choices that balanced a wide range of considerations. At the end of the exercise we undertook a group assessment of the various options open to the inhabitants of the two villages and produced rankings that significantly shifted the priorities of the government officials present who had previously focused on irrigation and reforestation (Table 11.1).

Chambers came away from the Wollo exercise convinced it was possible to move from top-down AEA and conventional RRA to an approach where rural people took the lead. In the ensuing months, experiments in villages in several countries demonstrated the capacity of rural people to produce their own diagrams, often in ways that showed great ingenuity and depth of knowledge. *Maps*, it was discovered, were readily created by providing villagers with chalk and colored powder and no further instruction other than the request to produce a map—of the village, the watershed, or a farm. A threshing floor or a cleared space in the village square was sufficient for villagers to produce excellent maps, often of considerable complexity. Sometimes the maps turned into models, and I remember one such in India where the farmers had collected soil from the various fields in the village and constructed a representation that faithfully laid out each field using its own soil. Maps, however, were only a beginning. *Seasonal calendars* could be constructed by people who were illiterate and barely numerate, using pebbles or seeds. A row of twelve pebbles is laid out on the ground to indicate the months of the year and then for various items—rainfall, availability of food, labor demand, risk of illness—the relative level of activity in each month is indicated by placing a number of seeds next to each pebble.

This new approach was embraced with enthusiasm, particularly by leaders of NGOs who were eager to increase local participation. Individuals such as Sam Joseph of Action Aid, Jimmy Mascarenhas and Aloysius Fernandez of MYRADA (Mysore Relief and Development Agency), and Parmesh Shah of the Aga Khan Rural Support Programme continued the experiments, training their own staff to facilitate the exercises. The range of diagrams quickly expanded; all those originally developed under AEA were accessible to farmers, and new ones were added. Farmers constructed pie diagrams—pieces of straw and colored powder laid out on an earthen floor—to indicate relative sources of income. In the process, villagers often revealed more complex, detailed knowledge than would have occurred through discussion or simple question-and-answer sessions.[13] In Chambers' words, the diagrams "have often astonished scientists, and farmers themselves, with the detail, complexity and utility of information, insight and assessment they reveal."[14]

Although this was in itself encouraging, the practical usefulness of the approach soon became apparent. Farmers were encouraged to use scoring matrices to evaluate different crop varieties in breeding programs.[15] Farmers in many parts of the world have proved adept at constructing and scoring matrices in which they compare different crop varieties according to their own varied and often numerous criteria.[16] Maps, seasonal calendars, and pie diagrams not only revealed existing patterns, but pointed to problems and opportunities, and were seized on by rural people as a means of making their needs felt and as a basis for collective planning. The relationship between "expert outsiders" and village people began to change. Productive dialogues replaced the traditional one-way flow of information and instruction. I remember a group of villagers in Haryana, northwestern India, constructing a map of their watershed on the ground and using four colors to indicate the degrees of degradation they observed. Present was a conservator of forests for the state, noting and comparing this classification with that of his department, which only recognized three classes. But the most enduring image was of the conservator and the villagers, down on their haunches on the ground, engaged in lively discussion about the watershed and what should be done. It was a liberating experience both for the villagers and the conservator.

The approach has now spread to most countries of the developing world and been adopted by government agencies, research centers and universities, and NGOs. As a deliberate policy, no central manual has been produced, although much has been written and there is an extensive network of practitioners. The methodologies, which are described by a bewildering variety of names—participatory rural appraisal (PRA), participatory analysis and learning methods (PALM), méthode accélére de recherche participative (MARP), to list only a few—have evolved according to local needs and customs and reflect local ingenuity.[17] Participatory Learning and Action (PLA) notes, produced by the IIED in London and distributed to several thousand individuals, disseminate good practice and new ideas, so that innovations in the approach reported from an African village are being tried out in an Asian village only a few weeks later.

In many situations PRA has become integrated into development projects as well as used in a wide diversity of local assessments ranging from biodiversity surveys, to identifying people with disabilities, to measuring the occurrence of livestock diseases.[18]

Participatory Extension

AEA and PRA have an obvious role to play in extension; although to date there have been few examples of successful application. In Cambodia, Ian Craig, who was one of the pioneers of the original Chiang Mai AEA, has helped the government include AEA and PRA in the National Extension System programs for the country's communes.[19] The analysis begins by delineating and describing the agroecological zones within each commune. A list is then drawn up of problems and opportunities for each zone as identified by local farmers. This in turn acts as a guide for extension and development activities aimed at solving the problems. The results of the process are presented at a meeting of the commune council where further participation in the shape of feedback shapes the final AEA report. This then feeds into the development of a commune agricultural plan (CAP).

A detailed AEA of Sna Ansar commune, which borders on the great Tonle Sap lake, was carried out by a team consisting of both government officials and local farmers.[20] Among the key problems were declining paddy rice yields and fisheries. An innovation assessment suggested a number of solutions to these and other problems; for example, finding new and more appropriate rice varieties and exploring the use of composting in paddy production, while putting into force protection of the flood forest and grassland areas that provide hatcheries and sanctuaries for the fish. These and other plans were then adopted by the commune for collective action. To date AEA has been conducted in an estimated five hundred communes, accounting for approximately one third of Cambodia.

Farmer Field Schools

The experience of working in participatory extension reveals the considerable knowledge possessed by farmers, derived from direct observation or from the experience handed down over the generations. But it also becomes evident that there are gaps—blind spots—in their knowledge (see Chapter 12). A complementary approach to participatory extension is to engage farmers in a form of adult, nonformal education and extension practice that not only teach farmers biology, agronomy, and land management but enables them to continue to share knowledge and to organize these activities within their community.[21] The technique, known as Farmer Field School (FFS), was developed in the Philippines in the 1970s, initially focused on integrated pest management (IPM; see Chapter 12). By the 1990s an estimated two million farmers had been trained through FFS in South and Southeast Asia. FFS programs are now operating or being piloted in over twelve African countries.[22]

A study of FFS in Kenya, Tanzania, and Uganda showed it to be especially beneficial to women (females constituted 50 percent of participants) and people with low literacy levels. Participation in FFSs increased income by an average of about 60 percent and led to increased production, productivity, and income in nearly all cases. In Kenya crop production increased by 80 percent, and in Tanzania by over 100 percent for agricultural income.[23] The costs per farmer for FFS training in several East African programs vary between $9 and $35 depending on whether farmer facilitators or extension agents are used. Various means of reducing costs are now being developed.[24]

Farmers as Experimenters

Farmers have been experimenters since the beginning of agriculture.[25] Hunters and gatherers had long ago learned to use fire to stimulate the growth of tubers and other food plants and of grass to attract game. Plant selection began when people found they could encourage favored fruiting trees by clearing their competitive neighbors, but the first steps toward intensive plant breeding were taken when an individual, probably a woman, deliberately sowed a seed from a high-yielding plant somewhere near the dwelling and observed it grow to maturity. Besides producing new varieties, experimentation also resulted in whole systems of agriculture—shifting cultivation; rice terracing; home gardens; irrigated agriculture; the Mediterranean trio of wheat, olives, and vines; the Latin American multiple cropping of maize, beans, and squashes; and various forms of integrated crop-livestock agriculture.

The Romans, as is evident from their writings, analyzed the structure and functions of agricultural systems in a scientific manner. They also described the process of experimentation. Marcus Terentius Varro, whom I quoted at the beginning of Chapter 6, urged farmers to both, "imitate others and attempt by experiment to do some things in a different way. Following not chance but some system: as, for instance, if we plough a second time, more or less deeply than others, to see what effect this will have."[26] The great agricultural revolution of Britain in the late eighteenth century was led by farmers. Jethro Tull is famous for his invention of the corn drill, Charles Townsend for the introduction of turnips, Thomas Coke for the Norfolk four-course crop rotation system, and Robert Bakewell for the selective breeding of livestock.[27]

As Jules Pretty of the University of Essex points out, however, these were well-educated landowners who had read the Latin texts and understood the basic principles of sustainable agriculture and set about popularizing; they were not the real innovators.[28] Over the previous hundred years, numerous "unknown" farmers had been developing and propagating new techniques through an informal process of rural tours and surveys, farmers' groups and societies, agricultural shows, training, and publications. The professionalization of agricultural research began only in the nineteenth century, notably following the creation of the Rothamsted Experimental Station in England in 1843 and of the land-grant colleges in the United States, although, even then, farmers were well represented on the boards of management, and research programs were highly responsive to farmers' needs.

Research in the developing countries, under colonial rule, was, inevitably, of a top-down nature, with a strong emphasis on export crops. Little changed after independence. The top-down tendency was reinforced by the Green Revolution, despite the shift in emphasis to food crops. Although the early work in Mexico, in the 1940s and 1950s, on the breeding of new disease-resistant wheat varieties, was concerned with local adaptability, the realization by western scientists of the enormous potential of the dwarfing genes present in East Asian wheat and rice germplasm led to a quest for varieties that would perform well, irrespective of local conditions. Inevitably, the richness of farmer's indigenous knowledge and their capacity to experiment was downplayed.[29] Nevertheless, observant field scientists continued to absorb knowledge from the farmers they worked with. Indeed, an anthropologist at IRRI who compiled a list of technologies on offer from the institute found that 90 percent of those being promoted had been derived from Asian farmers.[30]

Farmers as Breeders

Farmers usually hold strong, and often insightful, opinions on the qualities of crop types and livestock breeds. New varieties are often adversely compared with traditional varieties, with which they have long familiarity. In northern Thailand, highly commercial farmers grow as a first crop, for their own consumption, many varieties of the traditional sticky (nonglutinous) rice that has been part of their culture for hundreds of years, although later in the year they may plant new varieties for sale.[31] These are rarely simply prejudices, however; if asked farmers can identify attributes, compare positive and negative features, and rank varieties placed in front of them.

Scientists at CIAT (Centro Internacional de Agricultura Tropical) in Colombia, aware that the varieties they had developed according to research station criteria were often not accepted, asked farmers to rank the grain from bush beans and explain their reasons.[32] The resulting rankings were very different from those of the breeders and also revealed a difference in preferences between men and women, the latter choosing smaller, better-flavored grains, while the men preferred the larger grains that command good prices in the market (Table 11.2).

Breeders usually try out their selections in farmers' fields to determine acceptability. But this has customarily occurred at the end of the breeding process when many of the key decisions have already been made. What has changed in recent years is a greater involvement of farmers earlier in the process, eliciting not only

Table 11.2   Group evaluation of bush beans in Colombia: farmers' characterization of preferred type of bean variety

Men's preferences	Women's preferences
High yielding	*High yielding*
Long pod with 6 to 7 grains (related to yield)	Quick cooking
Tall erect plant (not sprawling) appropriate for planting higher density	*Grain swells quickly, increasing total portion size when cooked*
Adaptability to different soil fertility conditions, or fertilization	*Flavor (sweet, not bitter)*
Large grain size	Soft skin
Deep red grain color ("radical" type)	*Resistant to storage pests*
Shorter season (not longer than 85 days)	Pod opens easily for threshing
Disease resistant (one or two spraying adequate, not more)	*Short season*
Resistant to storage pests	
Pod that does not split open in the field causing grain loss at harvest	
Flavor	
Soft-skinned when cooked	
Stability of yield over at least three production seasons	

Source: Ashby, Quinos, and Rivers[33]
Note: Priority criteria in italics

reactions but positive inputs into the determination of breeding goals. The next stage at CIAT was to encourage farmers to take seed away and grow them in trials on their own land. At the end of the trials, they produced overall rankings on both grain quality and the performance of the plants. Yield was not the dominant criterion; farmers placed much greater emphasis on marketing, resistance to pests and disease, and labor requirements.

In Rwanda, a five-year experiment conducted by CIAT and ISAR (Institut des Sciences Agronomiques du Rwanda) progressively involved farmers at even earlier stages in the breeding process.[34] Beans are a key component of the Rwandan diet, providing 65 percent of the protein and 35 percent of the calories, and are grown by virtually all farmers. There is an extraordinary range of local varieties—over 550 identified—and farmers (mostly women) are adept at developing local mixtures that breeders have difficulty in bettering. In the first phase of the experiment, teams of expert farmers were asked to evaluate some 15 varieties, two to four seasons before normal on-farm testing. This revealed new criteria—for example, the ability of varieties to perform well when grown under bananas—and also made breeders more aware of the range of expertise among the farmers. Some women were particularly astute at distinguishing among different criteria.

Between 2000 and 2010 the partnership between ISAR and CIAT developed 35 new bean varieties through their participatory breeding program, including beans that are high yielding, disease resistant, climate change tolerant, and climbing beans to maximize land use. In 2010, 15 new bean varieties that could potentially triple yields were released at a farmer field day at the Rwerere research station and neighboring farmers' fields. The farmers involved in the selection process examined the seeds and gave them names, including one called Gasilida named after Cansilida Mujawarmariya, a local female farmer who supplied the source of the land race to breeders. The participatory breeding program is part of Rwanda's strategy to increase food security and the sustainability of the farming sector. Between 2007 and 2009, exports of beans from Rwanda increased 30 percent.[35]

Mother–Baby Trials

The Rwandan experience and others has led the international agricultural research centers to pay greater attention to farmers' capacity to experiment. There are now several approaches that engage the farmers more intimately in the analysis of the trials. An example is the technique known as mother-baby trials which has been developed in southern Africa (Box 11.2).

Box 11.2 Mother–Baby Trials[36]

These participatory breeding trials were devised by Sieglinde Snapp of ICRISAT and adopted by CIMMYT researchers, Marianne Bänzinger and Julien de Meyer, to study the performance of new maize varieties resistant to stresses of various kinds—tolerance of drought and low soil fertility, resistance to parasitic weeds—and to aid in their dissemination among farmers.[37]

While breeding can be done in the greenhouse or on an experimental farm, it will tend to focus on only one stress at a time. An advantage of carrying out the breeding trials on farmers' fields, with farmer participation, is to assess the tolerance of the varieties under a range of simultaneous stresses and to harness farmers' own knowledge and responses.

The technique involves a network of "stress breeding sites" with two main components:

- "*Mother trials*," involving up to twelve varieties, located close to the community and managed by schools, colleges, or extension agents
- "*Baby trials*" involving four to six varieties, located in the fields of farmers who used their own inputs and equipment

Out of this program have come some fifty new maize varieties, cu____ planted on over one million hectares in sixteen countries (as of 2005) in souther__ eastern Africa.[38]

The Participation Revolution

In some ways we have witnessed a revolution—a set of methodologies, an attitude, and a way of working that has finally challenged the traditional top-down process that has characterized so much development work. Participants from NGOs, government agencies, and research centers rapidly find themselves, usually unexpectedly, listening as much as talking, experiencing close to firsthand the conditions of life in poor households, and changing their perceptions about the kinds of interventions and the research needs that are required.[39]

Well known is the experience of World Neighbours in Guinope, Honduras, which has worked in partnership with the Ministry of Natural Resources and a Honduran NGO, ACORDE (Association for Coordinating Resources in Development).[40] Initially maize yields in the project area were low (400 kg/ha), poverty and malnutrition were widespread, and out-migration was common. The program started slowly and on a small scale, involving the local people in experiments with chicken manure and green manures, contour grass barriers, rock walls, and drainage ditches. Extensionists were selected from the most adept farmers, and they progressively involved others so that eventually several thousand farmers participated. Maize yields tripled on average, and the farmers began to diversify into coffee, oranges, and vegetables. Labor wages have risen from $2 to $3 a day and out-migration has been replaced by in-migration, people moving back from the slums to the homes and land they had abandoned.

Other, more recent programs have begun to make explicit use of the new analytical tools. An example is the Aga Khan Rural Support Programme (AKRSP), with whom I worked in the 1980s, in the Hunza Valley and neighboring valleys of northern Pakistan.[41] An arid, mountainous region, it is not naturally well endowed, but the inhabitants are skilled in the use of the local natural resources. The development program consists of a series of interactive dialogues through which the villagers, acting as a community (VO or village organization), identify, plan, and implement a key infrastructure project in each village. In many cases the projects are impressive feats of engineering that bring irrigation water from the glaciers to open up new land for agriculture. Several hundred such projects were completed in the 1980s, and the program was subsequently engaged in realizing the potential of the new infrastructure through a variety of initiatives funded by the villagers' savings and using diagramming techniques to determine options for innovation.[42]

Another program, facilitated by AKRSP (India), has made explicit use of PRA techniques in developing soil and water conservation in Gujarat in western India.[43] In the first stage, the villagers produce maps of their watersheds, detailing the problem areas, planning appropriate soil and conservation works, and choosing trees for planting using a technique of group ranking. This process takes one to six months. Next, village institutions are formed. They nominate extension volunteers, paid by the villagers, who are given training in PRA methods, in the necessary technical skills, and in project preparation and accounting procedures. They are responsible for managing teams of individuals who then implement the plans. Yields have grown by 20 to 50 percent, yet the costs of the watershed treatment are 1,340 rupees/ha compared with 3,000 to 7,000 rupees/ha on nearby government programs.

Being Innovative

Innovation—developing new ways of doing things—can take many forms. Sometimes it may be the process of discovering something entirely new; more often it involves finding novel applications for existing technologies. A classic example of technology adaptation was the worldwide response to the introduction of new technologies for potato storage in the 1970s (Box 11.3).[44]

Box 11.3   Adapting Potato Storage

Storing potatoes is a challenge because the potatoes are likely to sprout excessively, resulting in severe losses. In the 1970s, Jim Bryan, a seed specialist at the International Potato Center (CIP) in Peru, observed that farmers were successfully storing potatoes using diffused light, contrary to the normal practice of storing potatoes in the dark at low temperatures.[45] CIP then produced a package of methods using natural indirect light, including insulated roofs, translucent walls, and adequate ventilation, which was introduced to some twenty-five countries.[46]

But adoption did not proceed as expected. Virtually all of the farmers changed the technology; although the principle of diffused light storage (DLS) caught on, it was modified on each farm according to the local conditions, the household architecture, and the budgets of the farmers.

A further step in innovation going beyond adaptation, pioneered by Jacqueline Ashby and her colleagues at CIAT, was to establish "innovators' workshops" in which farmers design and evaluate experiments. One trial tackled the problem of a lack of stakes for climbing snap beans. The farmers suggested growing the beans after tomatoes to use the tomato stakes and the residual fertilizer, and they chose two snap bean varieties as appropriate for the new system.[47]

David Millar, who works for the Tamale Archdiocesan Agricultural Programme, asserts that there is no farmer in northern Ghana who is not in some way experimenting.[48] Some are pursuing curiosity experiments. One farmer, Dachil, had traveled to southern Ghana and brought back cocoyams, which naturally grow in the forest. He planted them in his yard under the shade of a mango tree. "If the results are good, my next step will be to set up a small garden on my farm. . . . I am just curious to find out everything I can about the crop." Other farmers are trying to solve problems. Millar describes the trials designed by Nafa and his brothers using different forms of crop rotation to eliminate a notorious weed, *Striga*. And large numbers of farmers in the region are engaged in adapting introduced technologies.

System of Rice Intensification

One of the most controversial and extraordinary examples of agricultural innovation in recent years has been the development of the system of rice intensification (SRI).[49] The system was invented in Madagascar by the Jesuit priest Father Henri de Laulanié with the input of local farmers in the early 1980s (although there is evidence of farmers using similar practices in the Indian state of Tamil Nadu a century ago). Laulanié wanted to investigate methods of increasing the low rice yields obtained by Malagasy farmers (less than 2 tons/ha) and he saw that two practices—transplanting seeds singly and maintaining moist rather than saturated soil—when employed by farmers, increased production. To this Laulanié added the transplanting of younger seedlings, their planting in a square formation, and the use of a rotary-hoe to weed the land in a perpendicular fashion in two directions. This markedly improved plant growth, and he named the system SRI.

At the core of SRI is the minimization of competition between the rice plants. Typically wet rice fields (paddies) are established by transplanting seedlings that are 3 to 4 weeks old, usually in clumps of three to four seedlings. Under SRI the seedlings are 8 to 15 days old, and only one seedling is planted per hill. Planting younger seedlings increases the time spent in the main field, encouraging the production of more tillers (shoots that grow from the parent stem, involved in vegetative propagation and in some cases seed production).[50] Most important the spacing is 25×25 cm between the hills, which gives the roots and leaves more space to grow. The effect is to reduce the density of seedlings required for planting from about 200m^2 to 16m^2. In addition, the fields are subject to intermittent watering to keep the soil moist rather than continuously flooded; organic manures are preferred; and increasing use is made of mechanical weeders.

SRI is controversial because so much of the evidence, until recently, has been anecdotal and some of the claims seem exaggerated.[51] Average yields in Madagascar increased from 2 tons/ha to an average of 8 tons/ha in the first two years of application and up to a phenomenal 20 tons/ha on some farmer's fields by the sixth year. There have been numerous comparisons between SRI yields and conventional yields. On-farm comparative evaluations conducted over nine seasons between 2002 and 2006 across eight provinces in Eastern Indonesia by the Directorate General for Water Resources showed the average yield increase was 78 percent, or 3.3 tons/ha higher than for non-SRI systems.[52] Similar results were achieved in several other countries.[53] Grain yields reported from field experiments carried out in different parts of India showed yield increases from SRI ranging from 9.3 percent to 68 percent when compared with conventional practice.[54] However, some claim the yields are not necessarily greater when compared with best management practices.[55]

A problem in assessing the technology is that it is difficult to separate out the effects of the different interventions that make up SRI. However, what is clear is that the increased spacing and greatly reduced competition increases grain yield per unit area. In the words of Amir Kassam, an FAO agronomist, Willem Stoop previously of the Africa Rice Centre (WARDA), and Norman Uphoff, a political scientist at Cornell, who have been major champions SRI, the plant spacing "minimizes

Box 11.4   Innovations Developed by SRI Farmers[60]

- Using raised beds to give the soil more aeration and growing seedlings in sand (Philippines)
- Raising seedlings in plastic trays (Eastern Indonesia)
- Marking fields for transplanting using wooden rake-markers with teeth at 25-cm intervals (Madagascar and India).
- In India, marking with a simple roller used to make rice flour patterns around windows and doorways during the Diwali festival.
- Using zero-tillage farming and raised beds (Sichuan Province, China)
- Adapting SRI to rain-fed conditions (Cambodia)

plant competition below and above ground, thereby encouraging greater root and canopy growth and distribution. Plant canopies have more uniform access to solar radiation while soil nutrients can be captured from a larger soil–root zone. At the same time, the photosynthetic process is prolonged and there is greater translocation to the panicles of the carbohydrates and nutrients stored in the rest of the plant."[56] Recent research has detailed the effects of the key elements of SRI on yields.[57] The spacing, the intermittent watering, and the early transplanting all have significant impacts on root systems, which are deeper and more active, while the intermittent watering also increases the amount, diversity, and activity of soil biota.

There are significant barriers to adoption: The labor requirement can be high initially, but this can decrease with time and as mechanical weeding replaces hand weeding. In some locations, accessing organic wastes may be difficult, as is effective water control.[58] It is not suitable for all rice-growing environments. However SRI is now being used in over forty countries by about one million farmers, with similar approaches being developed for other crops such as finger millet, wheat, teff, beans, and sugarcane. The most encouraging aspect of the technology may be its receptivity to innovation (Box 11.4).[59] Many farmers who have adopted SRI have proved highly innovative, significantly improving on the system's performance.

The Task Ahead

Far too often we assume that agricultural research and development programs are there simply to teach farmers methods of becoming more productive. But the practice of agriculture is intrinsically dynamic, requiring constant adaptation to changes in climate, in the resource and nutrient base, in pest and disease interactions, and the economy. Farmers know this and can articulate the challenges they face and be innovative in their responses. For the researcher and extensionist, the task ahead is to help further develop this capacity. When farmers experiment and adapt themselves, they often teach others in the process, enabling dissemination of knowledge from the ground up. One of the consequences of this innovation is productivity gains that are wider spread and more sustainable.[61]

In the next three chapters, I illustrate how similar participatory approaches to research and development are being applied to such diverse problems as integrated pest management, nutrient enhancement, the construction and management of small-scale irrigation, and aquaculture.

Part IV

12

Controlling Pests

Creative and ambitious measures must be taken to shatter the deeply ingrained
uncritical and dependent attitude towards pesticides which prevails at all levels
in developing countries, from ministries to the smallest farms.
Patricia Mattesson, Kevin Gallagher, and Peter Kenmore, Food and
Agriculture Organization (FAO) [1]

Pests, pathogens, and weeds are the most visible of threats to stable and resilient food
production.[2] Just how much crop and livestock loss they cause is largely guesswork;
estimates range between 10 and 40 percent. But in some situations the potential losses
can be considerably higher. Much depends on the nature of the crop: Where a premium is placed on the quality of the harvested product—for example, cotton, fruits,
and vegetables—even a small pest or pathogen population can cause the farmer
serious financial loss. Staple crops are not in this category, but the increasing intensity
of their cultivation can result in devastating pest and pathogen attack, on occasion
resulting in total destruction of the crop.

In this chapter I address the following questions:

- How extensive is pesticide use?
- What are the pesticide hazards for human health and the environment, as well
  as the reasons for their frequent ineffectiveness?
- What is the range of alternatives available for pest control?
- How can integrated pest management help?

Using Pesticides

Since World War II, the common approach to pest, pathogen, and weed problems
has been to spray crops with synthetic pesticides (insecticides, nematocides, fungicides, bactericides, and herbicides) produced by industrial processes. These compounds pose four major problems: (1) they may be harmful to human health; (2) they

may be damaging to wildlife and the environment; (3) they may be costly and ineffective at controlling pests; and (4) they elevate secondary, minor pests to major pests that can be more destructive than the pests the pesticides were originally targeting.

It is difficult to obtain up-to-date information on global pesticide use or even on sales—FAO does not have reliable pesticide statistics. Use in the developing countries grew rapidly in the 1960s and 1970s, reaching a total of over half a million tons (in terms of active ingredients) in the mid-1980s, about a fifth of world consumption.[3] Since 1990, consumption has been declining. By 1996, global pesticide consumption was at 2.6 million tons of active ingredient.[4] About half of developing country consumption is of insecticides—compounds responsible for the most serious health and environmental problems—and half the world's consumption of insecticides is accounted for by the developing countries.[5] Cotton growing accounts for the highest proportion (16 percent) of pesticide use, and yet cotton covers only an estimated 2.5 percent of cultivated land area across the globe.[6] In contrast, the developed countries account for 90 percent of herbicide use—compounds that, in general, are safer.

Pesticides give rise to many hazards: to humans causing morbidity and sometimes mortality; to the environment destroying wildlife directly or by disrupting their habitats.

Hazards to Human Health

Although pesticide use in developing countries is lower than in the developed countries, the relative impact on human health in the developing countries is probably greater. This is partly due to the high levels of insecticide use in the developing countries, but more important is the lack of appropriate legislation, widespread ignorance of the risks involved, poor labeling, inadequate supervision, and the discomfort of wearing full protective clothing in hot climates. These features greatly increase the risk of harm to both agricultural workers and the general public.

The hazards are made worse by the continued use of insecticides such as the organophosphates parathion, dichlorvos, and methamidophos, which are either banned or severely restricted in the developed countries. These and similar compounds are widely available in countries as disparate as China, Cambodia, and Bolivia.[7] A survey of farmers in the Andes of Bolivia conducted in 2002 showed 23 percent were using parathion.[8] While this practice is likely to produce harmful effects in developing countries, there may also be consequences for developed countries importing food products. According to the U.S. Food and Drug Administration (USDA), 5 to 7 percent of imported foods contain banned pesticides or have residues exceeding established U.S. tolerances, or are found on a commodity for which there is no established tolerance.[9]

However, the real extent of harm to human health in developing countries is difficult to ascertain.[10] Reporting is unreliable; the symptoms of pesticide poisoning are frequently confused with cardiovascular and respiratory diseases, or with epilepsy, brain tumors, and strokes. Some studies are confined to hospital cases; others confined to minor poisoning or pesticide exposure. Most studies have focused on a limited region and not been nationally representative; moreover, there is incomplete compilation of the data.[11]

The World Health Organization (WHO) estimates that, as of 1990, three million severe pesticide poisoning episodes occur globally each year. Of these, more than 250,000 die, with 99 percent of cases being in low- and middle-income countries. However, of the total number of deaths, only one million of these were estimated to be from accidental ingestion, most of the remainder presumably being suicides. Since this estimate is based only on hospitalized cases, it is likely to be a considerable underestimate.[12] Recent estimates continue to link most severe pesticide poisoning with suicides.[13] Part of the reason is the ease with which pesticides can be obtained and their lethality.

These data refer to acute poisoning; there are even fewer statistics for the effects of long-term exposure and chronic poisoning. The 1990 WHO report estimates nearly 800,000 cases involving long-term and chronic effects.[14] These are a very small proportion (less than 1 to 4 percent) of the several million cases of occupational injuries and ill health in agricultural workers worldwide.[15] Nevertheless, it is developing country farmers with lower education and greater poverty who are likely to be highly vulnerable.[16]

Field Data from the Philippines

Central Luzon in the Philippines was one of the Green Revolution lands where pesticide spraying on rice has been especially intensive. Following the introduction of the new varieties, farmers were spraying four to five times in a season with organochlorine compounds classified as extremely or highly hazardous (Box 12.1).

Box 12.1 Impacts of Pesticides on Human Health in the Philippines

In 1985, Michael Loevinsohn, a postgraduate student of mine at Imperial College (now at IDS), was investigating the ecology of rice pests in the Philippines. He became concerned when he saw how pesticides were being applied and decided to look at local mortality records. The results were startling. According to the records, over the period when pesticide use was doubling, death rates from diagnosed pesticide poisoning had risen by nearly 250 percent, and by 40 percent from other conditions that may have been related to pesticide use. The increased mortality was confined to men; for women and children it declined.[17] Moreover, the deaths peaked in February and August when spraying was most intensive after the introduction of double cropping in the mid-1970s.

Loevinsohn's evidence is circumstantial, but it was powerful enough to stimulate further investigation. Agnes Rola of the University of the Philippines and Prabhu Pingali of IRRI followed up with a detailed study that revealed a range of disease conditions in those exposed to pesticide use that were significantly higher than those not exposed (Figure 12.1).[18] These results seem to bear out Loevinsohn's findings and, together, they imply there are both high morbidity and mortality effects directly resulting from pesticide use in the Philippines. Loevinsohn estimated through extrapolation that at the time of his study there could be "many tens of thousands" of deaths each year.[19]

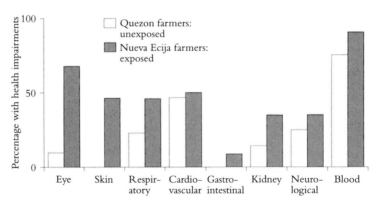

Figure 12.1 Health impairment of farmers exposed and unexposed to pesticide use in the Philippines.[20]

Pesticides and the Environment

Pesticide effects on wildlife are well documented for the developed countries.[21] Similar effects have occurred in the developing countries, although the evidence is not as complete. There is much anecdotal evidence; for example, many reports of deaths of cattle and other livestock. Pesticides are also responsible for the eradication of fish, shrimps, and crabs in rice fields—important sources of protein for the poor and landless, and in hampering pollination. For example, neonictonoids are highly toxic to bees.[22] In recent years, the highly toxic organochlorine and organophosphate insecticides have been partly, although not entirely, replaced by less widely toxic and persistent insecticides, such as the pyrethroids. Judging from experience in the developed countries, the harmful effects on the environment are less, but not insignificant.

Pesticide Inefficiency

Quite apart from these hazards, pesticides are frequently costly and inefficient. They have to be sprayed repeatedly if control is to be maintained, insect pests commonly become resistant to them, and, as ecological research has shown, they can kill off the natural enemies—the parasites and predators—that normally control pests.[23]

I first encountered the problems pesticides can cause when I was working as an ecologist in North Borneo (later the state of Sabah, Malaysia) in 1961 (Box 12.2).

Box 12.2 Pesticides and Pests of Cocoa in Borneo

In 1961, cocoa was a recently introduced crop being grown in large, partial clearings in the primary forest. When I arrived the crop was being devastated by pests: cocoa loopers and bagworms were stripping off all the leaves, cossid borers were destroying the branches, ring bark borers were killing whole trees, and a pest new to science—the bee bug—was damaging the cocoa pods.[24] At the time, cocoa fields were being heavily and repeatedly sprayed with insecticides, sometimes consisting of cocktails of organochlorines, such as DDT and dieldrin. They were having little effect. On the contrary, I believed they were making the situation worse. In their natural forest home, the pest species were probably being controlled by a variety of natural enemies, and, it seemed to me, the problem was being caused by the pesticides, which, being unselective in their action, were killing off the natural enemies.

At my recommendation all spraying was stopped. Two of the pests, the branch borer and the cocoa looper soon came under control by parasitic wasps. The bagworms

(continued)

Box 12.2 (*continued*)

continued to cause damage, and they were controlled by use of a highly selective pesticide before eventually being naturally controlled by a parasitic fly. The ring bark
borer was largely eliminated by destroying a secondary forest tree that had remained in
the fields and was the borer's natural host. Very selective spraying kept the bee bug in
check. Within a year all the major pests were being satisfactorily controlled, and this
persisted for several decades (Figure 12.2).

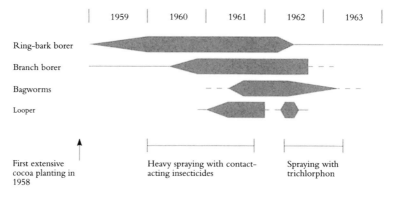

Figure 12.2   The control of cocoa pests in North Borneo, 1961.[25]

Modern Pesticides

Since the 1960s the broad-spectrum organochlorine insecticides so widely used in
North Borneo and elsewhere have been replaced by more selective compounds,
which also tend to be less damaging to wildlife and human health.[26] Increasingly
stringent regulations in the developed countries have forced manufacturers to engage
in exhaustive safety and environmental testing, both in the laboratory and in natural
field conditions.[27] New pesticides were discovered in the past by a largely random
process of screening thousands of synthetic compounds. Now, with the greater
understanding conferred by modern cellular and molecular biology, chemical companies have begun to search for tailor-made pesticides. For example, one group are
compounds that mimic the effects of juvenile hormones in insects, disrupting the
transition from one life cycle stage of an insect to another, preventing caterpillars
from becoming moths. They are valuable because they often affect only one species
of insect.

Natural Pesticides

There is also growing interest in natural plant compounds that have been tradition-ally used by farmers for pest control. They include custard apple, turmeric, croton oil tree, Simson weed, castor oil, ryania, and chili pepper.[28] The pyrethroids, based on the compound pyrethrum found in chrysanthemum plants, are effective against certain pests and are safe. One of the best known sources of a natural insecticide is the neem tree, which has been used against rice pests in India for centuries.[29]

Pesticides Used by Organic Farmers

Supporters of organic farming often claim organic growers do not use chemicals or pesticides. In fact certified organic crops can be treated, if there is no alternative, with a collection of various "natural or simple" chemicals. In the United States, the list includes sulfur, soft soap, copper and rotenone, some pyrethroids, iron orthophos-phate and paraffin oil, and certain bacterial and other microorganisms.[30]

Natural pesticides, however, are not necessarily environmentally friendly or less toxic than synthetic pesticides. In the words of Rebecca Hallett, an environmental scientist at the University of Guelph who has studied soybean cultivation in Europe: "It's too simplistic to say that because it's organic it's better for the environment. Organic growers are permitted to use pesticides that are of natural origin and in some cases these organic pesticides can have higher environmental impacts than synthetic pesticides often because they have to be used in large doses."[31]

Wild plants, from which some natural pesticides are derived, naturally abound with toxins that are produced to defend the plant against fungi, insects, and other animal predators. Bruce Ames of the University of California (who developed the Ames test for carcinogenicity) and his colleagues estimate that about 99.9 percent of the chemicals that humans ingest are naturally occurring.[32] Humans ingest roughly 5,000 to 10,000 different natural pesticides and their breakdown products.[33] Even though only a tiny proportion of natural pesticides have been tested for carcinoge-nicity, over half that have been tested are carcinogenic in rodents. Food is not neces-sarily made any safer by using "natural" pesticides.

Agronomic Methods

One alternative to pesticides, which is favored by organic farmers and increasingly by conventional farmers, is to rely on a variety of agronomic practices. These range from interplanting a crop with plants that deter pests to various techniques of cultivation.

Often the mechanism of control is subtle: Aromatic odors from the intercropping of cabbages and tomatoes repel the diamondback moth, a major pest of cruciferous crops; the shading effect of mung bean or sweet potato grown with maize reduces weed growth; and the liberation and spread of pathogen inocula can be reduced by growing cowpeas with maize.[34]

One of the most effective agronomic approaches is the "push–pull" system, which has built on ecological studies to create a polyculture agriculture that protects maize, millet, and sorghum from two devastating pests: an insect, the stem borer, and a weed, *Striga* (Box 12.3).

Box 12.3 The Push-Pull System of Pest and Weed Control[35]

Push-pull entails mixing plants that repel insect pests ("push") and planting, around a crop, diversionary trap plants that attract the pests ("pull"). The system was developed by collaboration between the International Centre of Insect Physiology and Ecology (ICIPE) and the Kenyan Agricultural Research Institute (KARI), both in Kenya, and Rothamsted Research in the United Kingdom. It is based on intercropping of the main cereal crop with the forage legume *Desmodium*. This plant both emits volatile chemicals that repel stem borer moths ("push") and attracts a natural enemy of the moths, parasitic wasps ("pull"). In addition, *Desmodium* secretes chemicals from its roots that cause "suicidal" germination of *Striga* seeds before they can attach to the maize roots. To ensure further protection, farmers can plant a "trap crop," such as Napier grass, around the edge of the field, which attracts the moths, pulling them away from the main crop (Figure 12.3).

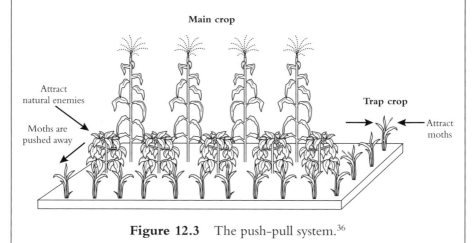

Figure 12.3 The push-pull system.[36]

As of 2010, 25,000 smallholders in East Africa are using push-pull. It allows them not only to control pests but also to increase soil fertility, protect against erosion, reduce pesticide use, and gain income from the *Desmodium* crop.[37]

Biological Control

Another nonpesticidal approach to pest control is to directly encourage the natural enemies of pests. Sometimes, although rarely, this can be spectacularly successful. An example is the biological control of the cassava mealybug in Africa, which was pioneered by Hans Herren, then at the International Institute of Tropical Agriculture (IITA).[38] The mealybug first appeared in the Congo and Zaire in 1973, but soon spread across a wide belt of central Africa, from Mozambique to Senegal, producing yield losses of up to 80 percent. Cassava originated in South America, and a search there for the mealybug's natural enemies identified a parasitic wasp in Paraguay. The wasps were released in Nigeria in 1981. The results were dramatic, with yield increases of up to 2.5 tons/ha; overall benefits are estimated in terms of billions of dollars.

Plant pathogens can also be controlled by their "enemies"; that is, organisms that act as antagonists or competitors.[39] One of the most serious pathogenic problems in Africa is the infection of crops such as maize and groundnuts by strains of the fungus *Aspergillus* that produce a very toxic, carcinogenic poison, aflatoxin, which is also a potential biological warfare agent.[40] High levels of aflatoxin are found in many African countries. In a serious outbreak in eastern Kenya in 2004, food samples had aflatoxin levels of 139–200 parts per billion (ppb) accompanied by many deaths.[41] African countries lose up to $1.2 billion each year due to rejection of infected foodstuffs. Contamination is common in hot and humid conditions, especially when crops are under stress; for example, from drought or insect attack. A current project conducted by IITA and the USDA Agricultural Research Service under the umbrella of the African Agricultural Technology Foundation (AATF) is investigating the use of benign strains of *Aspergillus*. When broadcast onto the soil, these outcompete the toxin-bearing strains. Initial field trials in Nigeria showed reduction in aflatoxin production of 50 to 99 percent.[42]

Often natural enemies of pests can be encouraged by creating a more diverse agroecosystem; for example, few pest problems occur in the Javanese home gardens described in Chapter 5 (see Figure 5.1). The diversity of plants in each garden encourages a diversity of insects which, in turn, supports a large population of general predators—spiders, ants, assassin bugs—that keep potential pests under control.

Sometimes even growing a mixture of two crops is enough.[43] In the Philippines, intercropping of maize and peanuts helps to control the maize stem borer. The predator is a spider that, as an adult, feeds on the stem borer caterpillars. But the young spiders feed on springtails, which they find in the leaf litter under the peanut plants. The simple intercrop is sufficient to create a complex, beneficial food web. The move toward large areas of monoculture has been one of the reasons why pest and disease outbreaks grew in the wake of the Green Revolution.[44]

Pest Evolution

Pests and pathogens, like all organisms, have the ability to adapt to new situations through natural selection. Michael Loevinsohn, working in the Philippines, has shown the remarkable capacity of rice pests to evolve in response to variation in the timing of rice cultivation. Within a few years genetically different populations of pests have arisen, each adapted to different rice cropping patterns that are only a few kilometers distant from each other. At Mapalad, at the base of the Sierra Madre Mountains in Luzon, where a single rain-fed crop is grown, populations of the yellow stem borer have a shorter generation time than do populations at Zaragoza, 10 km away in the center of the irrigated plain where two crops are grown. They also lay more eggs and have a lower survival rate. Planting is carried out more or less at the same time at Mapalad, and the crop matures uniformly; in these conditions there is a selective advantage for pests that mature quickly and increase rapidly in numbers. By contrast, under irrigated double cropping, the planting is asynchronous and the pests are more heavily attacked by natural enemies—predators and parasites. In these circumstances it is advantageous, for the pests, to mature more slowly but have a higher survival rate.

Not surprisingly, pest and pathogen populations responded very rapidly to the continuous cropping of the new varieties of wheat and rice. The first ten years of double cropping of rice at IRRI in the Philippines resulted in dramatic growth in pest populations. The numbers increased directly because of the introduction of a dry season rice crop, but there were more pests and more damage in the wet season crop as well. Thirteen percent of the wet season crop was lost under single cropping, but this rose to 33 percent when double cropping was introduced.[45] Under triple cropping, the numbers and damage were even higher. Only where there is a break in cultivation, such as a fallow, or where a cereal is alternated with a dissimilar crop, are pests and diseases held in check.

Pesticide Resistance

Pests, pathogens, and weeds are also capable of evolving rapid resistance to threats and adverse circumstances, in particular to the use of pesticides.[46] Today over 550 insect pest species in the world are resistant to one or more insecticides.[47] There are fewer fungi and bacteria resistant or tolerant to fungicides and bactericides, but it is a significant problem.[48] Nearly 200 weed species are resistant to herbicides.[49] Several important insect pests are resistant to all the major classes of insecticides: The diamondback moth, a pest of cabbages and other crucifers, is resistant not only to the older organochlorines and carbamates but also to the newer organophosphates and pyrethroids.[50]

Pesticide resistance is a classic example of natural selection. Resistance occurs as a result of natural mutation or genetic recombination in the absence of the pesticide, but usually at a very low level because it is generally disadvantageous in other respects. However, it will rapidly spread through the pest, disease, or weed population from generation to generation in response to the pressure of pesticide use. Survivors tend to be resistant and pass on the resistance through their genes to their descendants. It is almost impossible to prevent this from occurring, but the risk can be minimized in the following ways:[51]

- Using compounds that are slow to elicit a resistance response
- Using different compounds in rotation from year to year or as mosaics (i.e., different compounds for different fields or farms)
- Avoiding opportunities for cross-resistance by establishing links between resistance to different compounds
- Closely targeting pesticide application

Breeding for Crop Resistance

Pesticides are often very effective at controlling pests of various kinds, but as we have seen they have their drawbacks. Agronomic and biological approaches are to be preferred on environmental grounds, yet they tend to require high levels of skill and labor. The ideal solution is to put resistance to pests into the seed. It is by far the easiest and most convenient—and often the cheapest—approach, providing of course that the resistance is not too rapidly overcome.

Plant breeding for disease resistance goes back to the earliest days of modern agriculture. As I indicated earlier, plants naturally contain toxins that protect them against a variety of insects, fungi, bacteria, and weeds. Breeding aims to (1) find those plants, either in cultivation or in the wild, that produce the appropriate toxin or other defense mechanism; (2) identify the genes responsible; and (3) either select for them or transfer them to desired cultivars through conventional breeding or genetic engineering technologies. This effort is aided today by the many important collections of crop varieties maintained in various parts of the world.

Of course, just as crop plants will evolve and fight off attack, so are pests and pathogens equally adept at evolving ways of overcoming the defenses that naturally occur in crop plants or are bred into them by plant breeders. In 1950, a new race of wheat stem rust exploded in the United States and southern Canada and was carried by high winds into Mexico.[52] This was only the first of a series of epidemics. New races continued to arrive, and by 1960 a group of virulent races had almost completely replaced the existing, relatively weak forms. The wheat breeding program was able to keep pace with this changing pattern of disease but only by virtue of having new resistant varieties quickly available when each change in race occurred.

There has been little breeding for rust resistance in recent years because the problem seemed to have been overcome. But in 1999, a highly virulent strain of black stem rust appeared in Uganda and spread through the highlands of East Africa. In Kenya losses were as high as 80 percent. There are some fifty rust resistance genes, and the new strain, named UG99, is resistant to all but ten of these. Since 1999 the rust has spread beyond East Africa to Ethiopia, Sudan, Yemen, and Iran.[53] In the short term the urgent need is to identify the small number of wheat cultivars that remain resistant to UG99. Recently some new cultivars have been bred from some of these sources by Rajiv Singh and his colleagues at CIMMYT.[54] They contain a number of genes, each of which has a low level of resistance but when combined are very effective.

The ability of the fungus to mutate and evolve, however, means that protracted resistance will be unlikely. The discovery that rice is resistant to the entire taxon of rust fungi is a cause for optimism. If the mechanism of resistance can be identified and the genetic information translocated to wheat, it may be possible to create a range of durable resistant varieties.[55] The gene transfer technology is currently available to make this process feasible, but the severity of the current threat means researchers are racing against time.

Bt *Genes*

In recent years breeding for resistance has benefited from recombinant DNA techniques based on the genes that code for the toxins produced by a bacterium, *Bacillus thuringiensis (Bt)*. This bacterium occurs naturally and is widespread in the soil. During its life cycle it produces a crystalline protein toxin, known as a *cry* protein.

If the bacterium or extracts of the cry protein are sprayed on crop leaves or other parts of plants, they are ingested by insects feeding on the plants, particularly the caterpillars of moths such as cotton bollworms. Once ingested, the protein kills the insect. Since useful insects such as honey bees and natural parasites and predators do not feed on the sprayed plants, they are unaffected. The toxins do not harm humans.

These features make the cry proteins ideal insecticides. I found them useful in controlling the bagworms attacking cocoa (see Box 12.2) because they did not upset the emerging control by natural enemies. Not surprisingly the *Bt* proteins (mentioned in Chapter 9) are approved for use by organic farmers.

The genes are ideal targets for genetic engineering and have already benefited cotton production in developing countries (see section on *Bt* cotton in Chapter 9). The genes encoding for the proteins were first isolated in the early 1980s and have now been transferred to a wide variety of crops. Farmers in the United States are regularly growing maize, potatoes, and cotton containing the *Bt* gene. It has also been transferred to rice and maize where its potential lies in the control of stem borers, an important pest in developing countries against which conventional plant breeding has made little headway.[56]

A further potential for *Bt* is to gain control of the diamondback moth (*Plutella xylostella),* a devastating pest of cabbages, cauliflower, kale, mustard, and other brassica crops, important food plants across Asia and Africa.[57] Losses are as high as 90 percent. Even when sprayed with insecticides, often every other day, with 12 to 24 applications normal in a three-month season, there can still be up to 35 percent loss. The moth has developed resistance to almost all insecticides. One consequence of the heavy and frequent spraying is the serious health risk posed to farmers as well as to consumers and the environment.

GM technologies have also been used against crop pathogens. Indeed, one of the first and most successful GM crops was a papaya with resistance to ringspot virus, a devastating disease of papaya in Brazil, Taiwan, and Hawaii.[58] Another example is the current development of bananas resistant to banana wilt in Uganda (see Chapter 9).

Resistance to **Bt**

Inevitably, the use of cry proteins will, like synthetic chemical insecticides, lead to selection for resistance. Bruce Tabashnik, an entomologist, and colleagues at the University of Arizona have studied *Bt* resistance and found a number of insect species showing resistance in the laboratory. Resistance to the sprayed *Bt* in the field is known, but the first case of resistance to genetically engineered *Bt* in the field was in *Helicoverpa zea* (common names include cotton bollworm and corn earworm, depending on which crop the larva is feeding on) in Arkansas and Mississippi and not until 2003–2004.[59] This occurred despite the fact that *Bt* cotton and *Bt* corn have been grown on more than 162 million ha worldwide, "generating one of the largest selections for insect resistance ever known."[60] Monsanto confirmed a second case of field resistance on *Bt* cotton in the Indian state of Gujarat in 2009.[61]

One approach to slowing selection for resistance is to set up refuges of host plants without *Bt* genes.[62] The rare resistant insects surviving on *Bt* crops will mate with abundant susceptible pests from the refuges. If inheritance of resistance is recessive, the hybrid offspring will be killed by *Bt* crops, markedly slowing the evolution of resistance. In the case of *Helicoverpa,* where resistance has occurred, the resistance is dominant. So far it has not occured for pests with recessive resistance.

Fortunately, there are a number of different cry proteins produced by different *Bt* genes. An approach being widely adopted is to produce crops with at least two different genes in the expectation that the insects cannot readily develop resistance to both toxins simultaneously. It is also possible to develop plants with multiple toxic genes, each with a different mode of action. A promising companion to the *Bt* gene is a gene that encodes for a proteinase inhibitor contained in the tropical giant taro plant. This confers high resistance to insect attack. The protein probably inactivates the proteases of an insect so starving it to death, a very different insecticidal mechanism from that of the *Bt* gene. Another potentially useful gene makes the baculovirus NPV (nuclear polyhedrosis virus) more lethal to bollworms. Since the toxin is expressed only in cells where the virus is actively growing, there is little or no danger to nontarget organisms.

Integrated Pest Management

Clearly no one form of pest control is ideal. All have their downsides. Not surprisingly, pest and pathogen control has been—and to some extent still is—a hit-or-miss affair.[63] Usually the first response is to try a pesticide; if this does not work or causes further problems, a different pesticide or an alternative method is tried. And so the process continues. Often what works for one pest on a crop will not work for another, or the same intervention makes the other pest problem worse. Professionals in crop protection have long recognized the problem and since the 1950s have been developing a systematic approach to pest control that goes under the name of *Integrated Pest Management* (IPM).[64]

IPM looks at each crop and pest situation *as a whole* and then devises a program that integrates the various control methods in the light of all factors present. As practiced today, IPM combines modern technology, the application of synthetic, yet selective, pesticides, and the engineering of pest resistance with natural methods of control, including agronomic practices and the use of natural predators and parasites. As was demonstrated by two of the first applications of IPM in the developing countries—my control of cocoa pests in North Borneo (see Box 12.2 above) and Brian Wood's control of oil palm pests in Malaya—the outcome is sustainable, efficient pest control that is often more effective and cheaper than the conventional use of pesticides.[65]

Controlling the Brown Planthopper

One of the most extensive examples of IPM is for the control of the brown planthopper (BPH). The BPH is a sucking bug that feeds on the rice plant and, when present in high numbers, causes distinctive hopper-burn of the rice plants and loss of yields. It also transmits viruses that attack the rice plant. The BPH was virtually unknown as a pest before the introduction of the new rice varieties; at a major symposium on "The Major Insect Pests of the Rice Plant" at the International Rice Research Institute (IRRI) in 1964, BPH was hardly mentioned. Yet it soon began to cause severe damage. By 1977 the losses in Indonesia were over a million tons of rice.

At about this time a young scientist at IRRI, Peter Kenmore, now at FAO, uncovered the reason why insecticides, far from controlling the pest, seemed to be associated with the outbreaks.[66] He showed that BPH is naturally held in check by natural enemies in the rice fields: parasites that destroy the eggs and young nymphs and several species of wolf spider that prey on the adult planthopper. In North Sumatra, the population density of the pests rose in direct proportion to the number of insecticide applications; farmers were treating fields six to twenty times over a 4- to 8-week period without any success.[67] The pesticides were not only ineffective but also being heavily subsidized by the government, at 85 percent of their cost. Indonesia had become self-sufficient in rice production but, in the process, was accounting for 20 percent of the world's use of pesticides on rice. In 1986, the government acted on the basis of the mounting evidence implicating pesticides in the BPH outbreaks. A presidential decree banned fifty-seven of the sixty-six pesticides used on rice and began to phase out the subsidy.

The Indonesian BPH outbreak of 1977 was also tackled by introducing a new resistant rice variety, IR26, but within three seasons it had failed, and losses in 1979 were again very severe. Next to be introduced was IR36. It was more successful and rapidly adopted. By 1984 Indonesia had become self-sufficient in rice production. But IR36's resistance was also short-lived. By 1986 the planthoppers were back in force, threatening over 50 percent of Java's rice land; losses in 1986–1987 were estimated to be nearly $400 million.[68] One explanation for this sequence of events was that BPH exists in a number of different biotypes or races:

1. The original resistance in the variety IR26 was to biotype 1.
2. Biotype 2 emerged, to which IR36 was resistant.
3. In 1983, a new biotype (biotype 3) invaded Sumatra and attacked IR36.
4. IR56 was introduced about the same time and found to be resistant to all the biotypes.[69]

However, Peter Kenmore and his colleagues believed this was too simplistic an explanation.[70] In their experience, BPH populations are extremely variable and can rapidly evolve to local circumstances, as Michael Loevinsohn has clearly shown. They argued that heavy pesticide spraying accelerates the adaptation of BPH to new rice varieties. In their view, plant breeding approaches are sustainable only if they are a part of an integral strategy.

Under the IPM program developed for BPH, farmers are trained to recognize and regularly monitor the pests and their natural enemies. They then use simple, yet effective, rules to determine the minimum necessary use of pesticides. The outcome is a reduction in the average number of sprayings from over four to less than one per season. In the early years of the program, rice yields grew from 6 to nearly 7.5 tons per hectare. Rice production increased by 15 percent, and pesticide use declined by 60 percent, saving $120 million a year in subsidies. The total economic benefit in 1990 was estimated to be over $1 billion.[71] The farmers' health has improved and fish, traditionally consumed by rice farmers, have returned to the rice fields. On the IRRI farm where IPM has been continuously practiced since 1994, planthoppers have remained at low densities, less than five per rice hill, and the rice fields are rich in biodiversity (Table 12.1).

Table 12.1 Increasing arthropod diversity on the IRRI farm between 1989 and 2005

	Species richness	
Type of arthropod	1989	2005
Herbivores	14	36
Predators	38	65
Parasitoids	17	38
Detritivores	6	30

Source: Heong[72]

IPM in Farmer Field Schools

Over the past fifty years, IPM has grown into a sophisticated approach to pest control and has had notable successes.[73] A review of IPM in the developing countries in the 1990s, by Jules Pretty of the University of Essex, identified several programs where annual savings are in the range of $1–10 million.[74] But IPM has not been as widely adopted as might be expected. Why not? Part of the reason is that, despite its grounding in ecological principles, it has remained until recently a traditional top-down approach in its implementation. IPM programs have been worked out by specialists and then instructions passed on to farmers.

IPM is a more complex process than relying on a regular calendar of spraying, and some have argued that farmers cannot understand some of the technicalities involved. However, in recent years, this view has been effectively challenged. At the Escuela Agrícola Panamericana Zamorano (generally known as Zamorano) in Honduras, training programs have been discovering what farmers do and do not know about pest control.[75] For instance, they know much about bees but are unaware of the existence of solitary wasps that prey on insects, or of parasitic wasps that, as larvae, live inside other insects. They are familiar with many aspects of the ear rot disease of maize but not with how it reproduces. They are aware that pesticides are toxic, but equate this with the smell of the pesticide and take few precautions when they spray. Farmers in the training course look at fungi under the microscope, watch parasitoids emerge from pests, and, in the field, observe wasps and ants preying on pests.

A most rewarding result has been the farmers' readiness to experiment with their newfound understanding, integrating it with their traditional knowledge. One farmer, for example, intercropped amaranth among his vegetables to encourage predators; another placed his box of stored potatoes on an ants' nest; a third took parasite cocoons from his farm to a neighbor's farm.

The most extensive involvement of farmers in IPM has been the Indonesian rice program.[76] By 1993 over 100,000 Indonesian farmers had attended farmer field schools where they used simple Agroecosystem Analysis diagrams to understand and discuss the relationships between various pests and the rice crop. The life histories of pests and their predators and parasites are explained using an "insect zoo" and dyes used in knapsack sprayers to demonstrate where the insecticide sprays end up. The schools themselves have become the basis of farmer IPM groups where farmers continue to meet to discuss their problems and to organize villagewide monitoring of pests and predator populations. In 1990, an outbreak of white stem borer threatened to undermine the success of the program, but the calls to revert to spraying were successfully resisted. Through the schools, farmers were taught to recognize the egg masses of the stem borers and in a massive campaign searched for and destroyed them. Only a handful of rice fields were infested a year later.

Since 1990 some 20 percent of the farmers' training has been paid for by the farmers themselves. Observers are convinced this accounts for the very considerable savings on pesticide applications and the attainment of higher yields. As one graduate of the field school put it: "[Now] I have peace of mind. Because I now know how to investigate, I am not panicked any more into using pesticides as soon as I discover some pest damage symptoms."[77] Farmers in other Asian countries are now learning IPM in farmer field schools.

IPM and Bt Genes

The *Bt* gene is not a silver bullet. *Bt* crop varieties are very good at dealing with a range of insect pests, notably stem borers and bollworms, but they are ineffective against most other pests. Now that the *Bt* gene is controlling the bollworms, these "secondary" pests have exploded.[78] In northern China where 95 percent of the farmers use *Bt* to control the bollworm *Helicoverpa armigera,* the crop acts as a trap crop, killing bollworms and so reducing the infestations on other crops, maize, peanuts, soybeans, and vegetables. However, mirid bugs, which were previously a minor pest, have attained outbreak status because of the absence of insecticide spraying, and indeed the *Bt* cotton has become a regional source of the mirids.

For this and other reasons, including the prevention of resistance to *Bt*, the contemporary challenge for IPM is to see it as a vehicle for, and a complement to, pest control with the *Bt* gene. This should include the promotion of the use of (1) natural enemies of the pests, (2) as many possible forms of resistance as possible, and (3) traditional and modern pest control practices.

In some respects IPM is the most severe test of the theory of change of the Doubly Green Revolution. While heavy spraying with synthetic pesticides was instrumental in producing the high grain yields of the Green Revolution, it brought many undesirable effects, including severe health and environmental hazards. In many instances the compounds used were ineffective, for a variety of reasons including pesticide resistance and the destruction of natural enemy control. The development of IPM has provided the potential for an approach to control that is not only effective but environmentally friendly and that could increase both resilience and equitability. But the potential has yet to be fully realized. IPM has yet to make much headway in Africa, apart from its use on some high-value, export crops where pesticide use is already common. Its application to staple crops requires a rethink of the strategy.[79]

A fundamental challenge to IPM implementation is that it takes a relatively high level of skill and is often labor intensive. Not surprisingly pesticide spraying appears to be an easier approach. Alarmingly, the marketing of the new hybrid races has been accompanied by companion marketing of pesticides. There is now evidence that this is undermining the IPM rice programs so painstakingly put in place over the past three decades, with resultant upsurges of pest populations.[80]

Finally, we need to bear in mind that pests will nearly always be able to develop resistance to pest control methods of whatever kind. Moreover, there will always be trade-offs between one set of outcomes and another—and there will always be undesirable consequences. The task ahead is to minimize these trade-offs.

In Chapter 13 I consider the importance of healthy and nutritious soils for plant growth and the potential for integrated nutrient management as an analogue of IPM.

13

Rooted in the Soil

Once the natural vegetation is cleared, "the trees, cut down by the axe, cease to nourish their mother with their foliage." However, "we may reap greater harvests if the earth is quickened again by frequent, timely, and moderate manuring."

Lucius Columella, *De Re Rustica*[1]

Lucius Columella, writing in the century after Marcus Varro (quoted at the beginning of Chapter 6), was another Roman landowner who clearly understood the basis of sustainable agriculture: the soil.

It is relatively easy to sow a good seed in a pot of well-structured, organic soil, place it in a greenhouse, protected from pests and pathogens, water and fertilize the growing plant when necessary, and be rewarded with a phenomenal crop. Needless to say, conditions on a farm are far from this "ideal" environment. Individual farmers can do much to improve their situation—by buying high-quality seed, applying fertilizer, manure, or crop residues, incorporating organic matter in the soils, weeding the crop, and adopting an integrated approach to pest and pathogen control. But the biggest challenge lies in achieving a better soil environment for their crops.

In this chapter I address these questions:

- What is the nature of soil and of nutrient cycling?
- How are fertilizers used?
- What are the potential hazards of fertilizers to human health and the environment, and how can they be used better?
- What is the extent of land degradation and soil erosion?
- What are the various physical, biological, and agronomic approaches to soil conservation?

The Nature of Soil

A good soil has two aspects, one qualitative and one quantitative. Structurally, soil is made up of rock or mineral particles, organic matter, air, and water. Typically around 2 to 5 percent of soil solid matter is organic matter (also known as *humus*), which is inhabited by the millions of living organisms that break down organic compounds and make them available to plants. The rest of the solid matter consists of rock or mineral particles. The texture of the soil is determined by the type and size of mineral particle, ranging in size from clay at the smaller end of the scale to silt and sand, the largest. The size determines the amount of space around each particle; clay particles are small and thus tightly packed, meaning that a clay soil will hold water and nutrients but may suffer from poor water drainage and a lack of air needed by plant roots. On the other hand, large sand particles have pockets of space around them, leading to good aeration and drainage but also a tendency for soils to dry out and for nutrients to be leached.

Compacted soil such as compacted clay can restrict both soil organisms and root growth because there is little movement of air and water; hence they are less productive. Good soils have a mixture of small and large particles to enable water drainage and storage (particularly important in arid and semi-arid areas).[2]

Soil has a quantitative aspect: It needs an optimal amount and mix of nutrients if crop growth and yields are to be maximized on a resilient and sustainable basis. The primary nutrients are nitrogen, phosphorus, and potassium, but micronutrients are also usually essential. In the Chiang Mai valley described in Chapter 6, for example, declining yields were eventually traced to lack of boron in the soil. Other micronutrients include zinc, copper, iron, chloride, manganese, and molybdenum.

Nutrient Cycling

In natural ecosystems, the nutrients cycle. Nutrients are collected from the soil by the roots of plants, contribute to the growth of stems, leaves, and fruits, and when the plants die are returned to the soil as the vegetation rots. A similar cycling supports animal populations: Nutrients, such as nitrogen, are ingested as the animals graze on grass and other plants, are partly returned in the excreta and urine, and partly returned when the animals eventually die and decompose (Figure 13.1).

As all farmers recognize, when plants are treated as crops and animals as livestock, the process of harvesting removes the nutrients from the ecosystem. Some soils are naturally richer in nutrients than others and can be mined, at least for a period, but

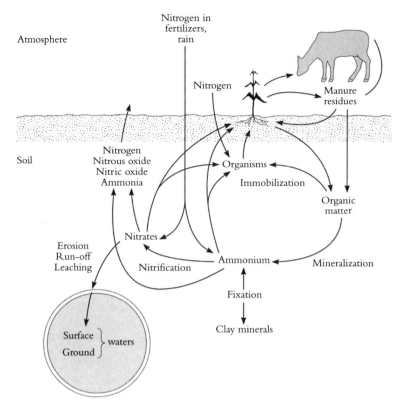

Figure 13.1 The nitrogen cycle in crops, livestock, and the soil.[3]

eventually for all soils the lost nutrients have to be replaced. Without nutrient replacement there is no agricultural sustainability.

Synthetic Fertilizers

Until the twentieth century, the only means of nutrient replacement available to the farmer was to apply composts or manures or, as the Romans recognized, to grow nitrogen-fixing legumes (such legumes contain bacteria in their root nodules that take up nitrogen from the atmosphere). Agriculture was transformed by the invention, at the beginning of the century, of the Haber-Bosch process for synthesizing ammonia from the atmosphere. Ammonia is the primary ingredient in synthetic fertilizers, and today, the manufacture of synthetic, inorganic fertilizers is highly efficient and, until recently, relatively inexpensive, relying as it does on atmospheric nitrogen and fossil fuels, such as methane and coal, for its basic ingredients. How-ever, as I indicated in Chapter 1, rising energy prices have significantly increased the costs of fertilizer production and hence resulted in rising prices of cereals and other food crops.

The other key nutrients, phosphorus and potassium, are also fairly plentiful in mineral forms, as phosphates and potash, respectively. However, they are geograph-ically narrowly confined (50 percent of the deposits are located in the Middle East and another 25 percent in South Africa), and there is controversy over whether the reserves of phosphorus are peaking.[4] Estimates of these reserves and availability of exploitable deposits vary greatly. High-grade phosphate ores, particularly those con-taining few contaminants, are being progressively depleted, and production costs are increasing. One review concludes that within some sixty to seventy years about half the world's phosphate resources will have been used up.[5]

The development of synthetic fertilizers opened up the potential for very high yields, far higher than were achieved by natural nutrient cycling. It was this potential that Norman Borlaug and his colleagues sought to exploit in the Green Revolution. Today about half of total fertilizers are applied to cereals, about 15 percent each for rice, wheat, and maize.[6]

In those countries that experienced the Green Revolution, fertilizer consump-tion is not much lower than in the developed countries.[8] In China, fertilizer use is now higher than in the developed countries (Figure 13.2). But, there are consider-able downsides to such heavy use. Crop plants rarely make efficient use of nitrogen, whether in synthetic fertilizer (i.e., produced by industrial processes such as Haber-Bosch) or in crop residue or manure. Of the 100 million tons of nitrogen fertilizer (N) produced industrially in 2005, only 17 percent was taken up by crops. Moreover, efficiency has been declining: between 1960 and 2000, the efficiency of nitrogen use for cereal production decreased from 80 to 30 percent.[9]

To make up for the losses, farmers tend to apply far more N than is needed, even taking into account the losses. A great deal is wasted. This is understandable. There is a massive variation in the response of crops to fertilizer applications—the soil type, the rainfall, and the kind of fertilizer all affect the efficiency with which the nutri-ents are taken up and converted to grain or other harvested product. It is difficult, even with sophisticated analyses, to determine the appropriate level of fertilization. The very high levels of fertilizer subsidy are a further incentive to overfertilize.

Nitrogen fertilizer overuse is ubiquitous. China now manufactures and uses a third of global nitrogen fertilizer. The percentage overuse in the Chinese provinces ranges from 50 to 100 percent, and it is estimated that China could halve N use without impacting yield or yield growth. But the government continues to push for high productivity and self-sufficiency and believes high N applications are essential.[10]

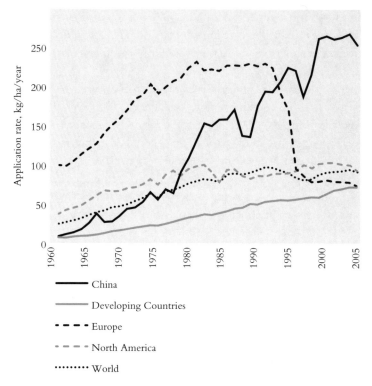

Figure 13.2 Growth of synthetic fertilizer use in arable land (nitrogen, phosphate, and potash).[7]

Possible Health Risks

These high rates of synthetic fertilizer use raise a variety of concerns, although some of the initial fears have proved unfounded. Application of nitrogen fertilizers, whether synthetic or organic (derived from plant wastes or animal manures), results in growing levels of nitrates in run-off waters and, through percolation, in subsoil water and aquifers. The consequent nitrate levels in aquifers and drinking water can be in excess of the World Health Organization (WHO) upper limit of 50 mg/ liter of nitrate and are likely to pose a health hazard. Such contamination is not confined to the Green Revolution lands, however. A survey of water from 3,000 dug wells in Indian villages revealed that about 20 percent contained nitrates in excess of the WHO limit.[11] But in most of these cases, the nitrates were derived from human or livestock waste rather than from fertilizers.[12]

250

Studies conducted over the last three decades have shown often conflicting and inconclusive results, and all point to the need for further study.[13] Yet it is important to recognize that the risk may be much greater in developing countries, where ground-water sources tend to be less protected, and drinking water is not treated to achieve the nitrate standards that are in place in developed countries. The WHO limits are intended to be applied worldwide, but they are unlikely to be implemented in the developing countries in the near future.

Environmental Pollution

Fertilizers are not directly toxic to wildlife, but they can damage wild plants by caus-ing excessive growth and can severely disrupt natural ecosystems. Perhaps the most damaging consequence of fertilizer contamination is the nutrient enrichment—*eutrophication*—of rivers, lakes, and coastal waters. Eutrophication can be caused by untreated sewage effluent or by run-off of fertilizers from agricultural land. Nitrates and phosphates that have run off the land are responsible for generating dense blooms of algae and of surface plants, such as water hyacinth and water cabbage. These plants shade out the underlying aquatic plants and, when they die and decompose, remove oxygen from the water, causing extensive fish kills. A thriving fish-pen in-dustry in the Philippine lake of Laguna de Bay in Luzon is frequently damaged in this way and although the bulk of the nutrients entering the lake comes from human waste, a significant in-flow is the fertilizer nitrogen from the surrounding intensively farmed ricefields.[14]

Eutrophication is a serious and growing problem in developed and developing countries alike. Thus, nitrogen fertilizer use in the developing countries may some-time be subject to the same kind of restrictions currently in force in the developed countries. In some respects, this should be welcomed. While modern agrochemi-cals, such as fertilizers and pesticides, are undoubtedly instrumental in raising yields, there is a great deal of waste, in part caused by high levels of subsidies but also be-cause of the lack of awareness by farmers in developing countries of the problems of overuse.[15]

Reducing Wastage

Less than half of nitrogen applied to crops ends up in the harvested product. In the tropics, even under highly controlled conditions and with the best agronomic prac-tices, the recovery rate is seldom more than 30 to 40 percent for wetland rice.[16]

The unique anaerobic conditions of paddy fields result in heavy losses of nitrogen, through volatilization as ammonia and some by emission of nitrous acid. Much of the remaining nitrogen runs off to surface waters or is leached to underground aquifers. In the seasonal tropics, leaching is encouraged by the alternation of the extremes of the wet and dry seasons. There is a slow buildup of nitrates in the topsoil during the dry season due to mineralization of organic nitrogen. This is followed by a rapid and short increase at the onset of the rains and then a decline as nitrates are flushed into surface- and groundwater.[17] On large-scale irrigated rice lands, which are mostly located in the seasonal tropics and are heavily fertilized, well over half the nitrogen is lost either to the atmosphere or via the irrigation outflows.[18]

Part of the answer to this waste is to improve the optimality of fertilizer application. In the developed countries, fields are regularly tested and farmers routinely provided with precise recommendations on application rates and timing, dependent on the soil type and the previous crop. There is an urgent need for similar advice in the developing countries.

Nitrogen utilization can also be improved by novel methods of application such as precision farming or soil-specific farming. In Chapter 7, I described the use of soda bottle caps to deliver small amounts of fertilizer to the soil pits where maize seed is sown. A similar principle has been applied to fertilizing rice paddies in Bangladesh. Urea super granules (USG), inserted in the middle of every four rice plants in the paddy fields, result in an extra ton of paddy with a reduction of fertilizer by over a third.[19] Coating urea fertilizer with sulfur produces a slower, more controlled release of nitrogen and also reduces methane, ammonia, and nitrous oxide losses.[20]

Breeding for Nitrogen Use Efficiency

A more fundamental approach to reducing wastage is to make crops more efficient at taking up and utilizing nitrogen. Conventional breeding can play a role; some varieties are more efficient than others, but nitrogen use efficiency is a complex trait with many components.[21] Attempts are being made to engineer improved efficiency.[22]

Nitrogen-fixing bacteria convert atmospheric ammonia to nitrogen using an enzyme, nitrogenase. To accomplish this bacteria need energy, which the symbiotic forms, known as rhizobia, obtain by living in nodules on the roots of certain plants. The plants furnish some of the products of their photosynthesis, and the bacteria reciprocate by supplying nitrogen. Virtually all such symbiotic relations are with legumes, such as peas, lupines, clover, and alfalfa.

The most promising approach is to create such symbioses with wheat, rice, and other cereals.[23] Rhizobial bacteria already colonize the rhizosphere of cereals, producing improvements in plant growth. This may have been the origin of root nodule symbiosis in legumes, which evolved into a full symbiosis in a gradual fashion. The way forward may be to reconstruct this process, but it will take some time to identify the genes involved and engineer them into cereal varieties.[24]

Land Degradation

Land degradation broadly refers to a decline in the capacity of the land to supply human needs, whether of food or other products and services such as supporting biodiversity. This is the sense in which it has been applied by the Millennium Ecosystem Assessment described in Chapter 6. But given such a wide definition, it is not surprising that there is much controversy about its extent or about the appropriate remedial measures. As Rattan Lal, a soil scientist formerly at IITA and now at Ohio State University, and his colleagues point out, there is a division between agronomists, ecologists, and soil scientists, who argue that land degradation is a serious concern with grave consequences for food security, and others, notably economists, who believe the problem is exaggerated and easily solvable by appropriate market forces.[25]

Not surprisingly estimates of land degradation vary widely. In Chapter 1 I referred to the *Global Assessment of Soil Degradation (GLASOD)*.[26] However, this was essentially a compilation of expert judgments without verified measurements. A new assessment, the Global Land Degradation Assessment (GLADA), is now underway under the auspices of the FAO.[27] This uses remote sensing to map vegetation as a measure of net primary productivity and to assess how this has changed as an indicator of land degradation. Preliminary results indicate that a quarter of the global land area has been degraded between 1981 and 2003, with the most severely affected areas being Africa south of the equator, Indochina, Myanmar, and Indonesia. Globally land degradation affects 1.5 billion people, who depend on these areas. Over 40 percent of the very poor live in degraded areas.[28]

Part of the assessment challenge is the need to incorporate a broad diversity of forms of degradation. Land can become degraded through the following processes:

- *Water erosion*: the principal cause of degradation, accounting for about two-thirds of the total (Box 13.1)
- *Wind erosion*: important in dryland areas, where it is responsible for much of the "desertification," accounting for a quarter of the total

The remainder is accounted for by:

- *Physical degradation*: crusting, compaction, sealing, de-vegetation, excessive till-age, impeded drainage, waterlogging, reduced infiltration and water-holding capacity
- *Chemical degradation*: salinization, alkalinization, acidification, nutrient leaching and depletion, removal of organic matter, burning of vegetative residues, agrochemicals, and industrial pollutants

These categories are clearly interconnected and feed on each other. For example, excessive tillage not only damages the soil but produces erosion and loss of nutrients.

Identifying land degradation on an individual farm is relatively straightforward. But there are difficulties in extrapolating from the individual farm or experimental plot to the whole catchment.[29] Often soil loss estimates do not match the levels of river or lake siltation in the same catchment. In many instances, the soil is simply moved from one part of the catchment to another and is not lost to the system.[30]

Assessing the extent of degradation at a national or regional level compounds these problems. The United Nations Environment Programme reported that "desertification threatens 35 per cent of the earth's land surface and 20 per cent of its population."[31] Yet as Jeremy Swift of the Institute of Development Studies (IDS) argues, much of the data is questionable.[32] In particular, it often relies on snapshot assessments, comparing drought with wet years, ignoring the often temporary nature of vegetation change, the capacity of dryland ecological systems to recover, and the ability of farmers and pastoralists to adapt to the climatic cycles. What may seem to be a desert one year is a productive tract of land in the next.

Nevertheless, in many parts of the world, the degradation of soils is clearly evident, and some of the most severely affected regions of the developing world—the uplands and highlands of Asia and Latin America, the semiarid lands of Sub-Saharan Africa, and the saline and waterlogged soils of South Asia—are where many of the rural poor and chronically undernourished now live. If we are concerned with their future, we have to identify, understand, and address soil degradation as it affects them and their livelihoods.

The Impact of Soil Erosion

In simple terms soil erosion consists of soil (from the top layers of the soil horizon but also from the subsoil in gully erosion) being transported from one place to another (Box. 13.1). It is not necessarily a bad thing. The soil may be fertile and deep and able to withstand losses; moreover, it may be transported somewhere else where

Box 13.1   The Nature and Consequences of Soil Erosion

Erosion consists of three sequential processes:[33]

1. *Detachment*—of particles
2. *Breakdown of aggregates*—by the impact of raindrops, shearing, or drag forces of water or wind, or chemical dissolution of cementing agents
3. *Transport and Redistribution*—by flowing water or wind
4. *Deposition*—when the velocity of the water or wind decreases because of the slope or ground cover

Although losses due to erosion can be considerable, in excess of 50 tons/ha per year, the severity of the consequences depends not only on the nature of the vegetative cover but also on the depth and intrinsic fertility of the soil. Rattan Lal and colleagues have estimated that erosion globally affects 1094 Mha and has caused yield reductions of 2 to 40 percent in Africa.[34]

it may be of benefit. For example, the erosion of soils from the fertile uplands of Java may be the source of the fertility and high productivity of the lowland rice fields.

Soil Conservation

For many years much time, money, and effort has gone into soil conservation measures in the developing countries.[35] The measures can be essentially physical or biological.

Physical Conservation

Physical structures of varying scale can check the surface flow of water, reducing water erosion and retaining soil and nutrients. The simplest approach is to throw up earth banks or bunds on contours, construct simple walls, or build terraces of differing degrees of complexity, which are more suitable for steeper slopes. But far too often the bunds are not properly installed or maintained and collapse or are broken. Why are such failures so common?

The techniques of soil and water conservation are well known.[36] But too often the technologies are inappropriate.[37] John Kerr and N. Sanghi of ICRISAT describe how contour bunds in India were rejected by farmers, even when heavily subsidized.[38] Among various faults: the bunds leave corners in some fields so that farmers risk losing their land to neighbors; the central watercourses provide benefits to some farmers but damage the land of others; and, if the facilities for dealing with surplus water are inadequate, the bunds readily breach and the water forms gullies. It was not uncommon for entire bunds to be leveled as soon as the project staff had gone to the next village.

Biological Soil Conservation

An alternative to physical conservation is to use biological techniques. One of the simplest is to plant the main crop along the contour, alternating with a protective crop such as a grass or legume. Water flowing down the slope meets with the rows of crops, is slowed down, and infiltrates the soil. This method is suitable for slopes of 3 to 8.5 degrees. Strips of grass will help to filter out particles and nutrients from the water and over time will build up into terraces. In Indonesian experiments, strips 0.5 to 1 meter wide of *Bahia* and signal grass were grown along the contours, alternating with 3- to 5-meter-wide strips of annual crops.[39] Erosion was reduced by 20 percent, and after four years natural terraces 60 cm high had been formed.

Conservation Farming

Conservation farming is an essentially ecological approach to soil conservation, which is gaining rapidly in popularity. In common with other new practices described in this book, it has several origins and a diversity of interpretations.

Over the years, especially in temperate climates, it has been common practice for farmers to till the soil in fields before seeds are sown in order to loosen and aerate the soil and to destroy weeds—either by hand with a hoe or with animal- or mechanically powered plows. Tillage breaks up heavy clay soils, and in winter the clods are broken down by snow and ice. However, on many soils prone to erosion or drought, tilling can harm soil structure and increase water loss.

In the United States in the 1940s, a few farmers began experimenting with ways to deal with the severe wind erosion affecting the Great Plains. Some started using mulching to control weeds without tilling the soil. However, it wasn't until the 1960s, with the development of effective selective herbicides, that farmers could begin practicing what is now known as conservation agriculture (Box 13.2).

Conservation farming experiments, ongoing today, were first conducted in Ohio in 1958. In Brazil and West Africa trials began in the 1970s, but the techniques have since become relatively widespread on erosion-prone fields on large farms in South America.[41] The development of herbicide-resistant crops that could be grown with the easier to apply broad-spectrum herbicides has further increased the popularity

Box 13.2   The Nature of Conservation Agriculture

Conservation agriculture includes various systems of reduced or no tillage. The advantages of these practices include the saving of labor used for plowing, protection of vulnerable soils from erosion by preventing topsoil from being blown or washed away, and improvement of soil fertility by keeping soil structure intact and by allowing more beneficial insects to thrive. It also keeps carbon and organic matter in the soil, both leading to a higher microbial content and sequestering carbon and thus reducing carbon emissions from agriculture (see Chapter 16).

Conservation farming requires learning new techniques, however, and can bring new challenges in weed control and drainage management. Increased used of herbicides can lead to negative environmental impacts, and farmers are now beginning to experiment with the use of more mulching and cover crops to control weeds without herbicides. In some situations farmers must also purchase new equipment such as special seed drills, which requires investment.[40]

in the United States and Latin America. In contrast, conservation-tillage practices are much less known in Europe, Asia, and Africa.[42]

In the Indo-Gangetic Plains of South Asia, where farmers practice a rice-wheat cropping system, minimum-tillage practices are rapidly being adopted thanks to the development and promotion of new techniques by the Rice-Wheat Consortium and the development of new no-till drills.[43]

There has also been growing interest in southern Africa. Experiments conducted by partnerships between local government bodies and the NGO Concern Worldwide, such as I recently saw being pioneered in western Zambia, are investigating the use of conservation farming as a replacement for the traditional long fallow system of the region. There the woodland is felled and burned before being plowed and sown to maize. Crops are grown for only a couple of years, and the land then takes several decades to return to a state where it can be felled and burned again. The alternative, conservation farming, is not to plow and instead sow the seed in small "pockets" in the soil to which have been added two cupfuls of manure and a bottle top of fertilizer. After harvest, the soil is covered with the stems and leaves of the maize and next year's seed is sown several months later in the same holes. Despite the need to hoe weeds, the labor is much less than in the conventional systems. Yields are high—some 4 to 5 tons of maize growing new drought-tolerant hybrids. In addition to building carbon in the cropped soil, such a system should allow tree or shrub cover to remain unburned more or less permanently, so increasing carbon sequestration and maintaining soil carbon levels, creating a more stable and sustainable farming system.

Returning Carbon to the Soil

Because under conservation farming the crop residues are left to rot down on the surface of the soil, there is a significant return of carbon to the soil. This reduces the likelihood of erosion and directly enhances yields, so creating a win–win situation. Other agricultural practices can also produce returns of carbon to the soil such as livestock rotations, the use of cover crops and composting.

Rattan Lal's analysis of several experiments has shown that an increase of 1 ton per ha of soil carbon in degraded croplands can increase maize yields by 200 to 300 kg/ha, wheat by 20 to 40 kg/ha, rice by 20 to 50 kg/ha, sorghum by 80 to 140 kg/ha, millet by 30 to 70 kg/ha, beans by 30 to 60 kg/ha, and soybeans by 20 to 50 kg/ha.[44] The more depleted the soil, the higher the increment in yield. Figure 13.3a shows this effect for maize in Thailand, while Figure 13.3b emphasizes the point that the addition of nitrogen adds to the carbon effect.

Organic Farming

In many respects conservation and organic farming have much in common. They both emphasize the importance of returning organic matter to the soil. There are strong overlaps: some conservation tillage is certified organic, and some organic

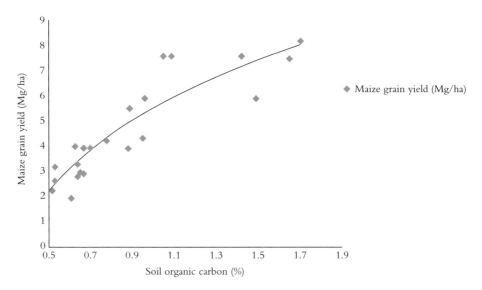

Figure 13.3a Effects of soil organic carbon in root zone on maize yields in Thailand.[45]

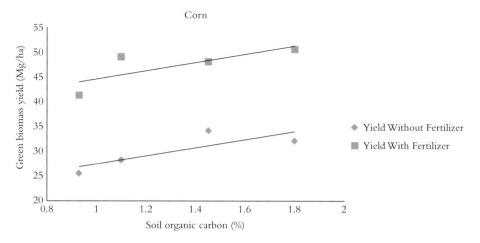

Figure 13.3b Effects of soil organic carbon and nitrogen applications on maize on a Russian Chernozem soil.[46]

farming makes use of minimum- or no-till practices. The difference lies in the strict exclusionary rules of organic farming: All synthetic fertilizers are banned and so are all herbicides and most insecticides and fungicides—the exceptions are various "natural or simple" chemicals (listed in Chapter 12).[47] The amount and origin of manure that can be applied is also restricted. Organic farmers have to rely on organic sources of nutrients, nitrogen fixing by legumes, natural forms of pest control, and a great deal of hand weeding. Farmers who adopt these and other restrictions can obtain certification to this effect from various national bodies and can sell their products as such.

The land under certified organic production has grown steadily over the past few decades. In 2007, it is estimated there were over 30 million ha of certified organic farmland worldwide, about 1 percent of total world production.[48] However, this does not include the millions of small producers who practice traditional agriculture that for one reason or another does not use inorganic fertilizers or synthetic pesticides. Such noncertified "organic" agriculture may be practiced on another 10 to 20 million ha in developing countries.[49]

Certified organic farming in developed countries has a well-established niche. While the costs of production relative to conventional production systems are high (largely due to the costs of the higher labor requirement), the products command premium prices. The question is whether an organic approach to farming can benefit developing countries. It certainly provides a profitable niche for the production of high-value crops for export to the developed countries. But can it, as the organic lobby maintains, provide a sustainable basis for growth, increasing incomes and helping to feed the world?

Although there is much controversy over the figures, organic agriculture produces significantly lower yields than conventional. (See the arguments between Catherine Badgley and colleagues at the University of Michigan and Keith Goulding of Rothamsted Research in the UK and colleagues.)[50] Comparative studies in developing countries are not thorough enough to generate firm conclusions, but there is extensive data in the developed countries. Thus long-term wheat experiments in the United Kingdom show comparable yields are obtained only with very heavy applications of manure, well above the amounts permitted under organic farming. Figure 13.4 shows the results from the famous Broadbalk experiment. Before synthetic fertilizers became widely available, there was usually a deficiency of nutrients, and yields of crops such as wheat were small and very variable. Yields with moderate amounts of fertilizers (144 kg N/ha) were two to three times those without fertilizers or manures. Modern pesticides further increased the yields. The best yields have been

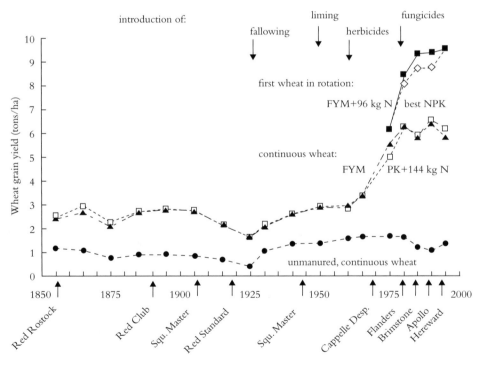

Figure 13.4 Yields of winter wheat varieties on the Broadbalk experiment at Rothamsted Research.[51] FYM=farmyard manure, NPK=nitrogen, phosphorus, potassium

obtained either with 250 to 300 kg N/ha (best NPK) or from 35 tons of farmyard manure plus 96 kg N.

Careful analysis of a wide range of other experiments suggests the typical ratio of organic to conventional wheat yields is 0.65 (i.e., organic cultivation yields 30 to 40 percent less), and this seems to be the approximate ratio for other crops.[52] However, the ratio could be an underestimate for developing countries. In drought-affected areas and under subsistence conditions, conversion to organic farming may well improve yields where the soils have been degraded over time.[53] Moreover, performance could change if organic crop varieties were bred to be more efficient at making use of scarce resources. The challenge is to breed organic varieties that are better at photosynthesis, at taking up nutrients contained in organic matter, at synthesizing their own nitrogen, at resisting pests and diseases, and at tolerating drought conditions. Some of this can be achieved by conventional breeding, but the process would be greatly enhanced by using recombinant DNA. At present certified "organic" has to be free of genetic modification (GM), but recombinant DNA uses naturally occurring DNA and enzymes. A couple of faculty at the University of California, Davis—Pamela Ronald, who works on GM rice, and her husband Raoul Adamchak, an organic agriculturalist—in their book *Tomorrow's Table* make an excellent case for bringing genetic engineering and organic agriculture together.[54]

Increasing Nitrogen

Good soil structure and organic matter are essential to producing good yields, but the key to high yields is the presence of nitrogen in the soil. Organic nitrogen can be boosted by encouraging the growth of certain microorganisms or, more directly, by applying plant and animal manures. Several kinds of bacteria, and other microorganisms such as blue-green algae, take up nitrogen from the atmosphere and convert it to ammonia, which can be used by plants. Some of these microorganisms are free-living in the soil, although they are often associated with the root zones of plants and their growth can be stimulated by certain crops. For example, in the presence of the rice variety IR42 they will produce up to 40 kg of N/ha per year. However, the best practical results have come from exploiting nitrogen-fixing microorganisms that live symbiotically in plants.[55]

There is a blue-green alga, *Anabaena azollae*, living in cavities in the leaves of a small fern, *Azolla,* that is a potentially phenomenal fixer of nitrogen—up to 400 kg N/ha per year, under experimental conditions. The fern will grow naturally in the water of rice fields without interfering with the growth of the rice plants. It quickly covers the surface, and after 100 days some 60 tons can be harvested per ha, containing

120 kg of nitrogen. However, the nitrogen is not directly available to the rice crop; the ferns have to be incorporated in the soil. Rice yields can be increased by a ton per ha or more, and the effect will carry over to a following crop; for example, if wheat is grown after *Azolla*-treated rice.[56]

Best known of the symbiotic, nitrogen-fixing microorganisms are the bacteria living in the root nodules of legumes, which can fix 100 to 200 kg N/ha per year. The fertilizing properties of legumes have been recognized for thousands of years. One of the earliest of the world's cropping systems—dating to soon after agriculture began in the valleys of Central America—was the interplanting of maize and beans; the seed of both crops was often placed in the same planting hole. It is a practice that, in various forms, continues today. For example, when cowpeas are cropped together with maize, the bacteria in cowpea root nodules can provide 30 percent of the nitrogen taken up by the maize.[57] Cowpeas and another legume, *lablab,* are particularly useful in lower-potential lands. Cowpeas are adapted to acid, infertile soils, while *lablab* is drought tolerant, produces good fodder, and can re-grow well after clipping.

Nitrogen from Legumes

Another way to capture legume nitrogen is by rotation of crops—inserting a legume such as alfalfa, clover, or a bean—between cereals. In the United States, a variety of alfalfa, known as *Nitro*, bred for this purpose can contribute up to 100 kg N/ha to a following maize crop.[58] There are also many bush and tree legumes that grow well in the tropics and can be interplanted with cereals and other food crops providing, under ideal conditions, 50 to 100 kg N/ha per year through their leaf litter or from intentional pruning. By carefully timing the pruning, it is often possible to ensure the nitrogen is available just when it is needed; for example, to coincide with maize germination.[59]

Green Manuring

The deliberate incorporation of legume crops in the soil, known as *green manuring,* is another practice of great antiquity, yet with considerable unexploited potential today. Varro, who I quoted at the beginning of Chapter 6, referred to some plants that are "also to be planted not so much for the immediate return as with a view to the year later, as when cut down and left on the ground they enrich it."[60] Most commonly used in this way was the lupine. In Bolivia today, a local lupine, *Lupinus mutabilis,* when intercropped or rotated with potatoes fixes 200 kg of N/ha per year,

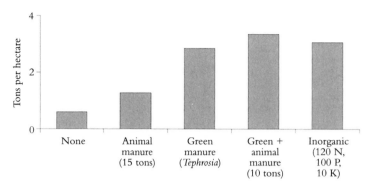

Figure 13.5 Maize yields in Rwanda under different fertilizer regimes.[62]

minimizing the need for fertilizers and, incidentally, reducing the incidence of virus disease.[61]

The best green manures in the tropics and subtropics are quick-growing legumes. In Rwanda, the shrub *Tephrosia vogelii* grows to three meters high in only ten months and produces 14 tons/ha of above-ground biomass. Maize grown after the legume has been incorporated in the soil delivers yields comparable with maize heavily treated with inorganic fertilizer (Figure 13.5). Usually the best approach is to combine green manures with small amounts of inorganic fertilizer, say half the usual rate of application.

An alternative is to grow a legume that doubles as a grain-yielding crop and a green manure. Cowpeas and lablab can be intercropped with upland rice and then, after the rice harvest, allowed to grow through the dry season. The peas are harvested and the vegetation plowed in before the next rice crop. Yields of 1 ton/ha can be nearly doubled with the added harvest of 0.5 to 1 ton of legume grain.[63] Legumes can also provide benefits in addition to increasing the availability of nitrogen and organic matter. Pigeon peas and cowpeas will access phosphate in phosphate-poor soils.[64] They are also deep rooting, which helps water infiltration.

Agroforestry

The interplanting of trees and agricultural crops is also a technique of great age—the trio of olives, vines, and wheat, grown in rows—was the mainstay of classical agriculture in the Roman and Greek empires. But contemporary interest lies in the tropical tree legumes because of their capacity to fix nitrogen. On experiment stations, for example at IITA, fast-growing species such as *Gliricidia sepium* and *Leucaena leucocephala* are grown in rows, with 4-meter-wide "alleys" in between for the annual crops.[65] The trees provide nitrogen, organic matter through their leaf fall and prunings, food for livestock, fuelwood, and timber, as well as conserving soil and water. The results on experiment stations and demonstration farms are often spectacular, but, for the most part, adoption has been poor. This is partly because alley cropping has been developed as a package, whereas farmers tend to be more willing to adopt various components, modifying their farms bit by bit.[66] Agroforestry also requires high labor input, the removal of land from crop production and carries the risk of introducing new species that may harm the growth, survival or reproduction of crop species through their biochemicals (allelopathy). It is also not enough for the tree crop to provide a soil and water conservation benefit. Farmers usually look for extra direct income as well.[67]

One of the main advantages of agroforestry is the provision of vegetative material that can be used as a mulch, a cover for the soil.[68] A particularly useful tree in China is *Paulownia,* a relative of the foxgloves, which is deep rooting and fast growing.[69] When mature it will provide up to 400 kg of young branches in a year and 30 kg of leaves.

Mulches protect the soil from erosion, desiccation, and excessive heating. They can also help reduce the spread of soil-borne diseases, by preventing splashing of water onto the lower leaves of the crops which spreads the fungal spores. Straw is a common mulch: in the Chiang Mai valley the second crop after the rice—cabbages, onions, and other high-value vegetables—is heavily mulched with rice straw.[70] One of the reasons why farmers in the valley retain the traditional rice varieties is because of the amount of straw produced. A short-strawed variety would produce a greater rice yield, but they would lose income on the second crop. A good mulch is thus much prized.

Manures and Composts

In addition to legumes, the other principal source of nitrogen on a farm is livestock. If properly integrated, crops and livestock can create one of the most sustainable forms of agriculture.[71] Livestock can make efficient use, not only of purpose-grown grain and forage crops, but also of straw and the by-products of other crops. Their excreta and manure returned to the soil maintains nutrient levels and enhances soil structure. In the developing countries, there is a long, and continuing, traditional use of manures, and the potential for greater utilization.

Ian Scoones, of the Institute of Development Studies (IDS), and Camilla Toulmin, of the International Institute for Environment and Development (IIED), describe the practice of farmers in the drylands of Africa who invite migrating pastoralists to pen their livestock overnight in the crop fields.[72] Farmers are sometimes willing to pay the herdsmen for the privilege. More recently, manures have become prized for their value in market garden cropping. [73]

Another long-standing practice is composting. Typically, composts consist of mixtures of animal manures, green material, and household wastes that are heaped together in such a way as to encourage anaerobic decomposition. The heat generated in the compost heap destroys pathogens and seeds and the roots of weeds. After the materials are sufficiently decomposed, the compost is applied, either in heaps around individual plants or worked into the soil. Composts are especially valuable in tropical climates since the high levels of organic matter help to store nutrients and protect them against leaching. The soil is made more friable, easier to plow, and maintains the moisture better.

Although composts are easier to make and apply in kitchen or home gardens, they can be used on a larger scale, as by the Wafipa people in Tanzania (Box 13.3).

Composting in such situations greatly improves productivity, but the process is labor demanding. As a consequence the Wafipa people moved toward use of the plow and inorganic fertilizer, despite its high cost. One answer may be to make composting more efficient and available commercially at prices that are lower than inorganic fertilizer. The Manor House Agricultural Centre in Kenya, which trains farmers in sustainable agricultural practices, has been successful in promoting the commercial use of composts through its cooperatives. The Pondeni Farmers Cooperative, for example, consists of a group of farmers who among other activities make and sell compost. It is sieved, mixed with bone meal to provide much-needed phosphorus, and packed in 90-kg bags that sell for $20 each.[75]

Box 13.3   Composting by the Wafipa in Tanzania

The Wafipa have long practiced a system of mound cultivation based on composts.[74] They begin by cutting grass and bushes, which are burned in small piles. Cucurbits are planted in the ash. Then, in between the cucurbits, mounds of about 90 cm high and 30 cm apart are constructed by piling up turves of grass, the grass facing inward. Legumes, such as beans and cowpeas, are planted in the mounds, which are left for the grass to rot. When the legumes have been harvested and the rains begin, the mounds are broken down and their contents spread over the land in preparation for sowing with millet or maize. After a couple of years of cultivation, the land is left fallow for four to ten years. The soils are poor, with little clay or natural organic matter and low water-holding capacity.

Participatory Conservation

Over the centuries, farmers in the developing countries have developed a great range of conservation systems, adapted to their local conditions.

Far too often, however, indigenous systems are overlooked in official programs. When Robert Chambers, Jenny McCracken, and I worked in the impoverished Wollo region of Ethiopia, we came across many examples of *gully plugs*—stone walls constructed across the gullies that trap silt, nutrients, and water—located high in the hills and supporting rich microplots of arable crops and trees.[76] They are, indeed, common in dryland areas in many regions of the world—in India, Pakistan, and Nepal, in Burkina Faso, and in Mexico, to name some countries.[77] The environments they create—small, flat, fertile, and moist fields—are quite unlike the surrounding countryside and can support high-value cereals and cash crops such as coffee or mango. Although construction of the gully plugs is expensive, they are relatively easy to maintain and produce an agriculture that is productive and dependable. Parmesh Shah of the Aga Khan Rural Support Programme reports that in Gujarat the gully plugs provide the most stable component of the household's food supply.[78] Yet they are often ignored in conservation programs, partly because they are not immediately obvious to the visitor.[79] In Ethiopia in the 1980s, the government was engaged in moving people from off the hillsides, and the gully plugs were in danger of being abandoned.

Integrated Nutrient Management

All these alternatives to synthetic inorganic fertilizers have distinct advantages. They are available, or can be created, on or near the farm and are generated from natural resources. Thus they tend to be relatively inexpensive. They can significantly increase yields, particularly on poor soils, and in some instances will perform as well or better than inorganic fertilizers. In nearly all situations they are good partial replacements, although they may not be less polluting, which bears remembering. Nitrates are liable to leaching whether they have an inorganic or organic origin. The main disadvantage of organic fertilizers is their high labor demand. It explains why they have fallen from favor in the developed countries.

Unfortunately, arguments about fertilizer use are often strongly polarized. On the one hand, many claim the only way to increase yields is to use large quantities of inorganic fertilizers and that promulgating the use of organic sources of nutrients will condemn many poor farmers to continuing low yields. Opponents of this view regard inorganic fertilizers as positively harmful and liable to trap farmers into high cost production. There is some truth in both arguments. Exclusive use of inorganic fertilizers is associated with long-term yield declines; yet where labor is a constraint and where hectarages are large, organic fertilization is insufficient to produce high yields. Putting more organic matter back in the soils is a high priority; at the same time we are going to need targeted, minimal use of synthetic fertilizers if we are to obtain the high yields we require.

As in pest management, I believe the approach for the future lies in integration, assessing each situation in agronomic, ecological, and socioeconomic terms, and then determining an appropriate mix of sources of inputs. This concept—of integrated plant nutrition systems—is still in its infancy and, like integrated pest management in its early days, is reliant on an expert, top-down approach.[80] But as some NGO programs have shown, it is amenable to a more participatory approach that will lead not only to greater efficiency but to more sustainability.

Most of this chapter has, ostensibly, been devoted to soil conservation but, as is clear from the examples I have cited, soil and water conservation are intimately connected. We turn to water in Chapter 14.

14

Sustained by Water

Increased pressure on already scarce water supplies in many parts of the world is a prime cause of famine, disease, and, increasingly, conflict.

Kofi Annan, World Water Day, 2011[1]

Water is as important for the productivity of plants as is the provision of a good soil structure and sufficient nutrients. A wheat grain may contain up to 25 percent water; a potato 80 percent. For rice, in particular, water is crucial; a gram of grain can require as much as 1,400 grams of water for its production.[2] Not surprisingly, water stress during growth results in major yield reductions for most crops.

In this chapter I address the following questions:

- What are the agricultural needs for water?
- What are the conflicts over water use and the damaging effects of overuse and misuse?
- What are the technological and community-based answers?
- How can aquaculture be better managed?

The Agricultural Demand

Crops require water, extracted from the soil through the plant root system, to grow, for cooling purposes, and to maintain turgor pressure (through which plants stay upright and direct their leaves toward the sun). Crop water use, known as *evapotranspiration*, is divided into two parts:

- *Evaporation:* the water lost from the soil and plant surfaces as well as that used for growth
- *Transpiration:* the water lost to the atmosphere through the stomata on the plant, used in cooling

The amount of water a crop needs is affected by the growth stage, the amount of water in the soil, weather, and crop rooting depths. For example, water needs are higher in a mature, fully grown plant than in a young seedling.

From the roots, water moves through the *xylem* (one of two types of transport tissues found in vascular plants; phloem is the other) to other parts of the plant, carrying minerals and chemicals such as glucose to where they are needed. Water is required in *photosynthesis,* the process of converting sunlight to usable chemical energy stored in the bonds of sugars. Light energy plus carbon dioxide (CO_2) and water (H_2O) are converted to sugars ($C_6H_{12}O_6$) and oxygen (O_2) in the reaction:

$$6CO_2 + 6H_2O \text{ (+light energy)} \rightarrow C_6H_{12}O_6 + 6O_2$$

If plants do not receive enough water to account for total evapotranspiration, they will yellow, their leaves will wilt, yield declines, and eventually, if they remain without adequate water, crops fail.[3] For rice, the hazard of not enough or too much water is especially great because of the short stature of the new varieties. The young, transplanted seedlings may die for lack of water in the first few weeks and will drown under excessive flooding. Ideally they need a constant flow of water at a depth of about 2.5 cm. Traditional rain-fed cultivation, which is subject to the vagaries of rainfall in the wet season, can rarely provide such exacting circumstances and high yields require supplemental irrigation in most situations (but see discussion of the system of rice intensification, Chapter 11). And for all modern cereal varieties, the potential to mature and produce grain irrespective of the season has placed a high premium on the provision of irrigation water in the dry season, when the potential yields are greatest.

The Hydrological Cycle

Agriculture sits in the middle of the global hydrological cycle, receiving water from rainfall, surface runoff, and groundwater and giving it up to the atmosphere through evapotranspiration (Figure 14.1).

Agriculture uses about 83 percent of freshwater withdrawals in developing countries (Africa uses 86 percent, Asia 81 percent, and Latin America 71 percent), and of that by far the largest part goes to irrigation, with livestock watering taking a small fraction.[5] Even though irrigated agriculture accounts for only about 18 percent of the cultivated area in the world, it produces about 40 percent of the value of its agricultural production.[6]

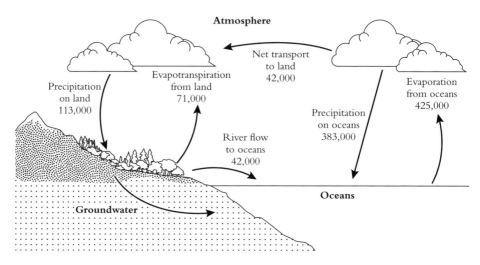

Figure 14.1 Hydrological cycle showing water transfer in cubic km per year.[4]

Today, much of Asia's grain harvest comes from the irrigated, annual double- and triple-crop, continuous rice systems in the tropical and subtropical lowlands of Asia and from the irrigated, annual rice-wheat, double-crop systems in northern India, Pakistan, Nepal, and southern China.

Conflicts over Water Use

The use of water has always been subject to conflict, but this is intensifying as a result of rapid urbanization and industrialization, as well as global warming (which I discuss in Chapter 15). Agriculture is also a major contributor because of excessive withdrawal of both surface- and groundwater. In many parts of the world, such as West Asia, the Indo-Gangetic Plain in South Asia, and the North China Plain, human water use exceeds annual average water replenishment. Freshwater shortage has been assessed as moderate or severe in more than half the regions studied in the Global International Waters Assessment (GIWA).[7] About 15 to 35 percent of total global water withdrawals for irrigated agriculture are unsustainable, and an estimated 1.4 billion people live in river basins with high water stress (Figure 14.2).[8] Rivers such as the Ganges and the Yellow River may not reach the sea during part of the year.

UN-Water estimated in 2007 that, by 2025, 3 billion people could be living in *water stressed* countries (water stress is calculated as the total water use divided by the

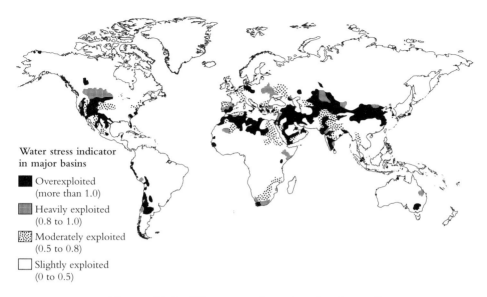

**Water stress indicator
in major basins**

■ Overexploited
(more than 1.0)

▨ Heavily exploited
(0.8 to 1.0)

▦ Moderately exploited
(0.5 to 0.8)

☐ Slightly exploited
(0 to 0.5)

Figure 14.2 Water stress in major river basins.[9]

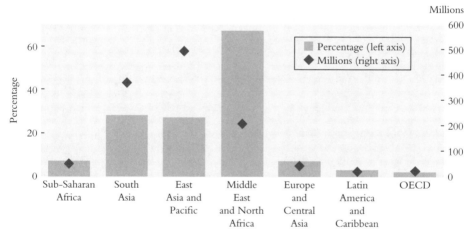

Figure 14.3 Population living in areas of absolute water scarcity.[10]

availability). The threshold below which the water requirements for agriculture, industry, domestic purposes, energy, and the environment are not met is 1,700 m^3/person/year. And of the 3 billion total, 1.8 billion people will be living in countries or regions experiencing absolute water scarcity, with only 500 m^3/person/year available (Figure 14.3).[11]

River Basins

Some of the most intense and problematic conflicts lie in river basins, especially where they cross national boundaries. Water shortages can be addressed through either increasing supply or decreasing demand. Usually both are needed, but each has its advantages and disadvantages, and the relative benefits of different options depend on local circumstances (Table 14.1).

For example, supply-side options will often have adverse environmental consequences, while demand–side measures, which rely on the cumulative actions of individuals, can be difficult to manage. A relatively successful resolution of some of these issues is the management of the Nile River basin (Box 14.1).

Large-Scale Irrigation

Over the centuries engineers have developed a wide range of proven technologies for increasing water storage and water flows. Not surprising, given the potential returns, the developing countries began to invest heavily in the late 1960s in large-scale, government designed and operated irrigation systems. The Upper Pampanga River Project in the Philippines, which covers some 80,000 ha, was completed in 1975 at a cost of over $100 million. The somewhat larger Muda River Project in

Table 14.1   Contrast between supply side and demand side of water management

Supply side	Demand side
Prospecting for and extraction of groundwater	Improvement of water-use efficiency by recycling water
Increasing storage capacity by building reservoirs and dams	Reduction in water demand for irrigation by changing the cropping calendar, crop mix, irrigation method, and area planted
Desalination of seawater	Reduction in water demand for irrigation by importing agricultural products (i.e., virtual water)
Expansion of rain-water storage	Promotion of indigenous practices for sustainable water use
Removal of invasive nonnative vegetation from riparian areas	Expanded use of water markets to reallocate water to highly valued uses
Water transfer	Expanded use of economic incentives, including metering and pricing, to encourage water conservation

Source: IPCC[12]

Box 14.1   The Nile River Basin

The Nile River basin supports the livelihoods of 160 million out of 300 million people living in ten countries: Burundi, Egypt, Eritrea, Ethiopia, Kenya, Rwanda, Sudan, Tanzania, Uganda, and the Democratic Republic of Congo. The population is expected to double in the next twenty-five years.

Given the high and increasing demand for water from the Nile basin, the increasing frequency of droughts in the area, and increasing pollution from activities on land, both water supply and water quality are threatened, which could lead to conflict among stakeholder countries.[13]

In 1999, the Nile Basin Initiative (NBI) was formed, and has inspired cooperation among countries. Through the NBI's Civil Society Stakeholder Initiative and the Nile Basin Discourse (NBD), civil society is engaged and, through forums, invited to air their opinions and concerns.[14] In 2010 to 2011, progress made by the NBI included $979 million worth of investment in such transboundary development projects as power interconnections, agriculture, and watershed management.[15] Since May 2010, six of the Nile Basin countries have signed up to the Cooperation Framework Agreement, which aims to set the legal basis for the equitable redistribution of Nile water between all parties. The agreement was to remain open for one year, but discussions with Egypt are ongoing as under Mubarak's regime they did not recognize the agreement. Egypt is now expected to ratify.[16]

Malaysia was finished five years earlier. By 1975 the proportion of rice land growing a second, dry season crop had risen in the Philippines and Malaysia to 60 percent and 90 percent respectively.[17]

For the developing countries as a whole, the amount of irrigated land and the proportion of arable land that is irrigated have more than doubled over the past fifty years (Figure 14.4). Nearly all of this increase has been in Asia; the amount of irrigated land in both Africa and Latin America is low. Although the amount of irrigated land continues to increase, the percentage irrigated has remained stable over the last ten years at around 26 to 27 percent, probably because the increase in arable land has been largely for unirrigated soybean and oil palm.

The rate of expansion of large-scale irrigation, however, has slowed partly because of increasing construction costs, but also because of considerable environmental and social costs. The siting of reservoirs invariably is a major source of contention. Villages and, sometimes, small towns are inundated, and the people have to be resettled; more often than not they end up on poor-quality land and receive few of the benefits from the irrigation scheme. Many case studies detail the consequences of large-scale irrigation systems.[19]

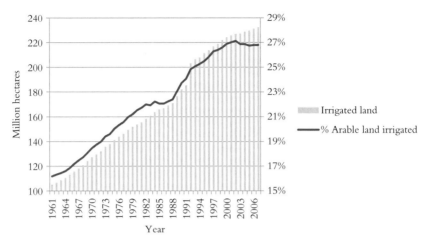

Figure 14.4 Growth in developing country irrigation.[18]

Often, in addition to the adverse impact on the local population, there are serious losses of forests and wildlife. It is now common to commission environmental impact assessments. These may run into several volumes of detailed information, but they are usually conducted by experts from outside the locality and can miss the crucial issues. By contrast participatory methods can ensure a better assessment of the likely effects and a more equitable and sustainable set of solutions to the problems that will inevitably arise. For environmental problems do not cease with the completion of construction. One of the consequences of high rates of erosion is the rapid buildup of sediments in reservoirs and irrigation canals, so shortening their expected life. A survey of seven reservoirs in India revealed reductions in their expected lives of 22 to 94 percent.[20]

Decreasing Groundwater

As an alternative to irrigation derived from rivers and reservoirs much investment has also gone into the provision of tube wells, which use long stainless steel pipes to extract water from the below ground water table. In India, the number of tube wells has increased from nearly 90,000 in 1950 to over 20 million in 2010.[21] They have several advantages over large-scale irrigation systems: they can be relatively easy and cheap to install and, because they occupy little land, create few environmental and social problems. But, and this is crucial, they will provide a sustainable supply of irrigation water only if the rate of extraction is below that of the rate of recharge to the underground aquifers from which the water is being obtained. If this is not the case, the water is effectively being mined.

274

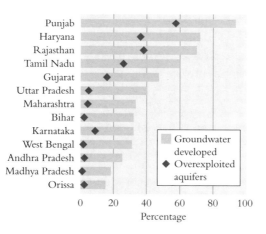

Figure 14.5 Depletion of groundwater aquifers in India.[26] Note: "Grandwater developed" is a percent of all available groundwater in a state. "Overexploited aquifers" is a percent of administrative blocks in which groundwater extraction exceeds recharge.

In many parts of West and South Asia, overpumping, encouraged by subsidized electricity, continues to cause an alarming fall in water levels—a problem compounded by falling rates of infiltration to the aquifers resulting from degradation of the upland watersheds. R. S. Narang and M. S. Gill of Punjab Agricultural University reported in the 1990s that over two-thirds of the Punjab water table was falling at 20 cm per annum.[22] The response has been to dig deeper wells, but this makes the problem worse. In the Indian state of Gujarat, water tables are falling by 6 meters a year, and wells 1-km deep are being dug in some places.[23] As I noted in Chapter 1, in China the groundwater overdraft rate exceeds 25 percent, and the rate is over 56 percent in parts of northwest India (Figure 14.5).[24] Groundwater levels there have declined by nearly 18 cubic km per year over the past decade.[25]

Waterlogging and Salinization

While in some areas water tables are falling, in others they are rising and creating serious problems of waterlogging and salinization. The common cause is a combination of excessive use of water and poor drainage. Cheap, subsidized electricity encourages profligate water use while growing capital costs have resulted in savings in drainage investment, sometimes eliminating drainage systems altogether from new projects. Salinization is usually due to rising water tables coupled with high rates of evaporation that bring toxic salts to the surface. In some coastal areas, acute salinization is created as a result of overextraction that leads to saltwater intrusions.

In the Indian state of Chennai, overextraction of groundwater has resulted in saline groundwater nearly 10 km inland of the sea.[27] Annual withdrawals from Mexico's coastal aquifer of Hermosillo of three to four times the recharge rate resulted in a 30 m drop in water tables and saltwater intrusion at the rate of 1 km per year, causing large agribusiness firms to relocate to other regions.[28] Similar problems can be found in populated coastal areas around the world.

Of course, salinization is not a new phenomenon. Agricultural collapses associated with the salinization of irrigated lands go back thousands of years; regrettably, little has been learned from this history.[29]

The extent of waterlogging and salinization in the developing countries is now considerable. Nearly 40 percent of irrigated land in dry areas of Asia are thought to be affected by salinization.[30] There are no exact assessments of the extent of global salinization, but one estimate suggests it seriously affects 20 to 30 million hectares (Mha) worldwide; that is, about 25 percent of the area under irrigation in arid and semi-arid zones and about 10 percent of all areas under irrigation.[31] Most of this is a legacy of large-scale water works developed from the 1950s. The spread of salinization in these old irrigation areas is now drastically reduced, and current rates of increase are in the order of 0.25–0.5 Mha per year.

Managing Irrigation Systems

Although technical solutions are available, the root causes of these economic, social, and environmental problems are bad irrigation design and poor maintenance and management. Irrigation, if it is to be effective, has to be reliable; otherwise much or all of the potential benefit will be lost. Reliability depends, in turn, on an efficient and responsive organization, and the question here is whether this can best be supplied under central government or local control. Many ancient irrigation systems, for example, in Mesopotamia, Sri Lanka, and China, relied on central government control. But other irrigation systems, such as the *subaks* in Bali established over a thousand years ago, are much smaller (about 200 ha), are built and operated by the local community, and have social and religious significance. Communal systems tend to be highly responsive because the supply and regulation of the water are an integral part of the traditional practices of resource management in the community.

Often the answer lies in creating a good partnership between government and local communities. For centuries, in the valleys of northern Thailand, for example, community-maintained dams constructed of stone and wood have been linked to irrigation systems governed by representative bodies known as *muang*. When, in the 1960s and 1970s, the government constructed large-scale diversion systems designed to provide year-round irrigation, in many places it grafted them onto the local systems. During our Agroecosystem Analysis of the Chiang Mai Valley (see Chapter 11), we discovered that triple-cropping was being practiced only in the areas of these joint systems where the water supply was reliable enough to risk planting the high-value third season crops.[32]

Corruption

Equitability in the sharing of water and other benefits is also a central and contentious issue in irrigation. Corruption in irrigation administration is widespread; direct action or bribery is a common means whereby farmers seek to redress perceived imbalances or to gain unplanned or illegal shares. During water shortages in the Minipe irrigation scheme in Sri Lanka, for example, farmers have been known to block the channels and divert water to their fields.[33] Open force is sometimes used. In South India farmers at the "tail end" of the irrigation scheme, who are usually those least likely to gain a fair share, have been known to "hire a jeep and budget for its costs out of their common fund raised for such purposes [to] patrol the main canal[,] lowering sluice gates and threatening violence. Occasionally a whole lorry load of farmers will go, brandishing sticks in a demonstration of force."[34]

In large-scale government irrigation systems, bribes are often paid to the irrigation engineers to ensure more reliable supplies, particularly by farmers at the tail ends of the system, and also by farmers wishing to grow unauthorized crops. The bribes may be in the form of an annual flat rate or individual payments. There may also be gifts of grain at harvest, mostly as rewards to the local field staff. In South India, the average cost of the bribes in the late 1970s were 4 to 10 rupees per ha over two seasons.[35] This was low compared with the 360 rupees net profit to be gained from a rice crop, but the costs can be much higher for tailenders and are likely to bear heavily on poorer farmers.

According to the Global Corruption Report, produced by Transparency International, corruption in the water sector overall can increase the cost of connecting a household to a central water supply by as much as 30 percent, which could mean an additional $48 million is needed to achieve the UN's Millennium Development Goals relating to water and sanitation.

In India, corruption adds an estimated 25 percent to irrigation contracts and exacerbates food insecurity and poverty because it is generally the more powerful farmers that can afford to "pay" for favorable rights to water. Indeed in Mexico, the top 20 percent of all farmers in terms of size as a producer received 70 percent of irrigation subsidies.

Irrigation engineers may also receive kickbacks from contractors undertaking system maintenance. Niranjan Pant describes the practice in the late 1970s on the Kosi Project in Bihar, northern India, where the contractor "has to spend about 30 percent of his bill on overseers and engineers, about 10 percent on office staff, about 10 to 20 percent for his own profit and thus only 40 to 50 percent is spent on the actual work."[37] Both engineers and contractors may make money by colluding in substandard work—so called "savings on the ground." Poor-quality cement is used, for example, and the savings are shared.

Investigations into corruption in the water sector led Transparency International to conclude that corruption must be prevented from the outset rather than addressed as an afterthought. To do so, it is important to understand the local water system. Approaches to tackling corruption need to be pro-poor; that is, actions should not threaten water sources most important to local livelihoods. Moreover while leadership from the top is important to drive change, bottom-up approaches including monitoring are equally important for accountability.[36]

Community Control of Irrigation

Corruption can be minimized by an institutionalized system of community control. One answer, in large irrigation schemes, may lie in delivering water to holding reservoirs before local distribution and placing the control over the reservoirs in the hands of local communities. For smaller systems, it becomes easier to institute effective community control that can extend to design and contract management (Box 14.2).

Box 14.2   Communal Rehabilitation of Irrigation

An example is provided by a program to rehabilitate small tank systems in the south Indian state of Tamil Nadu, where rainfall is less than 850 mm per year and is erratic.[38] The tanks are small natural, low-lying areas that are dammed so as to catch and store the monsoon rains. Subsequently, the water is used by the villagers to irrigate crop fields. The maintenance of the tanks and the irrigation canals has been the responsibility of government authorities, but the systems progressively fell into disrepair.

Normally rehabilitation is attempted by hiring contractors who work to a blueprint; not surprisingly, this produces inappropriate and excessively costly solutions. The alternative is for the District Rural Development Agency to give grants direct to the villagers, who with technical assistance from an NGO, PRADAN, and the Centre for Water Resources at Anna University, have formed water users associations and designed,

(continued)

planned, and managed the rehabilitation themselves. In one village, Panchanthangi Patti, the villagers contributed 25 percent of the costs in terms of labor, materials, and money. They determined the priorities and identified the work needed—including strengthening the earthen bund, partially desilting the tank and feeder channels, building check dams across the channels to prevent silting, and planting trees on the foreshore to prevent encroachment. The outcome is systems that the villagers feel they own and to which they are committed.

Over the last three decades, the number of tanks communally rehabilitated in three states of India—Tamil Nadu, Karnataka, and Pondicherry—totaled about 5,000 (out of 80,000) through assistance from external donors and the state.[39]

The general lesson from the experiences of the past thirty years is that small, community managed and designed irrigation systems are more likely to deliver sustainable water supplies. In many parts of the world, there is no other option. Although irrigation in Sub-Saharan Africa has grown steadily since the 1960s, it still amounts to only some 4 Mha (5 percent of total cultivated area)—compared with India's 66 Mha in 2009)—less than half a percent of the arable land.[40] Over much of the continent, the environmental conditions are not suitable for large-scale irrigation systems, and the future lies in small-scale systems (like those in Tamil Nadu) and, in the drier regions, in ingenious systems of water conservation and harvesting based on a microcatchment approach.[41]

Engineering Solutions

Engineering solutions to providing greater and more sustainable water supplies are part of the answer. Some are at considerable scale. While large-scale irrigation construction has fallen out of favor in some parts of the developing world, it is still being actively developed in some middle-income countries such as China.[42] The Three Gorges Dam on the Yangtze River in China, which began to operate in 2008, is primarily intended for hydropower generation, but it also helps to regulate the flow of the river to the 1.5 Mha of farmland in the Jianghan Plain and to minimize the frequently disastrous flooding in the middle and lower parts of the river.[43] The Three Gorges Dam is also one of three origins for a major undertaking to divert river waters from the south to the north of the country. When completed in 2050, it will link China's four main rivers—the Yangtze, Yellow River, Huaihe, and Haihe—and eventually divert 44.8 billion m³ of water annually to the drier north.

Box 14.3 Nanotechnology for Desalinization[44]

> Carbon nanotubes are sheets of carbon atoms rolled so tightly that only seven water molecules can fit across their diameter. Their small size makes them good candidates for separating molecules. At the same time they allow water to flow at the same rate as with pores considerably larger, reducing the amount of pressure needed to force water through, so saving energy and costs compared to reverse osmosis using conventional membranes.
>
> The nanotubes function as an array of pores, allowing water and certain gases through, while keeping larger molecules such as salt at bay.
>
> The new membranes could reduce the cost of desalinization by 75 percent, compared to conventional methods, and be brought to market within the next five to ten years.

In the least developed countries, more attention is now being paid to a variety of intermediate technologies, such as drip irrigation and treadle pumps, which will make more efficient use of available water (see Chapter 7). There are also many promising new technologies that can increase the supply of water, notably through desalinization of salt and brackish water (Box 14.3).

Water Harvesting

Even in very dry situations, there is usually some water available; the challenge is to harvest this in a cheap and efficient manner. Some of the most ingenious solutions are systems of so-called water harvesting, which aim to collect and utilize rainwater as it runs off the land.

Water harvesting from short slopes is relatively straightforward and cheap and can be highly efficient because of the distances involved. An example is the *zai* (or water pockets) technique pioneered by farmers in the dry, sun-baked, encrusted soils of northwest Burkina Faso decades ago as a way to create more arable land. It has now spread throughout similar climates in the rest of Burkina and in Mali and Niger (Box 14.4).[45]

Water harvesting from long slopes requires semipermeable, stone bunds along the contours, which will slow water runoff and encourage infiltration.[47] In the desert margins of West Asia and North Africa, harvesting targets the periodic flash floods through systems of barriers across the *wadi* floors. Under the Roman Empire, the combination of Roman engineering and local knowledge produced elaborate systems of water harvesting, producing large quantities of wheat and olives, in areas that are, today, largely desert. A modern equivalent in the Central Plateau of Burkina Faso consists of low semipermeable dams that concentrate and redirect

Box 14.4 The *Zai* System[46]

Farmers first dig medium-size (20 to 30 cm in diameter and 10 to 15 cm deep) holes (or *zais*) in rows across the fields during the dry season. Each zai is allowed to fill with leaves and sand as the winds blow across the land. Farmers add manure, which during the dry months attracts termites; these dig an extensive network of underground tunnels beneath the holes and bring up nutrients from the deeper soils. Stone earth bunds are constructed around the field to slow runoff when the rains come.

The rainwater is captured in the zais. Sorghum or millet seeds are sown in the holes where the water and manure are now concentrated. Water loss through drainage is limited by the manure and deep infiltration is made possible by the termites' porous tunnels. Thus, even in the drought-prone environment of the Sahel, sufficient water capture is ensured.

Farmers have consistently reported greatly increased yields using this technique. In a study conducted in Bafaloubé, Mali, between 2000 and 2003, sorghum and millet yields increased by between 80 percent and 170 percent, respectively.

The labor required to build the zais in the first year is quite high, but after that farmers may reuse the holes or dig more between the existing ones. In many cases, after around five years the entire land surface will be improved.

---

water flows. Natural terraces are formed as the sediment is deposited, on which yields of sorghum increase two- to threefold.[48]

The redirection of water is also a feature of the ancient Chinese "warping systems."[49] Storm and flood water is diverted around a series of obstacles with the aim of concentrating both water and nutrients. A system covering over 2,000 ha in Shanxi Province provides water in the dry season and helps build up high levels of nutrients and organic matter. Yields of maize, millet, and wheat are increased two- to fourfold.

Such systems are demanding of labor, which makes them particularly suitable for lower-potential lands with high population densities. Costs range from $100 to $1,000 per ha, but the returns are considerable. In all cases the intimacy of water and soil conservation is critical. Adequate water is a necessary but not sufficient condition; good crop growth also requires considerable supplies of nutrients and a supportive soil structure. The most successful systems are those where techniques are developed that promote all these requirements in a synergistic fashion.

Breeding for Water Use Efficiency

Despite the ingenuity of engineers and planners, water for agriculture (and for aquaculture—see below) is going to be increasingly in short supply. Perhaps the most sustainable solution is to breed into crop plants (and livestock) the capacity to make much more efficient use of limited water.

Breeding has been approached in two ways. First, new crop varieties are subjected to a range of stresses in the field, including drought, using such techniques as mother-baby trials (see Box 11.2 in Chapter 11). Second, the biological basis of drought tolerance is being investigated for all the staple cereal crops as a platform for improved breeding. Breeding for drought tolerance has so far been difficult given that it is not well inherited, the variety of effects of drought on a plant depend on the timing of the drought, and the limited understanding of drought physiology.

In the past decade, however, a suite of genes that regulate drought adaptation and/or tolerance have been identified, and their combination with transgenic approaches has led to rapid progress in improving drought tolerance.[50] One such gene is a so-called chaperone gene that can confer tolerance to stress of various kinds, including cold, heat, and lack of moisture.[51] The product of the gene helps to repair misfolded proteins caused by stress and so the plant recovers more quickly. One such gene found in bacterial RNA has been transferred to maize with excellent results in field trials. Plants with the gene show a 12 to 24 percent increase in growth in high-drought situations compared with plants without the gene. Field trials are now being carried out in Africa (Box 14.5).

Box 14.5   Water Efficient Maize for Africa

The WEMA (Water Efficient Maize for Africa) project was initiated in 2008 by the African Agricultural Technology Foundation (AATF)—a nonprofit NGO based in Nairobi. AATF aims to develop and make drought-tolerant maize available royalty free to smallholder farmers in Sub-Saharan Africa.

Using a combination of conventional breeding, marker-assisted selection, and genetic modification technology, a public-private partnership has been established between AATF, CIMMYT, Monsanto, and national agricultural research systems in the participating countries (Kenya, Mozambique, Tanzania, Uganda, and South Africa). The aim is to deliver maize varieties over the next decade that will increase yields by around 20 to 35 percent under moderate drought conditions in comparison to current varieties. This could result in an estimated two million tons of additional food, benefiting 14 to 21 million people.

The first conventional hybrids developed through marker-assisted selection should be available four or five years from now. The first transgenic trial was initiated in November 2009 in South Africa, and transgenic varieties are expected on the market between 2015 and 2017.[52]

Water and Fisheries

So far I have written about crops (and to some extent livestock) and their dependency on water. But I have ignored a key source of food for many rural communities: the fish and other aquatic animals whose dependency on water (fresh, brackish, and salt) is fundamental.

In some communities, fish are an important source of energy in the diet, but elsewhere their primary contribution is in terms of protein; they provided about 19 percent, on average, of worldwide animal protein consumption in 2007, about 6 percent of all protein consumed. In China and many other parts of the developing world, they contribute over half of the animal protein consumed. Fish are also an important source of vitamins, minerals, and fatty acids.[53] In 2008, around 80 percent of world fishery production was accounted for by developing countries, with 75 percent of fishery exports from developing countries going to developed countries. Without supplies of fish, it would be far more difficult to achieve food and nutrition security. Increasingly these supplies are in jeopardy.

Fish, like livestock, are harvested and managed along a continuum that extends from free-range, open access ecosystems to intensively farmed, industrialized systems. At one end are the marine fisheries of the high seas; at the other intensive forms of aquaculture. I will focus here on aquaculture.

For most of the world's wild fish stocks, the harvest is stagnant or declining. The global harvest captured in the oceans and inland waters has peaked in 2000 at 96 million tons and subsequently fallen to 90 million tons in 2003, remaining at that level until 2009. As a consequence, catch per person dropped from an average of 17 kg in the late 1980s to 14 kg in 2003.[54]

Aquaculture

Partial compensation for the reduction in marine fisheries has come from the growth in aquaculture, both freshwater and marine, which took off in the 1980s and continues to be the fastest-growing animal food producing sector, outpacing population growth. In 2008, aquaculture accounted for 46 percent of total food fish supply.[55] Production reached over 55 million tons in 2009, although the fish harvest per person appears to have stagnated between 13 kg and 15 kg since 1997.[56] Developing countries accounted for 92.5 percent of total food fish from aquaculture in 2008.

Much of the developing country growth has occurred in China, partly based on the long-standing tradition of raising herbivorous carp in conjunction with agriculture.[57] But there have also been large increases in shrimp production (where China produces 27 percent of global production) and mussels (where China produces 38 percent). Despite its ancient origins, large-scale aquaculture is still focused on only a limited range of species that have yet to benefit from intensive breeding programs.

In the rice-growing regions of Asia, there is a long tradition of raising fish in the wet paddy fields. It can be a highly symbiotic relationship: the fish are provided with a nutrient-rich and safe environment and, in return, eat weeds and insect pests and through their excreta increase the nutrient levels for crop production. Fish will graze on *Azolla* (a genus of seven species of aquatic fern) when present and convert it to available nitrogen for the rice plants. In the past, farmers have often exploited wild fish populations, constructing small ponds as refuges for the fish in the dry season. Today, high levels of production in rice fields are possible by purchasing good-quality fish stock, constructing nurseries to raise the fry before release, using supplementary feeding, and carefully controlling stocking rates.[58] The biggest obstacle is pesticide use, which, in most cases, either directly kills the fish or destroys the richness of the habitat on which they depend. Current efforts are demonstrating the combined benefits of integrated pest management and fish culture in rice fields. Under a CARE program in Bangladesh, the elimination of pesticides has not only increased fish yields but raised rice yields by some 25 percent.[59]

By contrast with the open sea where regulation of the fishery is difficult, farmed fisheries avoid problems of overexploitation. Ownership is usually not in question, stocks can be kept at an optimal size, feed is controlled, and harvesting timed to gain

Box 14.6   Aquaculture in Malawi

The WorldFish Center partnered with the Malawian Fisheries Department to present to farmers a set of technologies developed by the two institutions. Farmers were encouraged to provide constructive criticism of the technologies and to suggest improvements. If adopted, farmers were free to modify the technologies to fit their circumstances, and researchers often worked alongside these farmers evaluating the technologies' performance.

The technologies included using napier grass as a pond input instead of maize bran, which can be in short supply at certain times of the year; using a reed fence to harvest fish, as opposed to a net; developing a high-quality compost as a pond input; integrating vegetable and rice cultivation; using a smoking kiln to preserve fish; integrating chickens into the pond system as a source of manure; pond stirring to help recycle nutrients and food.[63]

Since 1990 over 200 smallholdings have been involved in farmer-led integrated resource management (IRM) and adopted many of the technologies trialed. The project has raised productivity and income of farmers and developed a successful farmer-participatory method of technology transfer.[64] IRM has now been adopted as the national Malawi strategy to develop smallholder aquaculture.[65]

the maximum return. However, the sustainability of aquaculture is being threatened by inadequate management and by pollution and conflicts over land use, particularly in coastal ecosystems. Clearing of land for shrimp farming in Thailand in the late 1980s destroyed over 17 percent of the country's mangrove forests in just six years and resulted in rapidly falling water tables.[60] Shrimp farming is especially subject to chronic disease problems. In several Asian countries, outbreaks of disease because of poor hygiene and quarantine, coupled with lack of control over water intakes and pond effluents, have contributed to irreversible crashes in shrimp production after only two to three years (in Taiwan shrimp production still has not recovered after a peak of 80,000 tons in 1987 collapsed to very limited production in 1991).[61]

Early attempts to duplicate the Asian aquaculture success in Sub-Saharan Africa have had little success because the conditions are different.[62] This prompted the WorldFish Center to begin a collaborative research project in Malawi (Box 14.6).

Improved Water Management

Finally, improved water use, whether for agriculture or aquaculture and whatever the source of water, will come about only through improved systems of management.[66] Traditionally water management has depended on probabilistic models of future hydrological changes.[67] But in a changing climate, the models are often invalid and such assumptions no longer apply. A more resilient, holistic approach—known as *integrated water resources management* (IWRM)—has been developed. This comprehensive approach encompasses basin or watershed-scale management of water resources; and integrates land and water activities, upstream and downstream areas, surface, ground, and coastal water resources, supply- and demand-side approaches, the various sectors and stakeholders involved in decision making, including water users and marginalized groups, and effective integration of different levels of policy, institutions, and regulation.[68] It is a highly ambitious set of goals; the challenge is to find a water management approach that is resilient to future climate changes, and reflects both the priorities of the poor and the biophysical complexities and uncertainties.

In Chapter 15 I examine the wider impacts of climate change on agriculture and food security.

15

Adapting to Climate Change

> Overcoming climate change will help to overcome poverty. . . . If we fail on one, we fail on the other.
>
> Lord (Nicholas) Stern[1]

Global climate change has been largely driven by the activities of the industrialized countries. Yet its most severe consequences will be and, indeed, are already being felt by the developing countries. Moreover, it is the poor of those countries who, in part because of their poverty, are most vulnerable. If left unchecked, climate change will increase hunger and cause further deterioration of the environmental resources on which sustainable agriculture depends.

In this chapter I will address the following questions:

- What do we know and not know about climate change?
- What will be the most serious consequences for agriculture and rural livelihoods in the developing countries?
- How well can agriculture adapt?

The Most Vulnerable

The developing countries are the most vulnerable to climate change.[2] They tend to be located in regions that are already subject to climatic extremes or where the extremes may become even worse. Their vulnerability to floods, drought, or sea-level rise may be very high. They tend to have a higher share of their wealth tied up in natural resources and environmental assets; anything that destroys the natural resource base will damage livelihoods, especially of the poor.

Whereas *resilience* (introduced in Chapter 5) is about the capacity to respond and adapt to the stress and shock created by climate change, *vulnerability* is a broader concept, encompassing how susceptible a system is to stress and shock. In other words, it includes the nature of the hazard.[3] In a survey of fifty-nine Sub-Saharan

286

African countries, thirty-three were classified as highly vulnerable or moderately highly vulnerable, depending on such factors as poverty, health status, economic inequality, and prices of particular crops. All the countries came out in the lowest quintile of a ranking on adaptive capacity of nations to climate change.[4]

Agriculture, as a sector, is particularly vulnerable to climate change. In Africa and parts of South Asia, subsistence farmers rely on natural rainfall, which means they are affected by small changes in rainfall patterns. Large areas of agricultural land are already classified as "dryland," and climate change is likely to change rainfall patterns and bring a shorter growing season in the future, expanding drylands over a larger area. As I indicated in Chapter 14, many parts of the developing world are already experiencing water shortages, and these may increase in scope and severity.

Moreover, infrastructure that can reduce the impact of climate hazards is either lacking or not fit for purpose. Developing country governments and institutions are often poorly resourced and unprepared; many people will have to cope on their own. Finally, most people in developing countries operate at low income levels with limited reserves and lack formal insurance coverage.

The Economic Costs

For the globe as a whole, a report produced by a team led by Nicholas Stern, an economist at the London School of Economics (LSE), and known as the Stern Report, projected that a 2°C rise in average global temperatures would reduce world GDP by an estimated 1 percent.[5] Regional estimates are higher: the loss to annual GDP in Africa and India caused by climate change could be 4 percent and 5 percent, respectively.[6] Some countries have started to assess costs, at least as a consequence of climate extremes. The Ningxia Hui Autonomous Region of northern China regularly suffers from a range of major shocks, the most serious being sandstorms, drought, and high temperatures. Although the region is prone to such disasters, the frequency and intensity appears to be increasing in recent years, possibly as a result of climate change. Over the past ten years till 2006, the total economic cost of these disasters rose steadily to about 1 billion renminbi ($150 million) a year, which is nearly 2 percent of Ningxia's GDP.[7] Statistics like these are beginning to bring home to developing country governments that climate change is already a significant threat to their development goals.

The Global Impacts

There is convincing evidence that global climate change is occurring and is the result of manmade emissions of greenhouse gases (GHGs)—primarily carbon dioxide (CO_2), methane (CH_4), and nitrous oxide (N_2O). The mechanism is relatively simple and increasingly understood: These gases form a layer over the earth's surface that traps an increasing proportion of the infrared radiation that would be otherwise radiated out to space, so warming the land and oceans beneath. As a consequence, the world as a whole is warming—so far by more than 0.7°C since the industrial revolution (Figure 15.1).

Since preindustrial times (around 1750), atmospheric CO_2 concentrations have increased by just over one third from 280 parts per million (ppm) to 385 ppm in 2008.[9] If we add three further fluorinated gases that deplete ozone, the six key GHGs produce total emissions of about 436 ppm of CO_2 equivalent. CO_2 equivalent includes GHGs in addition to CO_2 in terms of their warming potential relative to CO_2; for example 300 times in the case of nitrous oxide and 21 times in the case of methane.

Knowns and Unknowns

Alternative explanations for the rising temperature have been proposed. They include such phenomena as sunspot cycles and the effects of galactic cosmic rays, but, in my view and that of the great majority of the scientific community, none are convincing.[10] The Stern Report concluded: "It is now clear that, while natural

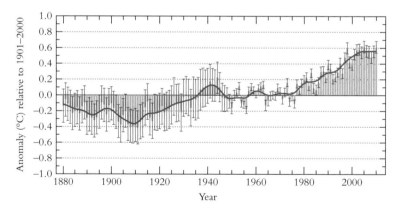

Figure 15.1   Global temperatures have risen by about 0.7°C since 1900 (Jan.–Dec. mean global temperatures over land and ocean).[8]

factors, such as changes in solar intensity and volcanic eruptions, can explain much of the trend in global temperatures in the early nineteenth century, the rising levels of greenhouse gases provide the only plausible explanation for the observed trend for at least the past 50 years."[11]

Much is still unknown, however, and there are many uncertainties generated by complicated feedback loops, potential tipping points in the climate system, and the need to take into account the following factors: the potential for irreversible damage to ecosystems; a very long planning horizon; long time lags between GHG emissions and effects; aerosols as well as multiple GHGs; wide regional variation in cause and effects; and the global scope of the problem.[12] Needless to say, the current and impending impacts of climate change as well as their scale and potential irreversibility mean we do not have the luxury of understanding many of these uncertainties prior to taking action.[13]

Despite the longer-term uncertainties, the following near-term consequences are highly likely. We can expect:

- Regions to be warmer
- Regions to be more prone to drought or flooding
- Higher sea levels
- More storm surges
- Greater variation in the weather and more intensive extreme events—hurricanes, tropical cyclones, floods, and droughts

The Global Drivers

Underlying these changes are global climate phenomena that interact in complex, and still not yet fully understood, ways.[14] Most of the developing countries are located in the tropics and subtropics; that is, they lie between 0° and 30° north and south of the equator. Within this latitudinal band are three critical processes: Two of these—*tropical convection* and the *alternation of the monsoons*—are relatively local processes that determine the regional and seasonal patterns of temperature and rainfall; the third, *El Niño–Southern Oscillation* (ENSO), although local in one respect, strongly influences the year-to-year rainfall and temperature patterns on a global scale (Box 15.1).

Although these three drivers are powerful forces, it is still not clear to what extent they are affected by climate change. As a general hypothesis, each of the drivers should be influenced by the rising sea surface temperatures resulting from global warming.[18] Higher surface temperatures should cause the land to heat up faster,

Box 15.1   The Global Drivers

Tropical Convection

Intense solar heating near the equator leads to rising warm, moist air and heavy rainfall. As it rises the air creates a surface low-pressure area, known as the Intertropical Convergence Zone (ITCZ). Each year the ITCZ moves north and south following the seasonal tilting of the globe toward the sun (wavy red lines in Plate 5). How far the ITCZ moves each year from the equator will determine the rainfall to the north and south. For example, when it migrates farther north than usual it brings heavy rain and floods to the Sahel, as happened in 2007; when it lies quite far south over the southwest Indian Ocean it will be very dry over southern Africa.

Alternation of the Monsoons

The monsoons are similarly affected by the seasons. Note the marked seasonal change in the direction of the surface winds, in January and July in Plate 5. The Indian monsoon is the most extreme form of monsoon with a 180° reversal of the wind. The southwest monsoon arises in spring and summer. As the air over northwest India and Pakistan becomes much warmer than over the Indian Ocean, it creates a low-pressure area drawing in warm, moist air from over the ocean. The air moves first northward and then, because of the effects of the Earth's rotation, is diverted northeastward. It begins to rise and cool and sheds its moisture as rain. In winter, the reverse occurs, the land cooling down more than the oceans, creating the northeast monsoon. These monsoon wind changes also affect lands far distant from South Asia; for example, along the eastern margins of Africa.

El Niño–Southern Oscillation (ENSO) of the Pacific

The third principal driver, characterized by a close coupling of the ocean and the atmosphere, is referred to as an oscillation because of the characteristic switch in the Pacific between two phases, La Niña and El Niño (Plate 6).

   Under "normal" conditions the trade winds move increasingly warm water westward from the Central Pacific's high pressure to the low pressure located over Indonesia. Very heavy and extensive rainfall occurs over the warm water of the western Pacific, while the eastern Pacific experiences relatively dry weather. La Niña is an extreme version of this condition. Sometimes the pattern is reversed, with wide-ranging consequences.[15] Every three to seven years El Niño sets in, changing the prevailing pattern of ocean surface temperatures and pressures and reversing the trade winds. The surface waters move eastward. Rain falls in the east, and droughts occur in Southeast Asia and Australia. In the Pacific, ENSO accounts for up to 40 percent of the variation in temperature and rainfall.[16]

   Although it is primarily a Pacific Ocean process, the effects are felt as far away as Africa and, indeed, in most regions of the world (see Plate 6). Thus six months after an

(continued)

El Niño phase, the global mean surface air temperature of the globe increases. After the severe El Niño of 1997–1998, it went up by nearly 0.2°C.[17] The effects on Africa and India can be profound. During an El Niño year, the December to February weather is usually wetter in eastern Africa but drier to the south, while La Niña produces the reverse effect. El Niño is associated with a drier Sahel, and La Niña is correlated with a wetter Sahel and a cooler West Africa. An El Niño event with strong warming in the central Pacific can also cause the Indian monsoon to switch into a "dry mode," characterized by significant reductions in rainfall leading to severe droughts.

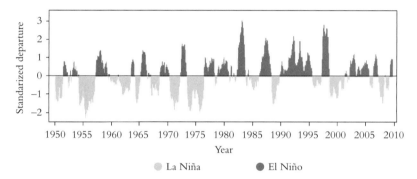

Figure 15.2   The increase in frequency of El Niño events since the mid-1970s. (The standard Multivariate ENSO Index is based on six variables measured across the Pacific.)[21]

leading to a greater contrast between the land and the ocean, and thus more intense monsoons.[19] However, some studies have found that the most vigorous monsoonal circulations have weakened, leading to decreased long rainy spells and increased shorter spells and, in general, a more erratic rainfall pattern.[20] But there are uncertainties about the data and how it is interpreted.

The ENSO phenomenon raises further, complicating issues because there is still no consensus on why the oscillation occurs in the first place. It may be as a result of a random "trigger," the El Niño phase appearing very approximately at three- to seven-year intervals. Nevertheless, this does not explain the shift in the pattern to a greater frequency of El Niño years that seems to have occurred in the mid-1970s (Figure 15.2).

Although these drivers are powerful global and regional forces, it is not yet clear whether their patterns are significantly altered by global warming. What we can be sure of is that global warming—expressed, for example, through higher sea and land surface temperatures—will affect their outcomes, increasing the incidence and severity of the droughts, floods, and other extreme weather events that they produce. What is also certain is that these changes have made it extremely difficult for farmers and others to predict the key seasonal rains. More advanced prediction tools and modeling will hopefully provide a more nuanced understanding of the drivers and their changes.

The Regional Changes

The uncertainties at the global level are repeated and magnified as we move to assessing regional impacts. The Intergovernmental Panel on Climate Change (IPCC) used a multimodel data set to consider various scenarios. The A1 storyline assumes rapid economic growth, a global population that peaks in midcentury, and rapid introduction of new, more efficient technologies. A1b (the scenario depicted in Plate 7) represents a balance of fossil-intensive and nonfossil energy sources.[22]

The potential changes to regional climates described in Plate 7 are large and wide ranging, with the capacity to affect many aspects of people's everyday lives. Even if we succeed in mitigating climate change by keeping the average global temperature increase to just 2°C above preindustrial levels, substantial impacts will still occur and require responses. If the temperature rise over the next few decades is significantly higher, the impacts will be extremely severe, and innovative adaptive approaches will be required.

The Impact on Crop Production

Plants are particularly vulnerable to high temperatures because of the damaging effects on certain enzymes central to the process of photosynthesis.[23] Higher temperatures will also cause *stomata* (pores in the leaves through which gases and water vapor are exchanged) to close to prevent water loss, so reducing photosynthesis, and there may be an impact on pollen viability and fertilization. Water availability is also critical; since photosynthesis uses the sun's energy to combine carbon dioxide and water to form the carbohydrates essential for plant growth, a lack of sufficient water leads to lower yields and, in the extreme, to death of the plant.

Many crops are already grown close to their limits of thermal tolerance. Just a few days of high temperature near flowering time can seriously affect yields of crops such as wheat, fruit trees, groundnut, and soybean.[24] In low-latitude regions, where most of the developing countries lie, even moderate temperature increases of 1 to

2°C can reduce yields of major cereals, and the effects of adaptation—such as changes in planting, changes in cultivar, and shifts from rain-fed to irrigated conditions—are likely to be limited (Figure 15.3).

David Lobell, an environmental scientist at Stanford University, and colleagues at the International Maize and Wheat Improvement Center (CIMMYT) examined data from 20,000 field trials of maize conducted in Africa between 1999 and 2007 and matched their yields with local weather station records. These revealed a non-linear relationship between warming and yields. More specifically, there was a yield loss of 1 percent under optimal rain-fed conditions and a loss of 1.7 percent under drought conditions for each degree day spent above 30°C. About three quarters of

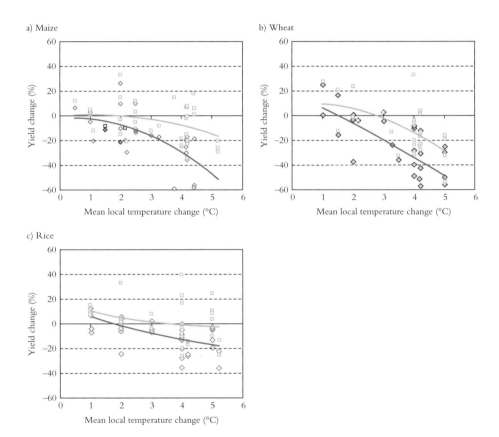

Figure 15.3 Cereal yield responses to temperature change. (Results of 69 published studies at multiple simulation sites. Responses include cases without adaptation [dark dots] and with adaptation [light dots].)[25]

293

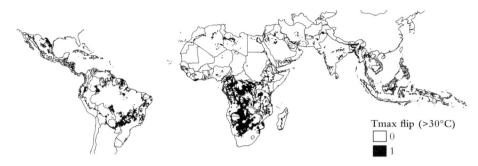

Tmax flip (>30°C)
☐ 0
■ 1

Figure 15.4 Areas where average annual maximum temperature will flip from <30°C to >30°C by 2050.[27]

Africa's maize crop area would experience a 20 percent loss for a 1°C warming. The results also indicate the importance of improving soil moisture and breeding drought-tolerant crops.[26]

A mapping exercise carried out by the CGIAR Research Program on Climate Change, Agriculture and Food Security (CCAFS) (using mean outcomes of four general climate models) reveals large areas of the developing countries where the annual maximum temperature will rapidly change to over 30°C (Figure 15.4).

In some instances moderately rising temperatures can be beneficial. The yields of some crops in temperate regions may increase, at least for rises of 1 to 2°C.[28] Higher minimum temperatures can also be beneficial. Milder winters in the north of China permit the northward extension of land devoted to winter wheat. Some crops, such as apples, however, need a winter cooling (a process known as vernalization) to initiate flower bud formation. The milder winters in the Western Cape of South Africa are already adversely affecting the apple crop there.[29]

The maize crop is particularly susceptible to lack of water: Some 20 million tons of potential tropical maize production is lost each year as a result of drought.[30]Maize crops in most parts of southern Africa already experience drought stress on an annual basis. The situation is likely to get worse with climate change; maize production in parts of Zimbabwe and South Africa may become impossible. Wheat yields in northern Africa are also likely to be threatened. Figure 15.5 shows that large areas of the developing countries already experience a high coefficient of variation for rainfall (a measure of the extent of variability in relation to the mean). This situation is likely to worsen, but by how much is not yet known.

The combination of changes in temperature and rainfall may also alter the length of the growing season. In China, both rising maximum and minimum temperatures will increase the length of the season throughout the country, and particularly in the

Figure 15.5 Areas where the coefficient of variation for rainfall is already more than 21 percent.[31]

Figure 15.6 Areas that will experience more than a 5 percent reduction in the length of growing period (LGP) by 2050.[32]

west, although this may be restricted by declines in rainfall in some areas. By contrast, in southern Africa and across western and north-central Africa, lower rainfall may cause the season to shorten, threatening the probability of getting a second crop in some areas and even the viability of a single crop in others (Figure 15.6).

Carbon Fertilization

Just how severe will be these various impacts on agriculture depends on the so-called carbon fertilization effect. Since carbon dioxide (CO_2) is the basic building block for plant growth, rising levels will, in theory at least, increase crop yields. There is also an indirect effect of higher CO_2 levels: the leaf stomata are able to partly close without sacrificing CO_2 intake and photosynthesis. Partial closure decreases evapotranspiration, increases leaf temperature, and improves water use efficiency that could lead to increased yields during times of drought.[33]

Greenhouse and field chamber experiments show the fertilization effect for plants with growth based on a so-called C3 metabolic pathway (such as wheat, rice, and soybean), but C4 pathway crops (such as maize, millet, sorghum, and sugar cane) are not responsive (C3 and C4 pathways use different enzymes to extract CO_2 from the air). However, the latest analyses of more realistic field trials suggest that the benefits of CO_2 may be significantly less than initially thought—only an 8 percent to 15 percent increase in yield for a doubling of CO_2 for responsive C3 species and no significant increase for nonresponsive species such as maize and sorghum, which are widely grown in Africa.[34] Hence this offsetting factor may be less than was previously assumed. Estimates produced for the IPCC of expected yield losses with a 3 to 4°C rise in temperature (with CO_2 fertilization) are 18 percent for wheat in northern Africa and 22 percent for maize in southern Africa.[35]

Livestock and Pasture Production

Some of the most serious impacts of climate change on livestock will be in the arid and semiarid regions of the world. Increased temperatures often lead to lower physical activity of livestock and declines in eating and grazing as well as an increase in body temperature and sweating, resulting in weight loss.[36] High temperatures reduce milk fat and protein content and put a ceiling on dairy milk yield irrespective of feed intake. In the tropics, this ceiling reaches between half and one-third of the potential of modern (Holstein) cow breeds. Increases in air temperature and/or humidity affect conception rates of livestock, particularly cattle that are not well adapted.[37] There is also a strong relationship between drought and livestock deaths.[38] In the 1980s, protracted drought killed 20 to 62 percent of cattle in countries as widespread as Botswana, Niger, and Ethiopia.[39] Increasing temperatures can also drastically increase water demand in livestock. For example, for zebu cattle (*Bos indicus*), water intake increases from around 3 kg per kg dry matter (DM) at 10°C ambient temperature to 5 kg at 30°C and 10 kg at 35°C.[40] Warmer climates may cause farmers to switch to different livestock mixes: poultry and cattle may be replaced by goats and sheep.[41]

Climate change will also affect the pastures and feed resources on which livestock depend. Pastures are likely to respond with rapid alterations in the species composition and diversity of forage plants. This may affect the quality of the forage and the grazing behavior of the livestock.[42] Land use change, as well as changes in primary productivity, species composition, and quality of plant materials will impact fodder production.[43] For example, higher temperatures increase lignification of plant tissue, reducing both the digestibility and rate of degradation of plants and resulting in lower nutrient availability for livestock.

Pests, Diseases, and Weeds

There is a strong likelihood that agricultural losses (in crops, livestock, and forests) will increase as a result of more frequent or severe pest and disease attacks stimulated by higher temperatures and/or humidity.[44] Such threats already have a major effect on crop and livestock production (Chapter 12). In some cases pests and diseases may extend their range as a result of climate change, moving into regions where the crops or livestock have no experience of them and hence little resistance.

Weed growth may also be directly affected by increased CO_2 levels. Fourteen of the world's most serious weeds are C4 plants that grow in fields of C3 crops. Here, the C3 crops may outcompete the weeds; however, where the opposite is the case (C3 weeds in C4 crops), the C3 weeds may become much more damaging.[45]

Water Supplies

The biggest impact on agricultural production will probably derive from the diminution of the water supplies on which agriculture depends. In Chapter 14, I showed the extent to which scarce water resources are already under pressure. In some places, global warming may make water more available, but for much of the developing world the effect will be to make water scarcity even more acute. The worldwide percentage of land in drought has risen dramatically in the last twenty-five years. In Africa, one-third of people live in drought-prone areas and will experience increasing water stress.[46] The IPCC estimates that, by 2080, the proportion of arid and semiarid land in Africa is likely to increase by 5 to 8 percent.[47]

It is difficult to predict with any certainty the future hydrological characteristics of major river basins.[48] However, we do know that in glacier- or snow-melt–fed river basins—which nurture one-sixth of the world's population—changes will be profound. The summer-season melts that provide much of the water at that time of year will decrease, and peak flows will move earlier in the year, with serious effects on agriculture. As the glaciers retreat, flows will increase, but then decrease over the next few decades as the amount of glaciated area is reduced.[49]

River flows elsewhere will be more dependent on changes in precipitation, rather than on snow or ice melt. A general conclusion is that seasonal flow will increase, with higher flows in the peak-flow season and either lower flows during the low-flow season or extended dry periods. Many semiarid and arid areas will suffer decreased river flows.

Extreme Events

Climate change will bring about gradual changes in the temperature and rainfall throughout the world with far-reaching consequences. Even more significant, perhaps, for agriculture will be the increasing variation about these trends and the greater frequency and intensity of extreme weather events—heat waves, heavy downpours, cyclones, and storms—leading to severe flooding and droughts. There is growing evidence that this is already occurring.

Disastrous floods will become more common in many parts of the developing world due to higher rainfalls, but even in drier regions there is likely to be a higher frequency of more intense downpours that may cause flooding. In 2007, for example, both eastern and western Africa experienced heavy flooding caused by heavy rainfall and thunderstorms within the rain belt of the ITCZ, which was farther north than usual. Much of the land was dry from years of drought, and the record rainfalls resulted in overwhelmingly high levels of runoff.

Several developing regions are also likely to experience greater cyclonic activity, bringing torrential rain, high waves, and damaging storm surges and resulting in inland and coastal flooding. The devastating supercyclone that hit the Indian state of Orissa in 1999 was one of the worst disasters of the last decade. It affected the livelihoods of 12.9 million people and resulted in the loss of 1.6 million houses, nearly 2 million hectares of crops, and 40,000 livestock.

The rising incidence of droughts, both short- and long term, may have the most significant impact on many developing countries. Most devastating will be situations where droughts occur for two or more successive years or when high temperatures coincide with drought spells. Such episodes of extreme weather are likely to become more frequent and will tend to force large areas of marginal agriculture out of production. Drought can have a catastrophic effect on rural communities. For example, in northeastern Ethiopia, drought-induced losses to crops and livestock between 1998 and 2000 were estimated at $266 per household—greater than the annual average cash income for more than 75 percent of the households.[50] The FAO reports that severe drought accounts for half of the world's food emergencies annually.[51] In 2003, the international community provided over 5 million tons of food aid to Sub-Saharan Africa.[52]

The Overall Effects

The overall consequences of these various effects are difficult to predict. Crop-climate models estimate a 3°C or more increase in temperature will cause crop losses in all regions. One study that combines agronomic and economic modeling suggests a 6 percent reduction in global agricultural production by 2080.[53] However in hotter climates, such as in parts of Africa, Asia, and Central America, losses may be higher. Where crops are already at the maximums of their heat tolerance ranges, yield losses for crops such as maize and wheat are likely to be very high—reductions of around 20 to 40 percent with an increase in temperature of 3 to 4°C (although theoretically carbon fertilization could reduce these declines by around half).[54] Parts of Africa and India are projected to suffer a 30 percent decline in food production under climate change (Plate 8).

Note that these estimates are for production potential and take no account of technology advances, the impact of adaptation and mitigation, changes in market prices for food, and the complex feedbacks between agroecosystems and the economy.[55]

Adaptation and Resilience

Adapting to climate change is as complex a process as the phenomenon of climate change itself.[56] The process of adaptation requires understanding of climatic impacts, the hazards they generate, where they occur, and with what degree of certainty predictions are made; as well as assessing the various dimensions of vulnerability and the appropriateness, including costs and benefits, of a range of potential options for action. In practice, adaptation is a collection of coping strategies, with each strategy focused on a particular threat. Some of these actions may be taken by autonomous individuals or communities reacting to climate change hazards as they occur; others may be more planned, depending for their initiation on government policies and institutions.[57]

The Responses of Farmers

Farmers throughout the developing world are already adapting to climate change. In a survey of eleven African countries, farmers were reported as growing different varieties and modifying planting dates and practices to account for shorter growing seasons.[58] When asked, farmers usually acknowledge that the climate in their area is changing and are ready to describe what is happening and how they are responding (Box 15.2).

Farmers faced with the threat of flooding may plant new flood-resistant rice varieties. They may also build protection around the rice field, ensure the flood water is quickly drained, or develop a more diverse livelihood so that other sources of

Box 15.2 Farmer Responses to Climate Change

Mozambique[59]

In the village of Nwadhajane in Southern Mozambique, the birthplace of the great Mozambique leader, Eduardo Mondlane, the villagers are aware of climate change affecting them and have already taken significant measures to counteract the worst features. They have two kinds of land: lowland and highland. On the former, the crops are very productive but are washed out by periodic floods; in the highlands, they produce good crops in the flood years but poor crops during the droughts. The villagers' response has been to create several farmer associations that have reassigned the land so that each farmer obtains a portion of highland as well as some lowland. The farmer associations are also experimenting with drought-resistant crops.

Morocco

In the Atlas Mountains of Morocco, the villagers cannot grow enough barley to feed themselves because of the continuing lack of rainfall. They are trying out drip irrigation as a possibility for high-value crops that they can sell in the markets on the coast. They are also harvesting some of the wild, typically drought-tolerant, plants growing on the hills around the villages; for example, the high-quality oil (similar to olive oil) from the Argan tree and the "honey" from *Euphorbia*. But the women are doing the harvesting, and they are getting little in return. The challenge is to process the oil and the honey in situ and thus derive some of the value added in the villages.

income will offset the losses from flooding. But not all farmers may be able to respond on their own; they may need governments to build suitable protective infrastructure or to develop specific policies that mitigate the effects of flooding. They may need help with insurance or from safety nets, through agricultural research and extension, new irrigation schemes, or dissemination of appropriate agrometeorological information. Some of these strategies will be technological, others social, economic, or political.

Hazards and Responses

Climatic and other hazards generally come in two forms, which may require different responses:

1. *Stresses*—gradual buildup of adverse events (e.g., increasing temperatures, rising sea levels, greater or lesser rainfall)
2. *Shocks*—usually dramatic, largely unexpected events (e.g., sudden floods, cyclones, earthquakes, tsunamis, disease outbreaks)

Unlike more gradual change, shocks can cause profound, sudden disruption to economies and communities. Stresses—although more predictable—can slowly build up to catastrophic outcomes.

A useful concept in adaptation is *resilience* (introduced in Chapter 5; see Box 5.3 and Figure 5.5). Figure 15.7 shows the range of responses to stresses and shocks.

Anticipation consists of surveys to determine the likely location and probability of potential disturbances. Such inventories can be depicted as hazard maps. Other smaller-scale surveys can be produced by local communities for their own planning. The advantage of these is that if a flood or other hazard arises, potentially affected communities can respond rapidly. Anticipation also involves producing long-range weather forecasts that can be used to put in place adaptive measures. They are made possible because of the relationship between the sea surface temperature (SST) and large-scale weather patterns. The slow changes in SST and the associated weather patterns can be predicted with some degree of accuracy up to six months in advance.[61]

The subsequent steps—prevention and tolerance, recovery and restoration—involve defining objectives, identifying the various options, and then appraising them in terms of their outcomes and the relevant costs and benefits.

Finally, building resilience is about learning. If a number of small-scale stresses and shocks are experienced by a country, or a community, it will learn to assess how it coped and how well its planned adaptations performed in practice. This means putting learning processes into place at all levels: in the household and community and at the district and national levels of government.

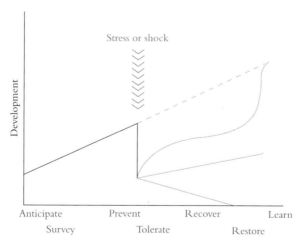

Figure 15.7 The range of responses to stresses and shocks.[60]

Coping with Stress

The biggest challenge for agriculture in the developing countries is coping with increased heat and drought stress, which often act together.[62] Many of the adaptive technologies have already been described in previous chapters. *Drought resistance* can be achieved in one of three ways: (1) drought-tolerant cropping systems, (2) drought-resistant crop varieties, and (3) provision of water resources. One of the most significant recent successes in water and soil management has been the adoption of "no-till" systems under the generic name of conservation agriculture (see Box 13.2). Advances in the use of conventional breeding and biotechnology to produce more water-efficient crops were described in Chapter 14.

The large-scale supply of water for irrigation was also discussed in Chapter 14. A number of other, small-scale technologies have been widely adopted and provide efficient water supply for cropping. One example is the combination of simple, plastic-tubed, drip irrigation with well-constructed treadle pumps for raising water from shallow wells (see Box 7.4). Such systems usually lie in relatively small watersheds and rely on an intimate knowledge of the temporal and spatial pattern of the water supply by farm communities or individual farmers.

Coping with Shocks

Shocks, such as floods and droughts, are best approached through the development of resilient forms of livelihood, utilizing the concepts described in earlier chapters. In principle, the answer is to rely on greater diversity of sources of income—whether of different crops or livestock, or of a range of nonfarm income, either on or off the farm. A personal anecdote: I once visited a small farm in the Sunderbans of India. The family had some rice fields, which both the husband and wife looked after. She also grew a range of root crops and vegetables in a small "market garden" plot. Her husband, with the help of their young son, raised fish fry for sale. He also had a bicycle with a small platform on the back that he used as a taxi, carrying goods and people about the village. On the thatched roof of their house was a small solar panel that they used to power light bulbs in the evening so the children could do their homework. If they graduated successfully from school, they could get a job in one of the nearby towns. If and when disaster struck (a cyclone hit a year after my visit), they could rely on one or more of the different sources of income on the farm and in due course could hope to receive remittances from their children in the towns.

Pursuing diversity in this way is common practice among poor households in the developing countries. In eastern Kenya farmers can fall back on a wide range of income sources in difficult times (Box 15.3).

Box 15.3   Kenyan Farmers Coping with Drought[63]

The Kati District of eastern Kenya suffered from poor rainfall in 1995 and 1996 and ran out of food between the July 1996 harvest and the next harvest in February 1997. Only 2 out of a random sample of 52 households had a maize crop that lasted them through this period. The farmers, when interviewed, listed a large number of activities that helped them endure this shock, including:

- Skilled work
- Selling land
- Collecting honey for consumption and sale
- Making bricks for sale
- Engaging in food producing or money-making group activities
- Business, such as selling snacks
- Burning charcoal for sale
- Salaries of householder or remittances
- Handicrafts for sale
- Selling or consuming exotic fruits from the farm
- Receiving credit
- Borrowing food or money from relatives
- Borrowing food or money from neighbors
- Engaging in casual labor
- Selling livestock
- Collecting indigenous fruit for consumption or sale
- Receiving food aid from government or other organizations

Each household averaged about six activities during the drought. After the drought this dropped to three, but diversity remains a common feature of their livelihoods.

There are clearly important lessons to be learned from people who have had to cope with frequent stresses and shocks. But such resilience is being steadily eroded in many places under the impact of migration, family breakdown, famine relief, poverty, and disease. One consequence is that people tend to become more dependent on outside aid.[64] A major challenge for governments, donors, and NGOs is to help communities rebuild their resilience mechanisms. Many of the activities in Box 15.3 rely on the informal sector; governments can help by creating links to the formal sector and by providing skills, knowledge, and access to markets. Women play a key role in creating resilient livelihoods. They may be primarily responsible for home gardens and for higher-value vegetable and fruit crops that help to diversify the agricultural production. Skills such as weaving and handicraft can provide a source of income when agriculture fails. Resilience is a family affair involving both men and women and, as they grow older, the children. Attempts to enhance livelihoods must take this wider holistic and more long-term approach.

In some respects the challenges seem special to the developing countries. Diverse livelihood strategies are not common in the developed countries but may become more prevalent as the consequences of climate change become more universally felt.

Building Resilience

The relatively poor state of knowledge of climate change highlights two key goals: (1) we urgently need more research into the dynamics of the global drivers and into the detailed consequences for agriculture at regional and local levels; (2) we must design adaptation measures to cope with both relatively predictable climatic stress *and* much less predictable but even more catastrophic extreme climatic events. Adaptation depends on developing resilience in the face of uncertainty.

Resilience is important at the national, regional, and local levels, involving not only technologies but also appropriate economic policies and institutional arrangements. It is the poor who will be most vulnerable to the effects of climate change. To some extent, the process of development itself will help them to adapt. If people are better fed and in better health, with access to education, jobs, and markets, they will have the capacity to be more resilient. Traditionally, poor people have developed various forms of resilient livelihood strategies to cope with a range of natural and manmade stresses and shocks. But these may be inadequate in the future or may have been lost in the development process. The urgent need is for governments, NGOs, and the private sector to work together with local communities to enhance the resilience of the poor.

I began this chapter with some estimates of the costs to agriculture of climate change. International donor investment in adaptation for developing countries currently stands at $150 to $300 million a year, falling short of the tens of billions required.[65] The International Food Policy Research Institute (IFPRI) estimates that to offset the negative impacts of climate change on human health and well-being—particularly declining calorie availability—additional investment of at least $7 billion per year is needed. Forty percent of this total is required in Sub-Saharan Africa, most notably in roads and rural infrastructure; whereas Asia and Latin America require a smaller share of the investment, mainly in irrigation efficiency and agricultural research, respectively.[66]

Finally, adaptation is only one component of the response to climate change. We also have to reduce greenhouse gas emissions, largely to ensure that global temperatures do not rise to levels where the consequences are truly catastrophic and the possibilities for adaptation greatly reduced. Here, agriculture as culprit rather than victim has a key role to play. I address these issues in Chapter 16.

16

Reducing Greenhouse Gases

It is impossible to say with certainty how bad the 21st century's heating will be, but there is a large chance of it getting hot enough to do harm, and a far from trivial chance of things turning catastrophic. This makes moving away from fossil fuels a global priority.

The Economist, July 10, 2010[1]

While agriculture is a major victim of climate change, it has considerable potential, albeit largely unrealized, to help mitigate climate change by reducing greenhouse gas (GHG) emissions and by providing low carbon energy. In this chapter I describe this potential, illustrated with examples of where this is beginning to be harnessed.

Over the past hundred years the bulk of GHG emissions have come from industrialization and urbanization. The developed countries have been mostly responsible with, in recent decades, a growing contribution from newly emerging countries such as China, India, and Brazil that are undergoing rapid industrialization. The less-developed countries, on a country-by-country basis, have contributed relatively small amounts of GHGs to total global emissions. Nevertheless, collectively they are a significant contributor through the processes of forest clearance and agricultural development.

In this chapter I address the following questions:

- What are the agricultural sources of GHG emissions and the estimates of their future growth?
- What are the theoretical and practical potentials for mitigation?
- How can mitigation be paid for?
- Does the cultivation of biofuel crops have a sustainable future?

Agricultural Emissions

The Intergovernmental Panel on Climate Change (IPCC) estimates that global GHG emissions generated by the agricultural sector (5,100 to 6,100 million tons of CO_2 equivalent) account for 10 to 12 percent of annual total global GHG emissions (45.4 Gigatons [Gt] CO_2 equivalent emitted in 2005 or translated into 12.4 Gt Carbon). When emissions from agricultural fuel use, fertilizer production, and land use change are included, this percentage increases to 30 percent.[2] These are very high levels given that agriculture's share in global GDP is only around 4 percent.[3] Lower estimates of total global emissions by, for example, The Global Carbon Project, put the figure at around 9.1 Gt Carbon per year between 2000 and 2008.[4]

The three main GHGs produced by agriculture are:

- *Carbon dioxide* (CO_2), which comes from microbial decay, fossil fuel combustion to supply power to machinery and construct grain silos, fuel for transport and to produce synthetic fertilizers and pesticides, and land clearing and the burning of biomass, plant litter, and soil organic matter.
- *Nitrous oxide* (N_2O), which predominantly originates in the soil following applications of manure, urine, and nitrogen fertilizer; it is around 310 times more potent (in terms of global warming) than CO_2.
- *Methane* (CH_4), which mainly originates in ruminant digestion, rice cultivation, and anaerobic soils; it has 21 times as much warming potential as CO_2.[5]

All the gases are combined in a single statistic—*CO_2 equivalent* (CO_2 eq.)—by multiplying each gas by its warming potential related to CO_2.

As can be seen from Table 16.1 most of agriculture's contribution to global warming (leaving aside the effects of land clearing) comes from N_2O and CH_4 emissions. The IPCC estimates that of the 5.1 to 6.1 Gt CO_2-eq. agriculture directly contributes to GHG emissions, 3.3 Gt CO_2-eq. is from methane and 2.8 Gt CO_2-eq. from nitrous oxide, with the carbon dioxide flux relatively small at 0.04 Gt CO_2-eq.[6]

Carbon dioxide becomes significant only when we take into account land use change, predominantly deforestation, and the considerable off-farm (beyond the farm gate) emissions that arise from storage, processing, manufacture, distribution, refrigeration, retailing, consumption, and waste disposal.[8]

Future Growth

Between 1990 and 2005, levels of agriculturally related N_2O and CH_4 increased by almost 17 percent, an increase originating predominantly from developing countries. By 2030 agricultural N_2O emissions will have risen by 35 to 60 percent mainly

Table 16.1 Sources of direct and indirect agriculture greenhouse gases (GHG)

Sources of agriculture GHG	Million tons CO_2-eq.
Nitrous oxide from soils	2,128
Methane from cattle enteric fermentation	1,792
Biomass burning	672
Rice production	616
Manure	413
Fertilizer production	410
Irrigation	369
Farm machinery (seeding, tilling, spraying, harvest)	158
Pesticide production	72
Subtotal	6,630
Land conversion to agriculture	5,900
Total	12,530

Source: Bellarby[7]

Note: CO_2-eq. = Carbon dioxide equivalent

due to an increasing use of N fertilizer and greater animal manure production.[9] The corresponding increase by 2030 for livestock-related CH_4 emissions is 60 percent. Combined CH_4 emissions from enteric fermentation and manure management will have grown 21 percent, and CH_4 emissions from rice crops 16 percent by 2020.[10]

The production of synthetic fertilizers and pesticides contributes relatively very small amounts of GHGs. For fertilizers, it is between 0.6 and 1.2 percent of global GHG emissions (0.3–0.6 Gt CO_2-eq/yr).[11] Part of these emissions is the CO_2 from the use of fossil fuel energy in fertilizer manufacture; part is N_2O from nitrate production. Pesticide production is associated with only 0.0003 to 0.14 Gt CO_2-eq/yr, and irrigation, on average, 0.05 to 0.68 Gt CO_2-eq/yr.[12]

Approaches to Mitigation

Opportunities for mitigation exist along the entire agricultural food chain. In this chapter I will focus on interventions with the greatest potential, especially those that will provide benefits additional to mitigation. As with the technologies discussed in Chapter 7, appropriate mitigation technologies and other measures need to be agriculturally productive and contribute to improved stability, resilience, and equitability. They also should improve adaptation to climate change. The goal is to find win-win technologies that help keep costs down and the cost-benefit ratios low.

Where Is the Carbon?

The soil contains 2,500 Gt of carbon (C), 3.3 times as much as the atmosphere and 4.5 times as much as the carbon in living plants and animals.[13] Of all land types, except deserts and semideserts, cropland stores the least amount of carbon. Wetlands store the greatest amount, but the extent to which they occupy the earth's surface is relatively small (Table 16.2).[14] Forests of various kinds store nearly half the total carbon and their preservation will do the most to reduce GHG emissions.

Preventing Land Use Change

When natural habitats are converted to agriculture, 75 percent or more of the soil organic carbon is depleted in tropical regions, compared with 60 percent in temperate areas. The Food and Agriculture Organization (FAO) estimates that 6 million hectares (Mha) of forestland and 7 Mha of other land (much in developing countries) were converted to agriculture and other uses each year in the last four decades, a trend predicted to persist in the future.[16] However, as I argued in Chapter 1, there is

Table 16.2 Global carbon stocks in vegetation and the top 1 m of soils

Biome	Area (10⁶ km²)	Carbon stocks (Gt)		
		Vegetation	Soils	Total
Tropical forests	17.6	212	216	428
Temperate forests	10.4	59	100	159
Boreal forests	13.7	88	471	559
Tropical savannas	22.5	66	264	330
Temperate grasslands	12.5	9	295	304
Deserts and semideserts	45.5	8	191	199
Tundra	9.5	6	121	127
Wetland	3.5	15	225	240
Croplands	16.0	3	128	131
Total	151.2	466	2,011	2,477

Source: IPCC[15] (data from WBGU, 1998)

little evidence of significantly increasing cropped areas over the past fifty years, except in the case of oil crops, such as soya bean and oil palm where the total area has grown from 100 to 300 Mha (i.e., about 4 Mha per year).

Despite the high estimates produced by FAO of available land with arable potential, I am not convinced that, in practical terms, this land exists (see Chapter 1). The exception is for oil crops, where lands such as the Cerrado of Brazil and the equatorial forests of the Amazon and Congo basins could be further converted. Estimates of forest loss are equally difficult to make. However, according to FAO, the rate of global deforestation has been slowing in recent years: between 2000 and 2010 the net reduction in forest area is estimated at 5.2 Mha per year, while in the previous decade it was 8.3 Mha per year. Areas suffering the largest losses of forest are South America and Africa, which, between 1990 and 2010 lost an estimated 4.1 Mha and 3.7 Mha, respectively, of forest per year. It is difficult to estimate what will happen in the future, but based on recent trends I would guess that these will be significantly less than 4 Mha per year, but still considerable in terms of carbon emissions.

Getting Carbon Back into the Soil

Since 1850 about 136 Gt of carbon has been emitted to the atmosphere from terrestrial ecosystems, just over half as a result of soil cultivation. Some soils may have lost as much as 20 to 80 tons of carbon per hectare (t C/ha). As organic carbon is depleted from the soil, soil quality, crop productivity, and water quality also decline, so that replacing the carbon is inherently a win–win proposition (Box 16.1).[17]

In addition, there is the potential for win–win situations where sequestration increases yields. The loss of soil carbon degrades soil quality; enhancing this pool, to the optimum range of soil organic carbon (SOC) concentration (1 to 2 percent) can improve yield through increasing available water capacity, improving plants' nutrient supplies, restoring soil structure, and minimizing the risk of soil erosion.[22]

Box 16.1   Soil Carbon Sequestration

Soil carbon *sequestration* is the process of removing carbon from the atmosphere and storing it in the soil indefinitely. Sequestration can occur at rates of 0 to 150 kg C/ha/yr in dry and warm climates and of 100 to 1,000 kg C/ha/yr in humid and cool climates. The sequestration process takes some time (a maximum of 5 to 50 years) to reach its optimum rate, and then continues until a new equilibrium is reached (when the soil sink capacity is full).[18]

(continued)

Putting the carbon back can be accomplished by a variety of means: no-till farming, cover crops, nutrient management, manuring and sludge application, soil restoration and woodland regeneration, improved grazing, water conservation and harvesting, efficient irrigation, agroforestry, and growing energy crops (Figure 16.1). Some of these have already been described in Chapters 13 and 14 and are further discussed below.

The technical potential to sequester carbon in the soils of terrestrial ecosystems as well as restored peatlands has been estimated at around 3 Gt C per year, the equivalent of 50 ppm of atmospheric CO_2 by 2100. This level of sequestration would be achieved by increasing the soil C pool by 1 ton/ha/year.[19] Collectively, this has the potential to offset 5 to 15 percent of global fossil fuel emissions and to increase annual production of food grains in developing countries by 24 to 32 million tons and of roots and tubers by six to 10 million tons.[20]

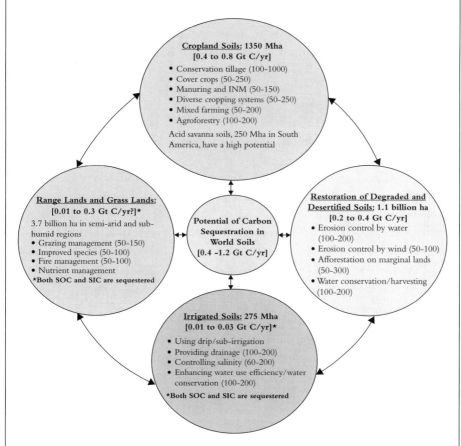

Figure 16.1 Potentials for soil carbon sequestration (individual rates are kg/ha/yr). SOC=Soil Organic Carbon; SIC=Soil Inorganic Carbon; INM=Integrated Nutrient Management[21]

An increase of 1 ton per ha of soil carbon in degraded croplands can increase wheat yield by 20 to 40 kg/ha, maize yield by 10 to 20 kg/ha, and cowpea yield by 0.5 to 1 kg/ha. In Indian alluvial soils, when SOC concentration in the root zone was increased by 1 percent, biomass production increased by 12.7 tons/ha for wheat, 7.9 tons/ha for maize and 1.6 tons/ha for cowpea. Given that the gap between cereal production and demand in Sub-Saharan Africa was around 16 million tons in 2001 and is predicted to rise to 52 million tons by 2015, soil carbon sequestration is significant for both mitigating climate change and reducing food insecurity.[23]

In practice, effective sequestration depends on the technologies and know-how available, the soil texture and structure, climatic conditions, the farming system, and associated practices of soil management.[24] While there are many successful small-scale projects, taking sequestration to scale faces many challenges. There are difficulties in measuring both the preexisting carbon in the soil and ongoing levels. As a consequence, technical and management skills are crucial. In some situations sequestration may also involve reduced yields. There are also likely to be significant transaction costs linked to such issues as determining land rights, monitoring, ensuring compliance, and, finally, paying farmers for sequestration—particularly if there are large numbers of small farmers.[25]

Measuring the potential soil carbon sequestration of soils is also difficult, although there are several new techniques, given the wide range of factors that can affect the process (see Clean Development Mechanism further on). Modeling of sequestration in China found that a combination of no-tillage and increasing crop residue return was the best option. Practicing 50 percent crop residue return and no-till farming on 50 percent of arable lands in China is estimated to give an annual soil carbon uptake of 32.5 million tons, "60 percent more than the current carbon uptake of all forests."[26] In some cases this can be much larger. It is a surprisingly large estimate, and needs further research.

Improving Productivity

Increasing crop yields can reduce GHG emissions. The more plants grow, the more they take in CO_2, part of which finds its way into the soil. Greater productivity should also lessen the demand for more land, so reducing the carbon emissions from land clearance.[27] More fundamentally, attention must be paid to breaking the vicious cycle of cultivation whereby falling nutrient levels in the soil produce lower yields and less crop residues, resulting in less organic carbon in the soil that, in turn, further lowers yields.[28] Crop yields can be increased through irrigation and use of fertilizers, but this will be energy intensive and lead to greater N_2O emissions. The alternative, virtuous cycle builds on improved varieties, use of perennial crops, cover or catch crops grown in between seasons, or crop plants and no-till agriculture that can increase yields and the level of carbon stored in the soil.[29] I present examples of these in the following pages, reflecting on their respective advantages and disadvantages.

Conservation Farming

I discussed conservation farming (or minimum-tillage farming) in Chapter 13. This approach is gaining considerable advocacy as a way of tackling widespread soil erosion and depletion of organic matter that more conventional farming systems, reliant on the plow, can induce. Reducing or eliminating tillage will, in many circumstances, prevent degradation and help to restore carbon, particularly if it is combined with covering the soil with the residues from the previous crop. Because the crop residues are left to rot down as the next crop grows, there is a significant return of carbon to the soil.

Despite differences in the rate of soil carbon sequestration in different parts of the world, the adoption of conservation farming leads to an average soil carbon sequestration of 1.8 tons of CO_2/ha/year in the first decade. If conservation farming were to be implemented across five billion ha of agricultural land, current annual global emissions of CO_2 from fossil fuel burning (approximately 27 Gt CO_2 per year) would be reduced by around a third.[30]

Reduced or no-till agriculture also impacts N_2O emissions, but according to soil and climatic conditions this effect can be positive or negative.[31]

Agroforestry

Building carbon in the soil can also be achieved by the planting of crop or other plants with favorable characteristics. For example, the deep-rooted *Brachiaria* grasses planted in the South American savannas will increase rates of carbon storage (see Box 10.8).[32] Intercropping with trees—*agroforestry*—also has high potential as a win–win approach to reducing GHGs, especially if the trees are nitrogen-fixing legumes. Planting trees, alongside crops or livestock, can increase both the levels of C sequestered in the soil as well as improve the quality and productivity of the land and the nutrients available to the crops (from lower soil horizons) (Box 16.2).

Trees are also sources of food, timber, fuel, and other nontimber forest products (NTFP, e.g., oils and resins), and, since income can be generated from a wide range of sources, agroforestry systems tend to be not only profitable but stable and resilient.[35] Significantly, the IPCC believes agroforestry offers the greatest potential for carbon sequestration in developing countries than all other land uses (Figure 16.2).[36]

Box 16.2 *Faidherbia* Systems[33]

Faidherbia albida, a nitrogen-fixing acacia tree that is widespread throughout Africa, growing in a variety of soils and climates, has been lauded for its ability to both make large quantities of nitrogen available to nearby crops and increase the store of carbon above ground and in the soil. In one study in Malawi, planting *F. albida* provided 300 kg of fertilizer per ha and boosted unfertilized maize yields to 2.5 to 4 tons per ha. In Niger, similar benefits have led to its expansion of on almost five million hectares of land. The climate change mitigation potential for systems incorporating trees with fertilizing properties lies in their ability to sequester some 2 to 4 tons of carbon/ha/year (compared with 0.2 to 0.4 tons of C/ha/yr under conventional conservation farming systems).[34]

This tree species also has "reversed leaf" phenology—meaning it sheds its leaves in the wet season and retains them in the dry season. This makes it ideal in an agroforestry system because it does not compete with crops for light, nutrients, and water and provides shelter for crops in the dry season. *Faidherbia albida* takes six years to fully develop, however, making investments hard to justify, particularly if land tenure is not secure. At present *F. albida* is grown only on 2 percent of Africa's maize area and 13 percent of sorghum and millet area.

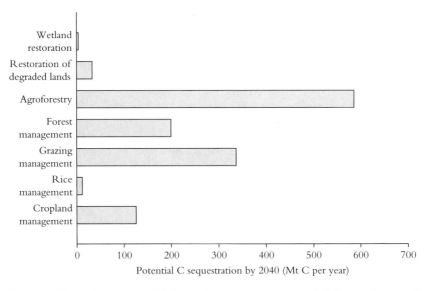

Figure 16.2 The potential for carbon sequestration of different forms of land use and management.[37]

Agroforestry can also assist agricultural systems adapt to greater climate variability. Deeper root systems allow trees to reach deeper sources of water and nutrients. Trees will also reduce runoff and soil exposure, helping conserve moisture in the soil. During floods, trees can increase soil aeration because of their high rates of evapotranspiration. Agroforestry, when practiced well, can deliver many wins.

Land Restoration

In the past, loss of carbon and organic matter from the soil was often seen as an unavoidable consequence of agricultural practices, despite the general recognition that healthy soils required high levels of organic matter. More recently, research on technologies that increase soil carbon sequestration and reduce nutrient loss, and that prevent disturbance, erosion, organic matter loss, acidification, and salinization have put concerns for healthy soils back into the mainstream of agricultural theory and practice.[38] As a consequence there is a growing realization of the considerable benefits of restoring areas previously cleared for agriculture or degraded through overuse and poor management.[39] One approach has been to take such marginal land and allow it to revert to natural or seminatural grass and woodland. Another has been to keep the land in production but under a carefully managed system that restores the soils; for example, using organic or conservation farming or integrated soil nutrient systems (see Chapter 13).

Reducing Nitrogen Emissions

As I discussed in Chapter 13, crop plants rarely make efficient use of nitrogen (N), whether in synthetic fertilizer, crop residue, or manure. Of the 100 million tons (Mt) of N produced industrially in 2005 only 17 percent was taken up by crops. Use of synthetic N can be avoided in various ways: by composting, through green manures, N-fixing cover crops, intercropping, or application of livestock manure.[40] Under some circumstances, these measures may improve efficiency of N use and reduce losses, enhance soil quality, and provide nutrients. However, they can also result in losses of N, although this is poorly understood. The alternative of breeding for higher N use efficiency is clearly going to take some time. Probably the most effective approach is to continue to apply synthetic fertilizers but using more appropriate timing, dosage, and location of the fertilizer to improve uptake and utilisation.[41]

Incorporation of urea fertilizers, formulated as briquettes, marbles, or supergranules, deep in the soil prior to planting is often more cost-effective than broadcasting the fertilizers on the surface.[42] Trials in Indonesia with deep placement of urea have produced a 10 percent yield increase with a 25 percent decrease in the amount of urea applied.[43]

Reducing denitrification also has potential. Two forms of nitrogen are important for crop growth: ammonium (NH_4+) and nitrate (NO_3-). Ammonium, being positively charged, is held by negatively charged soil particles and is not subject to leaching. But nitrate is negatively charged and is lost through both leaching and—through denitrification—as a gas to the atmosphere. If ammonium is not quickly taken up by plants, it is converted to nitrate through nitrification carried out by soil bacteria and can be further converted to nitrogen (N_2) and the GHG nitrous oxide (N_2O) in the soil. The process of nitrification can be slowed or prevented by using nitrification inhibitors that are incorporated at the fertilizer manufacture stage.[44] They delay or prevent nitrification and hence the conversion of ammonium to the easily lost nitrate.[45]

The production of N fertilizers is also a factor in GHG emissions. The energy needed varies widely; modern plants are efficient and in some cases generate negative energy balance. The storage and transportation of fertilizers and the machinery involved in their application are also sources of emissions. Total emissions from the production and use of N fertilizer were 1 to 2 percent of total GHG emissions in 2007.[46]

Reducing Livestock Emissions

Livestock produce GHGs, either directly, for example via enteric fermentation, or indirectly through the production of fodder and feed crops or the transformation of forest to grazing land. On the farm, methane (CH_4) and nitrous oxide (N_2O) are emitted from enteric fermentation and manure. In ruminants, microbial fermentation in the rumen converts fiber and cellulose into products that can be digested and utilized by the animals. Methane is exhaled during this process and is also generated by manure stored in warm, anaerobic conditions. Nitrous oxide is released to the atmosphere from manure during storage and spreading. Off the farm, the slaughtering of animals, processing, and the transportation of livestock products can all use significant amounts of fossil fuel energy.[47]

In contrast to ruminants, monogastric livestock such as pigs and poultry have greater feed conversion efficiency, require less land, and generate fewer emissions, particularly in terms of methane. Pig and poultry production generate half the amount of emissions per kg of product than sheep and beef (17 and 13 kg CO_2-eq. per kg meat for sheep and beef, respectively).[48]

Considering all the emissions together, FAO estimates that livestock contribute about 9 percent of total anthropogenic CO_2, 37 percent of CH_4, and 65 percent of NO_2. In combination, livestock production and associated land use change account for about 18 percent of total anthropogenic GHGs emitted (more than the transport sector). In the developing world, major reductions will depend on the transition from livestock extensively reared on low-quality grasses to more intensive systems (Chapter 10). This is already happening in emerging economies such as Brazil and China.

Extensively reared livestock produce more GHG emissions per unit output but often less in total than intensive systems. In some parts of the world, notably South America, the extension of large-scale grazing has been responsible for considerable deforestation.[49] Once created grazing lands can be managed to reduce emissions by adjusting the livestock intensity, for example, through rotations. In theory there is an optimal grazing intensity that produces the optimum carbon buildup, but this is difficult to determine given the great variety of grazing systems and practices. Moreover, managing manure in grazing systems is often challenging; once manure is excreted in the fields, much of the emissions have already occurred.[50]

It is easier to reduce emissions in intensive livestock production. The larger herd sizes mean more emissions overall but lower emissions per unit of product, lower land requirements, and a shorter length of time to rear an animal—leading to lower emissions over an animal's lifetime.[51] In the developed countries adding antibiotics and steroids to cattle feed reduces methane emissions, as does the use of bovine somatotropin (bST) to increase milk production in dairy cows.[52] There are also long-term possibilities of using biotechnology to modify the process of fermentation in the cattle rumen where the methane is produced. Breeding has the potential to improve feed efficiency, for example by reducing the amount of feed required per unit of output, beef or milk. Improving the health of livestock will also increase the efficiency of feed conversion.

Emissions from stall-fed animals can be reduced by altering the balance of carbohydrate and protein in the diet. Replacing fodder with concentrates and inclusion of high-quality forage in the diet can reduce overall emissions per kg of product. Supplementing dairy cattle diets in India with high-quality feed reduces methane emissions per liter of milk by a factor of three.[53] An alternative is to make greater use of crop wastes and of specially grown crops such as sugar cane: cane juice can be fed to pigs and poultry, and the forage to ruminants.[54] A wide range of dietary additives and supplements are being experimented with, ranging from antibiotics to probiotic strains that reduce methane production. Vaccines against methanogenic bacteria are also currently in development.[55]

Under intensive systems it is also easier to manage the manure. Emissions can be minimized by cooling or covering the manure, by separating solids from slurry, and by CH_4 capture and anaerobic digestion. Emissions are lower when the manure is drier and when animals are fed on lower-energy feed. Collected manure can be used as a source of methane for biogas, so reducing emissions and the dependence on fossil fuel.[56] The manure is broken down anaerobically in a digester (a large vessel where chemical or biological reactions are carried out) to produce the gas. However, the costs of digesters are relatively high, although exceeded by the long-term benefits, and hence require an appropriate microcredit financing.[57]

Changing Diets

Sustainable intensification is key to reducing livestock emissions and may have the potential to reduce pressure on grazing lands and protect conservation areas, but critics of intensive livestock production argue that animal health and welfare suffer and that intensive systems cause soil and water pollution. These problems are probably less serious in developing countries where livestock have traditionally been reared on large-scale grazing lands or in mixed farming systems (Chapter 10). The waste products often feed crop growth, while the waste products of the farm or farming household feed the livestock, so forming a closed loop of resources and preventing additional land being required to grow animal feed.

Livestock production is critical to the welfare of many if not most smallholders in developing countries. Although these closed loop systems conserve resources, they are ill-equipped to meet current and future meat demand. If the projected increase in demand is unavoidable, there will be an accelerated move to intensively reared livestock. The alternative is a significant change in eating habits toward vegetarian diets: then, there will be a place for a mix of livestock systems but with lower overall output.[58] But it is debatable whether such changes will come about unless there is a large increase in the prices of livestock products.

Because per capita consumption of meat is highest in the developed world, it follows that this is where the major dietary change should originate. But because populations are large in developing regions and their consumption patterns are changing toward more meat eating, reduced livestock consumption in the developed world will not be adequate in and of itself to meet GHG reduction targets. Developing countries will need to halt the shift in their patterns of consumption as well. But this will have implications for nutrition (see Chapter 13).

Reducing Emissions from Rice Fields

Methane is also produced from wetland rice soils where methane is generated by microbes in the water that help decay organic matter. Using less water can reduce methane emissions, although higher emissions of N_2O may result. Reduction of methane emissions from rice fields—as much as 88 percent without lowering yields—can be achieved by draining the fields at specific times.[59] A single draining of the rice paddy midseason can reduce methane emissions by almost 50 percent compared to employing continuous flooding.[60] The system of rice intensification (SRI), detailed in Chapter 11, does not require rice fields to be constantly flooded and hence is likely to result in lower methane emissions, but an in-depth analysis of the GHG balance of SRI is not yet available.[61] Other measures include the development and use of rice cultivars bred for their low secretion of methane, such as aerobic rice.[62] These tend to have few unproductive tillers, small root systems, high root oxidative activity, and high harvest index.[63] Methane emissions are also reduced by preventing waterlogging in the off-season; adding organic residues in the dry season rather than the wet, composting residues beforehand; and producing biogas as a by-product.[64]

Paying for Mitigation

So far I have discussed the technical issues involved in various approaches to mitigation. The next question is what financial incentives are needed for their development and adoption. They primarily take one of two forms:

1. *Indirect incentives*: Farmers may adopt mitigation measures because of other benefits—such as higher productivity or better adaptation to climate change—that the farmer perceives as beneficial even if the mitigation effect is not apparent. The planting of *Faidherbia* trees (see Box 16.2) is an example of such a "hidden" win-win effect. The incentives may or may not be accompanied by a subsidy for adoption.
2. *Direct incentives*: The incentive may be a direct payment of some kind tied to a price that is set for the sequestration of the carbon or the reduction in carbon or carbon equivalent emissions (Box 16.3).

Direct payments depend on carbon markets that were deliberately set up to shift economies away from GHG-producing activities. The UK Emissions Trading Scheme (UK ETS) was the world's first large-scale GHG emissions trading scheme, developed in 2002, but now superseded by the European Union (EU) ETS. This is the world's largest cap and trade scheme (Box 16.3), covering most of the high-emitting industries such as oil refineries, power stations, and factories in thirty EU countries, representing almost half of the EU's CO_2 emissions. There have been, and still are, operational problems with the EU ETS that have hampered its effectiveness in lowering emissions. For example, in the first phase of implementation (2005–2008) emitters were given permits to emit 130 million tons of CO_2 more than they actually did. The excess of permits caused the price to drop and because producers passed on the estimated costs of permits to their customers rather than

Box 16.3   Cap and Trade Carbon Markets

Cap and trade carbon markets are a form of property rights, giving "actors" the right to emit carbon in exchange for the purchase of permits. It does this by assigning a *cap*—the level of carbon emissions allowed overall that is set by a government or intergovernmental body—and then permitting companies to own a certain amount of emissions permits relating to the amount of carbon they can emit.[65] If companies can reduce emissions for less than the price of permits, they can sell them. Companies for whom the price of permits is less than the cost of measures they need to reduce emissions will purchase them. In this way companies reach emissions targets at the lowest cost.[66]

As the total number of emissions permits are reduced, their scarcity causes their price to increase, thereby making emissions reduction activities the cheaper option.[67]

the actual costs, they were able to make a substantial profit from the trading scheme without reducing emissions.

Carbon trading has many criticisms not only in terms of its practical implementation but in what trading can achieve. Steven Kelman of Harvard University believes that putting a monetary value on environmental goods and services legitimizes negative behavior, allowing those who can afford to pollute to continue to do so.[68] Carbon markets may also encourage short-term fixes while preventing the harder structural, political, and behavioral changes needed to sustain climate change mitigation.[69]

The questions are whether carbon markets are appropriate for agricultural activities and whether they can be applied in developing countries. Markets are being used to support soil carbon sequestration (e.g., no-till agriculture) in the developed countries (e.g., through the Chicago Climate Exchange), but the process is yet to be opened up to developing countries.[70] The World Bank's BioCarbon Fund was set up to purchase carbon from projects that sequester or conserve carbon in forests and agro-ecosystems. It is currently comprised of two tranches, one put into operation in 2004 and one in 2007, with a total fund of $90.4 million. A third tranche was announced at the COP17 in Durban in 2011. The aim of the fund is to open up carbon financing to developing countries. The World Bank is also proposing a new initiative called the Carbon Initiative for Development, which will aid low income countries in accessing finance for low carbon investments through carbon markets.[71]

The Mitigation Potential

If the price of carbon is high and there are no economic or other barriers, the IPCC believes agricultural emissions could be reduced or offset by around 5,500 to

6 Gt of CO_2 eq. per year by 2030.[72] The bulk (70 percent) of this potential for abatement in agriculture comes from developing countries, 20 percent from OECD countries, and 10 percent from emerging countries. Furthermore, 89 percent of the mitigation potential is linked to soil carbon sequestration, predominantly in the developing world, with 9 percent and 2 percent of the total potential associated with lowering CH_4 and N_2O emissions, respectively. Figure 16.3 shows the technical and economic potentials of different agricultural mitigation practices to reduce GHGs.[73]

Clean Development Mechanism

The alternative to a cap and trade system is the baseline and credit scheme established under the Clean Development Mechanism (CDM) of the Kyoto Protocol of the United Nations Framework Convention on Climate Change (UNFCCC), which came into force in 2005. The protocol commits countries to reduce their GHG emissions by binding targets, and the CDM enables developed countries to meet their targets by funding emissions reduction projects in developing countries. Developed countries are provided with certified emissions reduction (CER) credits, each equivalent to 1 ton of CO_2.[76] A country like Sweden, for example, can fund the construction of a hydroelectric dam in China and receive credits to offset its own emissions of GHGs.

The CDM has been moderately successful in delivering completed projects, but it is a complex process; the greatest beneficiary so far has been China. Moreover, carbon sequestration in agricultural soils has so far been excluded from the CDM. Some African countries have prepared national strategy studies for CDM, but there has been little progress in implementation.[77] Expansion of the activities covered by the CDM—to include forestation and reforestation, soil carbon sequestration, green biofuels, and agroforestry on agricultural lands—could speed adoption of mitigation measures.

But the challenge for all these carbon market mechanisms is to scale up and encompass a wide range of people, locations, and environments.[78] Monitoring and verification at this scale are problematic. Standard methodology for measuring soil carbon sequestration has been to take soil samples, determine the organic carbon concentration and calculate the SOC mass from bulk density. The need for quicker, cheaper and more accurate methods, however, has inspired the development of new tools such as laser-induce breakdown spectroscopy and inelastic neutron scattering.[79]

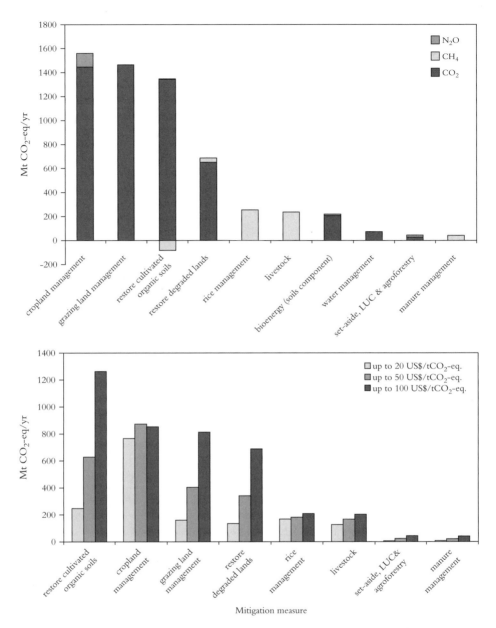

Figure 16.3 a. Technical mitigation potential of various approaches to GHG reduction by 2030.[74] b. Economic potential for GHG agricultural mitigation at various carbon prices given the social costs and benefits.[75] Both for the IPCC B2 scenario.

Trade-Offs

Finally, it is important to recognize that under the impact of climate change, adaptation and mitigation measures may involve trade-offs with each other and with food security.[80] For example, reduced tillage and use of organic residues can aid carbon storage and lessen the impacts of drought but may reduce yields. Conversely, breeding for higher yields under drought conditions may result in higher losses of carbon unless combined with forms of conservation farming. Interestingly, though, greater levels of microbial decomposition under global warming may reduce soil carbon sequestration potential.[81]

Both adaptation and mitigation will lead to winners and losers. There may be short-term losses followed by long-term gains.[82] A *whole systems approach*—incorporating both large-scale and locally specific economic development, and environmental conditions and their interactions with different mitigation measure portfolios—will be essential to minimizing the potential trade-offs.[83]

Using Biomass for Fuel

In addition to reducing greenhouse gases by modifying agricultural practices, there is a key role for agriculture in producing sources of energy that are either low carbon or even carbon free.[84] Green plant material is a valuable source of stored chemical energy, converted by the plants from the sun through the process of photosynthesis. Burning biomass (organic matter comprising plant materials and animal waste) releases that energy. Thus burning wood for heat and cooking was the first form of energy exploited by humans, and it remains an important component of global energy production today.

Bioenergy is any energy produced directly from biomass; it is considered a renewable energy source—as long as the materials are harvested sustainably, replaced, or grown for the specific purpose of energy production. Biomass fuels (biofuels) include the following:

- *Wood fuels:* wood, charcoal, and forestry residues (the most common type)
- *Crop fuels:* energy crops such as sugar and cereals; oil crops such as soya and jatropha (produced from the seeds of *Jatropha curcas*)
- *Agricultural and livestock by-products:* such as stalks, straw, and manure dung

These may be produced as solids (e.g., charcoal), liquids (e.g., alcohol, biodiesel), or gases (e.g., methane).

Currently, bioenergy accounts for about 9 percent of global primary energy supply, although there is a large variation between regions. As of 2001 developed countries obtained only about 5 percent of their energy from biomass, while developing countries averaged around 22 percent, with Asia at 19 percent and Sub-Saharan Africa at 60 percent.[85] In rural areas of developing countries, the figure is often as high as 90 to 95 percent.[86] Biomass is the largest used source of renewable energy both globally and in developing countries, dwarfing the still-growing hydroelectric, geothermal, wind, solar and tide renewable industries (Figure 16.4).

Using more biomass for energy holds potential for billions of people without access to electricity or improved cooking fuels. The challenge, however, is to make energy production from biomass sources more sustainable and efficient, causing less GHG emissions and, ideally, acting as an additional source of income and enterprise. Moreover, where crops are grown specifically for energy production, we have to ensure that they do not compete with food production or distort food prices in ways that harm the poor.[88]

A starting point is the improvement of the traditional practices of burning biomass for cooking using open fires. About three billion people rely on solid fuels such as wood, coal, and dung for cooking. Harvesting wood for stoves and for charcoal production degrades surrounding forests if it is not done sustainably. It can lead not only to a decline in environmental health and an increase in soil erosion, but it also increases the time that women must spend searching for fuel and poses human health risks. The smoke released by these fuels, when burned inside homes, more than doubles the risk of respiratory illnesses, such as bronchitis and pneumonia, leading to around 1.6 million deaths per year.[89] Rural women and children

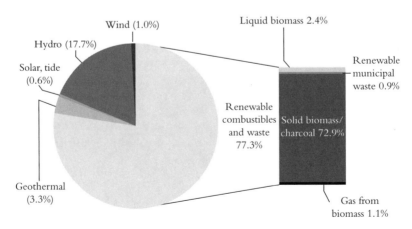

Figure 16.4 Biomass dominates global renewable energy supply.[87]

are most at risk. Interventions that can effectively reduce indoor air pollution include improving fuels, reducing the time needed for heating food, and improving
the cooking devices themselves.[90]

The Future of Biofuels

The alternative is to replace direct burning of biomass with its conversion to a fuel
that then can be used not only for cooking but for transportation and many other
energy uses. Both noncrop and crop plants can be grown specifically as feedstock for
conversion to biological fuels (biofuels). Depending on the nature of the feedstock,
either thermal or biochemical processes can be used for conversion; the number of
options is continuously growing. For example, wood or other solid fuels, such as
agricultural residues, can be partially burned at a high temperature to produce a synthetic gas. More commonly, manure (or other animal or vegetable waste) is converted
using the biochemical process of *anaerobic digestion*, where the organic materials are
broken down by microbial activity in the absence of air and at slightly elevated
temperatures. This produces biogas (60 percent CH_4 and 40 percent CO_2), which
can be used for cooking or lighting, fuel, or electricity generation using modified
gas engines or turbines.

Global production of biofuels has tripled from 18 billion liters in 2000 to around
60 billion liters in 2008, although their share in global energy supply is relatively
small, 3.5 percent in 2007. Production is centered around a handful of countries,
with the United States and Brazil accounting for around 80 percent of total biofuels
production. Global demand for biofuels is predicted to double between 2009 and
2015 to 184 billion liters.[91]

Of growing importance in the developing countries is the use of sugar crops
(e.g., sugar cane and sweet sorghum) and starch crops (e.g., cassava) for the production of bioethanol using fermentation. The feedstock is ground, the sugar is
dissolved out, and it is then fed to yeast in a closed, anaerobic chamber. The yeast
ferments the sugar to produce ethanol, which is then distilled to produce bioethanol. This can be used on its own in specialized combustion engines or mixed with
petrol. Drawbacks to bioethanol production include the competition it places on
land that could otherwise be used to grow food crops and the inefficiency of the
conversion process.

The alternative to bioethanol is to produce biodiesel from oils such as animal
fats, waste cooking oil, or rapeseed, sunflower, soya, palm, or coconut oil, which
can be converted through transesterification. A catalyst splits the oil into glyc-
erin and methyl esters so forming biodiesel that can be used as fuel in a diesel
engine. Engines can also be modified to run on less purified forms of oil, such as

Box 16.4   Jatropha Powered Electrification in Mali[92]

In the Garalo commune in southern Mali, thirty-three villages have successfully united in local energy production using oil from the seeds of the shrub *Jatropha curcas* to power a hybrid power plant.

Jatropha curcas, known as *bagani* in Mali, is an inedible perennial shrub that grows widely across the country, thriving in poor soils with little water. The small tree is often planted as a natural fence around crop fields, because its smell and taste repel grazing animals. Seed production starts within 6 to 9 months, and the plant reaches maturity in 5 years. It will remain in constant production for up to 25 years, and can live for up to 50 years.[93] The seeds contain over 30% of oil and 3 kg of seed will yield about 1 liter of high quality oil. The seeds are pressed using a locally manufactured machine, which itself is powered by jatropha oil (the energy used to press the nuts amounts to less than 10 percent of the oil obtained). The residues from the seeds are higher in nitrogen and phosphorous than cow dung, and so act as an excellent organic fertilizer that can be used on other food crops or sold.

Farmers are now intercropping *J. curcas* with their normal crops on land previously used for less profitable cotton production. The jatropha nuts are harvested by the farmers and processed by the local cooperative using the oil press; the oil is then sold to the local power company, ACCESS, which is running a 300 kW–capacity vegetable oil-diesel hybrid power plant. So far around 700 communities in Mali are using *Jatropha curcas* and local biodiesel generators to meet their energy needs. The Malian government has not approved export of the biofuel as it aims first to become energy self sufficient, powering all 12,000 villages in the country with affordable and renewable energy.[94]

straight vegetable oil or the oil from the seeds of plants such as *Jatropha curcas* (Box 16.4).

So far biofuel development in the developing countries has been mainly in Brazil and China. As of 2011 Brazil was producing 24 percent of the world's total biofuels, a share projected to increase to 27% by 2020. China is currently producing 6 percent of the global total but this share is expected to decrease to 4% by 2020.[95]

The production of biofuels is predicted to increase in Africa, with several international and regional initiatives' support. In Senegal, for example, the National Program for Biofuel Production was launched in 2006 with the goal of producing 1.2 billion liters of refined oil from 312,000 ha of *Jatropha* plantation. Many private biofuels projects are also occurring. In Zambia, the Zambian company Marli Investments initiated an outgrower *Jatropha* scheme in Kabwe, and, as of 2009, 12 million seedlings and seeds had been distributed to 25,000 outgrowers covering 18,500 ha. Marli Investments had planned to construct a biodiesel production plant by 2011, but the global recession has prevented its completion.[96]

The trade-offs between food security and energy security also need critical attention. There is a calculation—using World Bank and WHO data by Indur Goklany, a fellow at the Political Economy Research Centre of the World Bank—that the increase in poverty owing to growth in biofuels production over 2004 levels led to at least 192,000 excess deaths in 2010, a figure that exceeds the 141,000 deaths attributed to climate change.[97] Another study by the OECD found that if 2020 targets for biofuels were met, crop prices would increase by 30 percent, although the risk appears minimal if such targets are below 10 percent (i.e., the share of biofuels in a country's overall fuel use is 10 percent or less). Despite encouraging efforts, the lack of regional policies or strategies on biofuels leads to a lack of regulation and underinvestment into biofuels research and development.

The Challenges Ahead

If biofuel production is to move ahead, especially in the developing countries, several issues have to be resolved:

1. Accessible markets need to be created that will provide fair, stable, and relatively predictable prices.
2. Stable land ownership is required, bearing in mind the need for a balance between large and small farms.
3. Second-generation biofuel plants (i.e., highly efficient biofuel crops that do not compete with food crops [e.g., *Miscanthus*]) and their related conversion technologies need to be quickly developed.
4. Appropriate technical advice needs to be readily available.
5. The actual or potential competition between biofuel crops and food crops needs to be understood and minimized.[98]

As I mentioned in Chapter 1, there has been considerable argument over the impact of biofuel production on the food price spike of 2007–2008. While it may not be as high as first claimed, with increasing links between energy and food prices, biofuel policies will need to take food security into account. Joachim Von Braun of the University of Bonn, for example, suggests the removal of subsidies on biofuels during times of high food prices to prevent the diversion of crop production away from the food market.[99] It may be increasingly desirable to cultivate nonfood biofuel crops (e.g., *Jatropha*) or crops where the biofuel is a by-product (e.g., molasses from sugar cane).[100]

The Barriers

Despite the many examples of successful projects, overall there has been little progress in implementing large-scale mitigation measures given the potential. Barriers include limited access to technology and resources as well as lack of appropriate political, institutional, and economic policies.[101]

So far, international policies have not generally supported agricultural mitigation for developing countries. The agricultural sector has, to date, not been significantly included in climate change negotiations, such as meetings in Copenhagen in 2009 and Cancun in 2010. The large numbers of global stakeholders and the technical difficulties and economic costs of measurement, reporting, and verification are seen as formidable obstacles to large-scale mitigation efforts. However, limited progress is being made, and at the time of going to press, the preparation for the UN Climate Change Conference in Durban in December 2011 holds some promise.

17

Conclusion

Can We Feed the World?

To the people of poor nations, we pledge to work alongside you to make your farms flourish and let clean waters flow; to nourish starved bodies and feed hungry minds. And to those nations like ours that enjoy relative plenty, we say we can no longer afford indifference to the suffering outside our borders; nor can we consume the world's resources without regard to effect. For the world has changed, and we must change with it.

President Barack Obama, inauguration speech, January 20, 2009[1]

I am by nature an optimist (it is probably in my genes), so my answer to the question posed here and in the book as a whole is "yes." But it is a qualified yes. We will be able to feed the one billion chronically hungry and get to a food secure world in 2050, but only if we focus our efforts, provide sufficient aid and public and private investment, harness new technologies, remove trade restrictions, create appropriate enabling environments and governance, including efficient, noncorrupt and fair markets, and vigorously tackle climate change. This is a tall order, especially because in many directions our efforts have been lukewarm despite the promises. As U.S. Secretary of State Hillary Clinton puts it, "The question is not whether we can end hunger, it's whether we will."[2]

In this final chapter I review each of the qualifications to my "yes" answer.

We Can Feed the World If . . .

1. We Recognize Food Security Affects Us All and the Time to Act Is Now

There are about a billion people in the world who are chronically hungry. Nearly all of these are in the developing countries. Hunger particularly affects women and children. A third of all children under the age of five are stunted; that figures increases to as many as 50 percent in some African countries. Yet we know what to do. There are several interventions, medical and agricultural, that are evidence based and cost-effective.

At the same time, the poor of the developed countries are feeling the impact of higher food prices, and there is growing malnutrition in adults and children. Food insecurity—*a lack of access by all people at all times to enough food for an active, healthy life*—is becoming a global problem.

Political leaders, in both developed and developing countries have a new and growing commitment to agricultural development and combating hunger. Members of the African Union are committed to increase resources for agriculture and rural development. In Sub-Saharan Africa average annual growth in agriculture has been over 3 percent in recent years (between 2000 and 2008).[3] This combination is creating opportunities in domestic, regional, and international agricultural markets, especially where there are supportive, stable governments.

The time to act is now.

2. We Acknowledge the Challenges We Face Are Unprecedented and Require Concerted Action

There are three major, interconnected challenges for food security and agricultural development: (1) Food price spikes similar to those of 2007–2008 and 2010–2011 are likely to reoccur, partly because commodity markets have become more volatile and, at least in the medium term, extreme climatic events will be more frequent and severe. (2) About a billion people are chronically hungry in the world—more than at any time in history. (3) We have to approximately double food production if we are to feed the world by 2050.

We need to produce more food so that there is enough available both globally and in every community, recognizing that access to food is as important as its production. We also need to target those most deprived populations, whether in emerging economies or in the poorest countries. To feed the increased population by 2050 is on one level relatively straightforward (an increased annual production of about 1 percent per year) but is compounded by a variety of factors—changing diets, increased livestock consumption, high fertilizer prices, lack of good-quality land and water, and, most important over the longer term, the threat of climate change. It will not be enough to simply meet annual demand; we need reserve capacity to respond to shortfalls produced by extreme climatic events.

These are formidable challenges and must be tackled by concerted action. Far too often individual donor countries pursue favored elements of development that are often idiosyncratic. Donors need to actively support the UN Committee on World

Food Security (CFS) as the overarching strategic body and the Global Agriculture and Food Security Program (GAFSP), which is a channel for new multilateral funding.

Poor and hungry people around the world deserve better-focused and coordinated aid.

3. The World's Leading Donors of Aid Implement Their Commitments to Food Security

The food price spikes of 2007–2008 and 2010–2011 served to focus global attention on the scale of the problem of food insecurity. After decades of neglect, aid for agricultural development began to show significant growth (Figure 17.1).

The strongest commitments have come from the United States and the European Union. USAID has developed a new five-year program, Feed the Future, with an emphasis on supporting some twenty country-owned food security investment plans. The U.S. government provided $1.31 billion to this program in 2010, with allocations made on an annual basis by Congress. The U.S. administration requested $1.84 billion requested in 2011, and $1.55 billion in 2012.[5]

The European Union now gives about $1.75 billion a year to agricultural development with nearly 50 percent going to Sub-Saharan Africa. It created a €1 billion Food Facility to help developing countries combat the effects of the 2008 food price

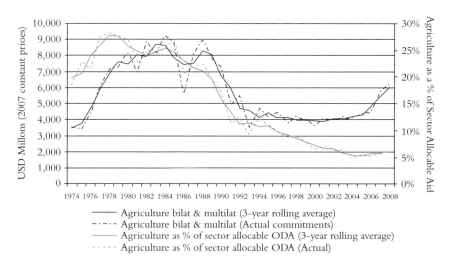

Figure 17.1   Long-term trends in global agricultural aid. ODA=Overseas Development Assistance.[4]

spike, and has an ongoing programmatic focus on agricultural development through the Food Security Thematic Programmes (FSTP).

But for most donors the G8 pledges at L'Aquila in 2009 have not been realized. According to a report commissioned by the ONE campaign group, donors have disbursed only 22 percent of their pledges to date, and most have not reported how they plan to reach the full pledged amount.[6]

The rural poor of the developing countries expect and deserve donors to honor their commitments.

4. The Doha Round Is Completed with Satisfactory Outcomes for Developing Countries

In economic theory a liberal trading system, free of subsidies, can lead to a more efficient allocation of resources, and food prices should be lower than they might otherwise be. The Uruguay round of negotiations (1986–1995) under the World Trade Organization made some progress, and many developing countries are better off as a result. But the round left much unfinished business.

The Doha Round of trade negotiations began in 2001, but for much of the past decade the talks have been stalled (Chapter 4). The negotiations are riven by arguments between different power blocks and even within individual groupings (e.g., Brazil, India, and China). Despite being in the majority, the developing countries have had little influence. The US and the EU will make concessions only if they can gain from greater liberalization in manufacturing, services, or investment or in new rules for intellectual property, labor standards, and environmental protection. Then they will provide freer access to developing countries, as is urgently needed; for example, in sugar and cotton where the distortions are severe and, for African countries, crippling.

As many have remarked, there is something deeply hypocritical about the developed countries giving, apparently generously, considerable sums in aid while simultaneously denying developing countries access to their markets. At the time of writing, the aim was to complete the talks by the end of 2011, but so far little progress has been made.

Satisfactory completion of the Doha Round is a critical test of the resolve of the global community to tackle the growing challenges of food security and agricultural development.

5. There Is Explicit Attention to the Creation of Enabling Environments

Where trade liberalization has occurred, the experience of developing countries has been mixed. The evidence is that critical to success is an enabling environment, since only then will farmers invest in new technologies and purchase fertilizers and seed in the confidence they will get a decent return (Chapter 8).

An *enabling environment* has been defined in many ways, but in simple terms embraces the macroeconomic policies that favor markets and trade, the provision of inputs and related physical infrastructure (e.g., roads and irrigation) and social infrastructure (e.g., education, research), together with the accompanying institutions and regulations. Which of these investments are needed and which are most cost-effective will vary from country to country. But in all environments, connecting farmers to rural and urban markers—either physically, for example through all weather roads or virtually through some form of information and communication technology (ICT; see Chapter 7)—is crucial.

Some of the needed investments are public goods and provided by governments, but others are the province of the private sector. Unfortunately many donor governments are selective about what they will fund. Primary education and agricultural research get considerable attention, but infrastructure such as roads and irrigation tend not to be in favor. The priorities have to be set and owned by developing country governments; for example, as part of the process of the Comprehensive Africa Agricultural Development Program (CAADP).

Enabling environments underpin sustainable agricultural development, and donors need to be explicit in their support for the necessary strategies and investments.

6. The Appropriate Governance for Food Security and Agricultural Development Is in Place

Appropriate enabling environments require appropriate governance. The necessary macroeconomic policies and physical and social infrastructure, together with the accompanying institutions and regulations, will be created only if there is a supportive governance structure. In turn this requires leadership that is explicitly committed to dealing with food security and promoting agricultural and rural development and is genuinely responsive to the needs of farmers and consumers.

A key precondition is political stability. Some of the most food insecure countries today are unstable. Seventeen of the forty-one countries in Sub-Saharan Africa are, or have recently been, categorized by the World Bank as "chronically, politically instable."[7] In many cases underlying the instability is conflict, either within or between countries. Political instability means that institutions do not function efficiently or fairly. Poor governance is frequently characterized by corruption and the misuse of public power. As a result the state is unaccountable and unpredictable, resulting in malfunctioning public services.

333

In contrast, good governance entails creating efficient institutions that promote rural development and attract private investment whether by farmers or the private sector as a whole. At the same time, good governance means creating policies, laws, and regulations that ensure markets and financial institutions function fairly, protecting women, children, minorities, and the poor from discrimination and exploitation.

Without good governance there can be no food security.

7. Fair, Efficient Output and Input Markets Are Created on a Countrywide Basis

A key component of an enabling environment is the creation of markets; only through access to markets can poor farmers increase the income from their labor and lift themselves and their families out of poverty. Yet, most poor farmers are not linked to agricultural markets. Smallholders, in particular, often have little contact with the market and hence a poor understanding of, and ability to react to, market forces.

Developing country markets are changing rapidly under the influence of a myriad of factors—urbanization, population growth, increasing per capita incomes, changes in consumer preferences, the modernization of food processing and retailing, and improvements in transport and communications infrastructure. These offer new opportunities for smallholder farmers, but at the same time create heightened risks and new and difficult barriers to surmount.

Part of the answer lies in helping to create countrywide networks of small and large markets supported by commodity exchanges. Also needed are countrywide networks of village-level agrodealers, small mom-and-pop stores, selling key inputs to farmers in small, affordable quantities, providing extension advice, and, most significantly, reducing the distances farmers have to go to get inputs and advice.

Access to input and output markets provides not only greater food security but a route out of poverty.

8. Greater Attention Is Paid to Gaining Increased Value for Farmers through Producer Associations and Widespread Availability of Microinsurance and Microcredit

If value chains are to offer proper opportunities for growth, then those markets in which smallholders can have a "comparative advantage" need to be identified and they have to be actively helped to capture a fair share of the improved value and equity. One way for smallholders to access the benefits from value chains is to be part of a producer association, such as a farmer cooperative or a farmer group engaged

334

in contract farming. These can, if carefully established, bring about higher incomes and reduced risks through economies of scale. So far, however, they have had mixed success. Getting better deals for farmers may require more aggressive measures, such as through trade unions, state intervention, or fair trade agreements.

Functioning markets inevitably require farmers to take the risk of purchasing inputs. This risk is lessened if they have some form of insurance should the crop fail. The most promising route is through the growth of the microinsurance industry that today reaches an estimated 15 million low-income people. In many instances farmers also do not have the cash available to buy inputs. Typically, appropriate loans are very small, yet are relatively costly to service and hence are not attractive to commercial banks. However, there is now sufficient experience to demonstrate the effectiveness of local, self-managed credit groups run by NGOs or private banks. Smallholders can also access credit from other sources; for example, warrantage and warehouse receipt systems (see Box 8.7).

Although these needs will be fulfilled essentially by the private sector, government support is necessary to establish smallholder-friendly regulatory and political environments, to encourage private-sector investment in rural producers, to help producers organize themselves, and to support rural infrastructure development.

Countrywide systems of farmer associations, microinsurance, and microcredit schemes offer the best route for smallholders to better gain from value chains.

9. We Acknowledge the Key Role of Agriculture in Development

Appropriate enabling environments will be created only if governments, both of developing countries and of donors, explicitly recognize the important role of agricultural development in the development process. Agriculture dominates the least-developed countries, typically accounting for over 80 percent of the labor force and 50 percent of GDP. While this may be seen as a sign of "backwardness," it has a potential strength since any individual small improvement, say in a crop yield, can be multiplied throughout the agricultural economy. Productive innovations can lead to fast and inclusive growth. A 1 percent gain in GDP originating from agriculture will generate a 6 percent increase in overall expenditure of the poorest 10 percent of the population; in contrast a 1 percent gain in GDP originating from nonagricultural sectors creates zero growth.

In effect agricultural development drives a virtuous circle. As agriculture develops—greater yields for both subsistence and cash crops—farmers become more prosperous, and the rural poor, whether landless or on smallholdings, also benefit through wage labor. Chronic hunger decreases. The rural economy also grows—through the creation of small rural businesses—providing more employment and improved rural facilities, especially schools and health clinics. Roads and markets develop so that the rural economy connects to the urban economy and to the growing industrial sector. Free trade provides opportunities for greater imports and exports. In particular high-value agricultural exports can accelerate agricultural development, further intensifying the virtuous circle.

Agricultural development is the most powerful means of lifting rural communities out of poverty.

10. We Recognize the Need for a New Doubly Green Revolution

In this book, I present a theory of change as a way of creating greater programmatic coherence and, hopefully, better outcomes. The theory begins with the history of the Green Revolution, which successfully used new technologies to increase the yields and production of the staple crops at a rate that exceeded the population growth rate and also brought down their prices. Yet it also recognizes that the Green Revolution had important limitations: Its impact on the poor was less than expected, it did not reduce (and in some cases encouraged) natural resource degradation and environmental problems, its geographic impact was localized, and there were eventually signs of diminishing returns.

My argument is that, given the nature of the current food security crises, we need another revolution, but not one that simply reflects the strategies and outcomes of the first. It must not only benefit the poor more directly but also be applicable under highly diverse conditions and be environmentally sustainable. In effect, we require a Doubly Green Revolution, a revolution that is even more productive than the first Green Revolution, even more "green" in terms of conserving natural resources and the environment, and even more effective in reducing hunger and poverty. The route is an agriculture that is:

- *Highly productive*—able to double food production and in an efficient manner by 2050
- *Stable*—less affected by the vagaries of the weather and the market
- *Resilient*—resistant to, or tolerant of, stress and shocks, especially those generated by climate change
- *Equitable*—providing accessible food and incomes, not just to the better off, but to the poor and hungry

Moreover it has to be:

- *Sustainable*—We must achieve a pattern of equitable growth that lasts from generation to generation and ensure we do not undermine the environmental and natural resource base on which agriculture depends.

The Doubly Green Revolution is the theory of change that will deliver sustainable food security and agricultural development.

11. There Is Explicit Recognition of the Critical Role of Smallholder Agriculture

The theory also acknowledges smallholders as the key agents of change in the Doubly Green Revolution. The vast majority of farmers in the developing world are smallholders; very approximately there are 400 to 500 million small farms (i.e., less than 2 ha in size) in the world. As has long been recognized, smallholders are in many respects highly efficient. Small farms produce more per hectare than large farms. Achieving a growth in total global production is important because it will ensure there is enough "to go round." But this is a necessary but not sufficient condition for agricultural development.

For poor farmers and the landless, it is labor productivity that is the most crucial since it determines whether they escape poverty. Labor productivity needs to be over $700 per worker per year to get them and their dependents out of poverty. For half the countries of Africa, the current average is less than $350. In many respects this challenge lies at the heart of the Doubly Green Revolution. How can we increase the returns so that smallholder farming becomes a route out of poverty? It is not enough to help smallholders achieve subsistence, even if it is sustainable. They also need incomes—not least to pay medical bills and schooling, as well as to pay for food when harvests are poor. The goal of the Doubly Green Revolution is not just sustainable existence but sustainable development.

Some analysts argue this can only occur through large-scale farming. In practice the way forward lies not in *either* small *or* large farms, but in making a deliberate choice of *both* small *and* large farms. They both have a role to play in agricultural development.

Smallholders, because of their sheer numbers, the total land area they occupy, and their efficiency will play the dominant role in agricultural development for several decades to come.

12. More Attention Is Paid to Agroecological Research and Development

There is little likelihood of significantly more arable land becoming available; for the future the only solution to the food security problem is to get more production out of the existing land, but to do it in a way that is sustainable. This will depend on human ingenuity, in particular in harnessing the benefits of ecological processes. Agronomy and husbandry comprise the tools whereby modern ecology is applied to agriculture. They are used to recognize and identify relevant environmental services and exploit them in a productive, sustainable manner. Some examples of modern agroecological practices that require more funding for research and for implementation are:

- *Intercropping*—growing of two or more crops simultaneously on the same piece of land
- *Rotations*—growing of two or more crops in sequence on the same piece of land
- *Agroforestry*—intercropping annual herbaceous crops with perennial trees or shrubs
- *Silvopasture*—combining trees with grassland and other fodder species on which livestock graze
- *Green manuring*—growing of legumes and other plants to fix nitrogen and then incorporating them in the soil for the following crop
- *Conservation systems*—minimum tillage or no-tillage before sowing
- *Biological control*—use of natural enemies, parasites, or predators to control pests
- *Integrated pest management*—use of all appropriate techniques of controlling pests in an integrated manner

Agroecological principles and practices underpin sustainable intensification.

13. There Is a Major Focus on Getting Poor Rural People Out of Poverty

Sustainable intensification is also about people. It depends crucially on how farm households (farmers and their families) take their environment—physical, biological, social, economic, and political—and wrest from it a sustainable living. Farm households are usually complex entities, often extending through brothers and sisters and their families to embrace a considerable number of people. While agriculture has sometimes been seen as essentially a male activity, in recent years there has been growing recognition among the development community of the critical importance of women in farm households in shaping rural livelihoods.

Farming may be at the core of a livelihood, but nonagricultural activities also play an important role, especially as farm households move out of poverty. A critical element is the diversification of income by establishing links with the urban economy. Households who have escaped from poverty report a member who has obtained a job, mostly in the private sector; some have established a craft or trade in a city, or built trading links. A significant number have established a small business in the neighborhood of the village. Almost invariably, households who have made a successful entry into the urban economy, whether the informal or formal sector, possess a privileged connection, for example an uncle or cousin, who could take a new entrant under their wing. These pathways out of poverty need to be explicitly recognized, and rural development designed to promote these necessary linkages.

The route out of poverty for the rural poor depends on establishing connections with urban communities.

14. Technologies for Agricultural Development Are Developed and Applied, Providing They Are Locally Appropriate, Whatever the Source

Whatever the source and wherever the application, the important feature of a technology is that it is *locally appropriate*; meaning it has to be:

- Effective
- Readily accessible and affordable
- Easy to use
- Environmentally friendly
- Serves a real need

This means seeking interventions that are productive, stable, resilient, and equitable. Depending on the circumstances, any of the following technologies may be appropriate:

- *Conventional:* Technologies from industrialized countries developed through the application of modern physical, chemical, and biological knowledge; delivered as products in a packaged form for a regional or global market (e.g., *synthetic fertilizers but placed with precision*)
- *Traditional:* Technologies that have been developed, usually over an extended period of time, by communities in developing countries to meet local needs (e.g., *selection and cultivation of land races of rice and other crops*)
- *Intermediate:* Traditional technologies that have been improved in appropriate respects by their integration with modern conventional technologies (e.g., *traditional treadle pump improved by engineers*)
- *New Platform:* New scientific "platforms" for innovation (nanotechnology, biotechnology and ICT), based on advanced sciences (e.g., *genetically modified, drought-, or pest-resistant crop varieties*)

Appropriate technologies and other interventions are necessary to achieve sustainable intensification.

15. We Accept That Biotechnology Is an Essential Tool in Attaining Food Security

For most agroecosystems, the only way to increase labor productivity is to raise yields. We urgently need new high-yielding varieties of staple crops (and of livestock breeds) that make more efficient use of sunlight, water, and nutrients and so increase the yield potential, but this will not be easy. At the same time we should be concentrating on decreasing the yield gaps for rain-fed crops. This will also be difficult, but the gains may be greater.

Conventional plant breeding—for example, in the production of hybrid maize and rice—has contributed a great deal to food security over the past 150 years. It still has much to offer and will remain a mainstay of breeding for the foreseeable future. Nevertheless, conventional techniques have practical limitations. The crossing of two parent plants, each with desirable characteristics, in the expectation of producing offspring with a new, improved combination of these characteristics is essentially a random process. There are also natural limitations to conventional plant breeding. Traits that a breeder wishes to incorporate in a plant or an animal may not be present in any variety, breed, or species with which a cross can be made, although they may occur in quite unrelated species. Conventional plant breeding is also a relatively slow process.

It is for these reasons that the techniques of biotechnology are already having a significant impact on plant and animal breeding, benefiting poor people as well as those who are better off. There are three practical techniques (discussed in Chapter 9):

- *Marker-assisted selection*—e.g., to cross high-quality maize with local varieties
- *Cell and Tissue culture*—e.g., hybridization of Asian and African rices; generation of disease-free bananas
- *Recombinant DNA* technology (also known as genetic engineering [GE] or modification [GM])—e.g., in producing pest-resistant cotton and drought-resistant maize

The potentials for genetic engineering are almost endless. But there are hazards: some apparent, others perceived. Concerns over adverse effects on human health exist, but these are largely unfounded (see Chapter 9). Perhaps of greater concern is the possibility of genes moving from GM crop plants to other crops, such as organic varieties or wild relatives, creating unforeseen consequences. Genes can indeed move in this way. Finally there is concern that the technology has been captured by large, multinational corporations and is consequently exploitive. However, many of the new GM varieties arise from research in developing country government laboratories.

Biotechnology is not a magic bullet, but it has much to offer where appropriate.

16. There Is More Funding for Improving Mixed Livestock Systems

The Green Revolution was not the only agricultural transformation of the twentieth century. Less well understood and celebrated was a revolutionary growth in livestock production. The popularity of livestock production and consumption is not too difficult to comprehend. Livestock contribute 40 percent of the global value of agricultural output and support the livelihoods and food security of almost a billion people. Crucially they contribute 15 percent of total food energy, 25 percent of dietary protein, and provide essential micronutrients. Livestock systems vary enormously in terms of scale and intensity of production but essentially comprise: grazing systems, mixed systems, and industrial systems (see Chapter 10).

A fundamental principle of mixed systems is the mutual utilization of waste—crop waste to feed animals and livestock waste to provide nutrients for crops—but there are many other cross-benefits. In recent years, there has been a progressive evolution from grazing toward mixed farming systems. Arable farmers need to increase production; the presence of animals on a farm can provide wastes for fertilization and a useful form of traction, allowing for cultivation of more land. Crop residues are increasingly collected from the arable fields and transported to the homestead to feed animals that are kept in kraals or stables. Eventually, farms adopt zero grazing, with animals fed on residues or grain, grown on the farm or purchased. In many developing countries, the most integrated mixed farms are those centered on the keeping of dairy cows or goats.

Arguably the biggest constraint on mixed livestock farming is the quality of the breeds and the availability of fodder. In recent years there have been a number of sophisticated cross-breeding schemes involving taurine and zebu cattle aimed at high milk production. There is an urgent need to develop better-quality fodder and feed and improved livestock breeds for mixed systems, if the growing demand for livestock products is to be met without increased pressure on land and environmental resources.

Mixed livestock systems can deliver multiple benefits on a sustainable basis.

17. We Recognize the Role of Farmers as Innovators

Far too often we assume that agricultural research and development programs are there simply to teach farmers methods of becoming more productive. But the practice of agriculture is intrinsically dynamic, requiring constant adaptation to changes in climate, in the resource and nutrient base, in pest and disease dynamics, and in the economy. Farmers know this and can articulate the challenges they face and be innovative in their responses.

One challenge is to replace the traditional top-down training and visit extension system with an approach that uses the range of participatory methods now available (see Chapter 11). A similar challenge applies to breeding. Breeders usually try out their selections in farmers' fields to determine acceptability. Customarily this has occurred at the end of the breeding process when many of the key decisions have already been made. What has changed in recent years is a greater involvement of farmers earlier in the process, eliciting not only reactions but positive inputs into the determination of breeding goals. Several approaches now engage farmers more intimately in analysis of trials. Farmer innovation can take many forms: sometimes it may be the process of discovering something entirely new; more often it involves finding novel applications for existing technologies.

Farmer experimentation and innovation needs to be recognized as one of the keys to sustainable agricultural intensification.

18. There Is Increased Support for Integrated Pest Management (IPM) Systems

Pesticides continue to be an unsatisfactory approach to pest control: First, they may be harmful to human health; second, they may be damaging to wildlife and the environment; and third, they may be costly and not very effective at controlling pests.

Integrated pest management (IPM) has been developed over the past fifty years as an effective tool that looks at each crop and pest situation *as a whole* and then devises a program integrating the various control methods in light of all the factors present. As practiced today, it combines modern technology, the application of synthetic, yet selective, pesticides, and the engineering of pest resistance with natural methods of control, including agronomic practices and the use of natural predators and parasites. It has a good track record, but success depends on a relatively high level of skill and is often labor intensive. Not surprisingly, pesticide spraying appears to be an easier approach.

In recent years, there have been dramatic results from using *Bt* genes in the control of serious pests (e.g., cotton bollworms) formerly subject to very high pesticide spraying. This technology is not a magic bullet, and there are potential problems of resistance to *Bt* and of stimulating secondary pest resurgence. Nevertheless, *Bt* and other forms of biotechnology-bred pest-resistant varieties are readily taken up by farmers.

A synergistic combination of biotechnology and IPM could be highly effective, relatively cheap, and environmentally friendly.

19. There Is Widespread Adoption, in Appropriate Environments, of Systems of Conservation Farming

One of the biggest and most difficult challenges facing smallholders is the achievement of a better soil environment for their crops. A good soil has two aspects: one qualitative, consisting of a good supportive structure, and one quantitative, comprising an optimal amount and mix of macro- and micronutrients. Estimates of land degradation vary widely; soil erosion is easier to define and measure, and in many developing countries is a serious problem.

Much time, money, and effort has gone into soil conservation measures, but too often the technologies are inappropriate. In recent years, an essentially ecological approach is gaining rapid popularity under the name "conservation farming." At its core is either the elimination or reduction of tillage. The advantages include the saving of labor used for plowing, protection of vulnerable soils from erosion by keeping topsoil from being blown or washed away, and improved soil fertility by keeping soil structure intact and by allowing more beneficial insects to thrive. It also keeps carbon and organic matter in the soil, leading to a higher microbial content as well as sequestering carbon and thus reducing carbon emissions from agriculture.

In practice, for example in western Zambia, next year's seed is sown in small "pockets" in the soil to which have been added two cupfuls of manure and a bottle top of fertilizer. After harvest, the soil is covered with the stems and leaves of the maize and next year's seed is sown several months later in the same holes. Yields are high—some 4 to 5 tons of maize growing new drought-tolerant hybrids, creating a more stable and sustainable farming system.

Conservation farming has all the elements of a truly win-win-win practice.

20. We Focus Our Investments on Small-Scale Water Harvesting and Community Water

The use of water has always been subject to conflict, but this is intensifying as a result of rapid urbanization and industrialization as well as global warming. Agriculture itself is a major contributor because of excessive withdrawal of both surface and groundwater. About 15 to 35 percent of total global water withdrawals for irrigated agriculture are unsustainable, and an estimated 1.4 billion people live in river basins with high environmental stress.

Tube wells have several advantages: They can be relatively easy and cheap to install and, because they occupy little land, create few environmental and social problems. But—and this is crucial—they will provide a sustainable supply of irrigation water only if the rate of extraction is below that of the rate of recharge to the underground aquifers from which the water is being obtained. Water tables are falling in some areas but rising in others, creating serious problems of waterlogging and salinization. The common cause is a combination of excessive use of water and poor drainage. Cheap, subsidized electricity encourages profligate water use.

Although technical solutions are available, the root causes of these economic, social, and environmental problems are bad irrigation design and poor maintenance and management. The alternative lies in exploiting the opportunities for small-scale water harvesting and small, community managed and designed irrigation systems. The lesson from the experiences of the past thirty years is that these are more likely to deliver sustainable water supplies.

Small-scale water systems, if communally managed, are the most promising approach to sustained water use.

21. There Is Significant Investment in Agricultural Adaptation to Climate Change

Global climate change has been largely driven by the activities of the industrialized countries. Yet its most severe consequences will be and, indeed, are already being felt by the developing countries. Moreover, it is the poor of those countries who, in part because of their poverty, are most vulnerable. If left unchecked, climate change will increase hunger and cause the further deterioration of the environmental resources on which sustainable agriculture depends.

Agriculture, as a sector, is particularly vulnerable to climate change. Many smallholders rely on natural rainfall and are vulnerable to quite small changes in rainfall patterns. Climate change is likely to bring shorter growing seasons in the future, expanding drylands over a larger area. Many parts of the developing world are already experiencing water shortages, and these may increase in scope and severity. Rates of photosynthesis are lowered by high temperatures and lack of water. Many crops are already grown close to their limits of thermal tolerance. It is estimated that a 1°C warming could result in a 20 percent loss of maize yield in three quarters of Africa's entire maize crop.

We must design adaptation measures to cope both with relatively predictable climatic stress and with much less predictable but even more catastrophic extreme climatic events. Drought resistance, for example, can be achieved in one of three ways:

- Drought-tolerant cropping systems; e.g., conservation farming
- Drought-resistant crop varieties, utilizing genes such as the chaperone gene
- Provision of water resources; e.g., small-scale water harvesting technologies

Traditionally poor people have developed various forms of resilient livelihood strategies to cope with a range of natural and manmade stresses and shocks. But these may be inadequate in the future or may have been lost in the development process. The urgent need is for governments, NGOs, and the private sector to work together with local communities to enhance the resilience of the poor.

Future food security depends crucially on appropriate adaptation to climate change, utilizing the best of science and the innovativeness of local communities.

22. There Is Urgent Attention to Financing the Reduction of Greenhouse Gas Emissions from Agriculture

While agriculture is a major victim of climate change, it also has considerable potential, albeit largely unrealized, to contribute to mitigating climate change by reducing greenhouse gas (GHG) emissions. The less-developed countries, on a country-by-country basis, contribute relatively small amounts of GHGs to total global emissions. Collectively, however, they are a significant contributor through the processes of forest clearance and agricultural development. Most of agriculture's contribution to global warming (leaving aside the effects of land clearing) comes from nitrous oxide and methane emissions. If the price of carbon is high, and there are no economic or other barriers, the Intergovernmental Panel on Climate Change (IPCC) believes agricultural emissions could be reduced or offset by around 5,500 to 6,000 million tons of carbon dioxide equivalent (CO_2 eq.) per year by 2030.

There is a long list of potential measures that can be taken, some more effective than others. The priority is to identify those that will provide benefits additional to mitigation, including improved adaptation to climate change. The goal is to find win-win technologies that help keep costs down and the benefit-cost ratios high. However, significant progress will not be made until systems are found that will directly reward smallholders for mitigation measures such as carbon sequestration and reduction in nitrous oxide and methane emissions. Existing approaches, involving carbon markets, do not seem to be enough. Scaling up mitigation also depends on finding satisfactory methods of monitoring and verification on a medium- to large scale.

Agriculture can contribute significantly to mitigating the emissions of greenhouse gases and hence reduce global warming, providing appropriate financial incentives are found.

23. We Invest in Scaling Up as a Route to Rapid Success

Finally if we are to achieve a Doubly Green Revolution, we need to find ways of replicating success rapidly. In simple terms the question is: If a local community, perhaps aided by the national government, by an aid donor, or by an NGO, has been able to build a productive, stable, resilient, and equitable system of technologies and/or processes that works and appears sustainable, can it be replicated on a much larger scale to benefit not just hundreds but many thousands of smallholders?

The challenge is to help take these local systems, especially those that significantly increase farmer and laborer incomes, and to scale them up so that the poor—both farmers and the landless—can benefit in a way that brings about the kind of virtuous circle of rural development I described earlier.

Although there is considerable experience of going to scale, there is no simple recipe. However, some principles are beginning to emerge:

1. The private sector has much of the necessary experience, skills, and processes to make scaling up work. This is primarily because any agricultural technology or process that significantly increases income is "marketable" and hence saleable.

2. Rarely can the private sector, whether indigenous or foreign, be left to itself to bring about the scaling-up transformation. In most cases, there has to be a public-private partnership. Sometimes this will consist simply of governmental action to provide the right kind of enabling environment for the private sector to operate. In others, more formal public-private-community partnerships are required that harness the different qualities and comparative advantages of the relevant actors.

3. Each value chain is likely to be different. For instance, scaling-up practices will vary for livestock, export high-value crops, and local staples.

4. If equitable benefits are to be derived, the value added needs to be biased to the lower levels of the value chain, e.g., to poor smallholders. Scaling up cannot benefit only the larger, better-off producers.

5. Related to the previous point, there is likely to be a significant role for farmer associations, cooperatives, and other bodies that will fight to ensure the benefits are widely shared.

346

6. Finally, much of the success of scaling up depends on the details of the pathways, processes, and deals struck between the partners.

Going to scale is the quickest route to achieving the goals of the Doubly Green Revolution.

24. We Recognize That Public-Private-Community Partnerships Are Crucial to Successful Scale-Up

Public-private-community partnerships can help link smallholders to markets. Essentially they are cooperative ventures between the public and private sectors and local communities designed to ensure that the expertise of each partner is used to meet commonly defined goals. The social welfare goals of the public sector can be complemented by the innovation and results- and demand-driven goals of the private sector. For example, private-public-community partnerships hold a great deal of potential for helping smallholder cooperatives gain access to input and output markets because these partnerships can sell and distribute seeds, fertilizer, and crop products at fair market prices. The private sector has also typically played a dominant role in offering financial services, but the government can help smallholders access credit through credit subsidy and by supporting the initial development of financial services into the rural economy. Creation of fair and sustainable partnerships between the public and private sectors and local communities is critical to achieving food security and sustainable agricultural development.

Conclusion

We can feed the world, but it will not be easy. There are no magic bullets. The answer lies in seeking win-win-win solutions where there are economic, social, and environmental benefits. These are often elusive and difficult to implement. Yet that has to be the direction of travel.

In summary, to achieve a Doubly Green Revolution, we need:

- To act *holistically* to tackle multiple interacting factors and outcomes and their trade-offs
- To act *now*—it's urgent
- To act *wisely*: We need all options on the table and the knowledge about how to use those options appropriately
- To act on all fronts:

- Creating enabling environments using public and private investments
- Utilizing modern laboratory and field knowledge in both the biological and socioeconomic sectors
- Developing appropriate technologies, whether conventional, traditional, intermediate, or new platforms
- Involving local people and community groups
- Using different modalities as appropriate (e.g., sustainable intensification, value chains, agrodealers, contract farming, cooperatives.)

Not only is it necessary, but it is right to act. We have a responsibility to ensure that the world's most vulnerable are not victims of our failure to act.

As Thomas Jefferson said in a letter to George Washington, "Agriculture . . . is our wisest pursuit, because it will in the end contribute most to real wealth, good morals, and happiness."[8] We have to combine this ancient wisdom with the new wisdom that derives from modern science and technology and the contemporary experience of creating markets that are efficient and fair, benefiting poor people, and especially women and children. If this book is successful, we will have helped to make the world a better place for our children, grandchildren, and great-grandchildren.

References

Front Matter

1. Conway, G., Lele, U., Peacock, J., and Piñeiro, M. 1994. *Sustainable agriculture for a food secure world*. Washington, DC: Consultative Group on International Agricultural Research and Stockholm: Swedish Agency for Research Cooperation with Developing Countries.

1. Acute and Chronic Crises

1. Bennett, G. 2008. *Haitians riot over sharply rising food prices*. National Public Radio (NPR). http://www.npr.org/blogs/newsandviews/2008/04/haitians_riot_over_sharply_ris .html.

2. Dictionary.com. 2010. "Hunger." Online Etymology Dictionary. Douglas Harper, historian. http://dictionary.reference.com/browse/hunger.

3. Smith, M., Pointing, J., and Maxwell, S. 1992. *Household food security, concepts and definitions: an annotated bibliography*. Development Bibliography No. 8. Brighton, UK: Institute of Development Studies, University of Sussex.

4. Maxwell, S. 1996a. *"Walking on two legs, but with one leg longer than the other": a strategy for the world food summit*. Mimeo. Brighton, UK: Institute of Development Studies, University of Sussex.
   Maxwell, S. 1996b. Food security: a post-modern perspective. *Food Policy* 21:155–170.

5. World Bank. 1986. *Poverty and hunger: issues and options for food security in developing countries*. Washington, DC: World Bank.

6. Conway, G., and Toenniessen, G. 2003. Science for African food security. *Science* 299:1187–1188.

7. Trostle, R. 2008. *Global agricultural supply and demand: factors contributing to the recent increase in food commodity prices*. Economic Research Service Report WRS-0801. Washington, DC: United States Department of Agriculture (USDA).

8. Food and Agriculture Organisation. 2011. *Food Price Indices*. http://www.fao.org/world foodsituation/FoodPricesIndex/en/.

9. World Bank. 2010. *Food price watch*. February 2010. Poverty Reduction and Equity Group (PREM). Washingotn, DC: World Bank.

10. OECD-FAO. 2009. *Agricultural outlook 2009–2018*. Paris: OECD.

11. Piesse, J., and Thirtle, C. 2009. Three bubbles and a panic: an explanatory review of recent food commodity price events. *Food Policy* 34:119–129.

349

Mitchell, D. 2008. *A note on rising food prices.* Policy Research Working Paper 4682, Development Prospects Group. Washington, DC: World Bank.

12. Piesse and Thirtle. 2009. Three bubbles.

   Trostle, R. 2008. *Global agricultural supply and demand.* Reprinted with permission from Elsevier.

13. Mitchell. 2008. *A note on rising food prices.*

14. Piesse and Thirtle. 2009. Three bubbles.

   Abbott, P., Hurt, C., and Tyler, W. 2008. *What's driving food prices?* Issue Report 75, Oak Brook, Illinois: Farm Foundation.

   Trostle, R. 2008. *Global agricultural supply and demand.*

15. Foresight. 2011. *The future of food and farming.* Final project report. London: Government Office for Science.

16. The Economist. 2010. *Wheat: straw man.* http://www.economist.com/node/16792838.

17. Bals, J. 2010. Weather woes spark fears for wheat. *Financial Times.* http://www.ft.com/cms/s/0/053717e6–9e64–11df-a5a4 00144feab49a,s01 = 1.html#axzz1LZ0ArdIV.

18. World Hunger Education Service Associates. 2010. *World hunger and poverty facts and statistics 2010.* http://wwwworldhunger.org/articles/Learn/world%20hunger%20facts%202002.htm.

   D'Emilio, F. 2010. *UN: number of hungry people declines.* http://www.google.com/hosted news/ap/article/ALeqM5gWSvZrY6HG3-SdhS1bdxfXofm3OAD9I7SGA80.

19. Short, R., and Potts, M., eds. 2009. The impact of population growth on tomorrow's world. *Philosophical Transactions of the Royal Society B* 364:2971–3115.

20. United Nations. 2010. *World population prospects: the 2010 revision, highlights.* New York: Population Division of the Department of Economic and Social Affairs of the United Nations Secretariat.

21. OECD. 2011. *Fertility rates.* Paris: OECD Family Database, Social Policy Division. www.oecd.org/dataoecd/37/59/40192107.pdf.

22. UN. 2010. *World population prospects*; UN Office of the High Representative for the Least Developed Countries, Landlocked Developing Countries and Small Island Developing States. 2011. *Least developed countries: about LDCs.* New York: UN-OHRLLS http://www.unohrlls.org/en/ldc/25/.

23. *The Economist.* 2009. Go forth and multiply a lot less: lower fertility is changing the world for the better. http://www.economist.com/node/14743589.

24. *The Economist.* 2009. Go forth and multiply a lot less.

25. Ezeh, A., Mberu, B., and Emina, J. 2009. Stall in fertility decline in Eastern African countries: regional analysis of patterns, determinants and implications. *Philosophical Transactions of the Royal Society B* 364:2991–3007.

26. Prata, N. 2009. Making family planning accessible in resource-poor settings. *Philosophical Transactions of the Royal Society B* 364:3093–3099.

27. UN. 2009. *World population prospects.*

28. Hopfenberg, R. and Pimental, D. 2001. Human population numbers as a function of food supply. *Environment, Development and Sustainability.* 3:1–15.

29. Foster, A., and Rosenzwieg, M. 2006. *Does economic growth reduce fertility? Rural India 1971–1999.* Delhi: NCAER India Policy Forum.
    Gertler, P., and Molyneaux, J. 1994. How economic development and family planning programs combine to reduce Indonesian fertility. *Demography* 31:33–63.
    Poston, D. Jr, and Gu, B. 1987. Socioeconomic development, family planning and fertility in China. *Demography* 24:531–551.

30. UN. 2010. *World population prospects.*

31. World Bank. 2010. *World development indicators.* http://data.worldbank.org/indicator.

32. World Bank. 2010. *World development indicators.*

33. Pingali, P. 2006. Westernization of Asian diets and the transformation of food systems: implications for research and policy. *Food Policy* 32:2881–298.
    Kumar, P., Mruthyunjaya, and Birthal, P. 2007. Changing composition pattern in South Asia. In: Joshi, P. K., Gulati, A., and Cummings, R., eds. *Agricultural diversification and smallholders in South Asia.* New Delhi: Academic Foundation.

34. Timmer, P. 2010. *The changing role of rice in Asia's food security.* ADB Sustainable Development Working Paper Series. No. 15. Manila, Philippines: Asian Development Bank.

35. Delgado, C., Rosegrant, M., Steinfeld, H., Ehui, S., and Courbois, C. 1999. *Livestock to 2020: the next food revolution.* Food, Agriculture, and the Environment Discussion Paper 28. Washington, DC: International Food Policy Research Institute.

36. FAO. 2006. *World agriculture: towards 2030/2050. Interim report. Prospects for food, nutrition, agriculture and major commodity groups.* Global Perspective Studies Unit. Rome: FAO.

37. FAO. 2009. *The state of food and agriculture.* Rome: FAO.
    von Braun, J. 2007. *The world food situation: new driving forces and required actions.* Food Policy Report 18. Washington, DC: IFPRI.
    Centre for World Food Studies. 2005. *China's rapidly growing meat demand: a domestic or an international challenge?* SOW-VU Brief No. 3. Amsterdam: Centre for World Food Studies.

38. Global Information and Early Warning System/FAO. 2007. *Crop prospects and food situation,* No. 6. http://www.fao.org/giews/english/cpfs/index.htm.

39. FAO. 2009. *The state of food and agriculture.*

40. Piesse and Thirtle. 2009. Three bubbles and a panic.
    Mitchell. 2008. *A note on rising food prices.*

41. FAO. 2006. *World agriculture: towards 2030/2050.*

42. Keyzer, M., Merbis, M., and Pavel, F. 2001. *Can we feed the animals? Origins and implications of rising meat demand.* Amsterdam: Centre for World Food Studies.

43. FAO. 2006. *World agriculture: towards 2030/2050.*

44. Ren21. 2010. *Renewables 2010: global status report.* Renewable Energy Policy Network for the 21st Century. http://www.ren21.net/Portals/97/documents/GSR/REN21 _GSR_2010_full_revised per cent20Sept2010.pdf.

45. Timmer, C. 2008. *Causes of high food prices.* ADB Economics Working Paper Series 128. Mandaluyong City, Philippines: Asian Development Bank.

46. Rosegrant, M. 2008. *Biofuels and grain prices: impacts and policy responses.* Washington, DC: IFPRI. http://ifpri.org/pubs/testimony/rosegrant20080507.pdf.
Mitchell. 2008. *A note on rising food prices.*

47. Timmer. 2008. *Causes of high food prices.*
Abbott et al. 2008. *What's driving food prices?*

48. Timmer. 2008. *Causes of high food prices.*

49. Zhang, W., Yu, E., Rozelle, S., Yang, J. and Msangi, S. 2009. *The impact of biofuel growth on agriculture: why is the range of estimates so wide?* Working Paper 2. Washington, DC: Biofuels and the Poor. http://biofuelsandthepoor.com/wp-content/uploads/2011 /10/The-Impact-of-Biofuel-Growth-on-Agriculture_Why-is-the-Range-of-Esti mates-so-Wide.pdf.

50. Biofuels and the poor. 2008. *Biofuels and the poor,* Washington, DC: IFPRI and Seattle: Bill & Melinda gates Foundation. http://biofuelsandthepoor.com/

51. USDA. 2008. Cost of production estimates. USDA Economic Research Service (ERS). www.ers.usda.gov.data.

52. Piesse and Thirtle. 2009. Three bubbles.

53. Brown, L. 2009. *Book bytes—the oil intensity of food.* http://www.earth-policy.org/index .php?/book_bytes/2009/pb3ch02_ss3#.

54. IFDC. 2010. *Fertilizer statistics.* Washington, DC: IFDC. http://www.ifdc.org/Media _Info/Fertilizer_Statistics.

55. Evans, L. 1998. *Feeding the ten billion: plants and population growth,* Cambridge, UK: Cambridge University Press.

56. FAO. 2003. *Production yearbook.* Rome: FAO. http://www.fao.org/economic/the -statistics-division-ess/publications-studies/statistical-yearbook/the-fao-production -yearbook/en/. FAO. 2007. *Digital soil map of the world.* FAO GeoNetwork. Rome: FAO. http://www.fao.org/geonetwork/srv/en/metadata.show?id=14116.
FAO. 2010. *Statistical Yearbook.* Rome: FAO. http://www.fao.org/economic/ess/ess -publications/ess-yearbook/en/

57. FAO. 2010. *FAOSTAT.* http://faostat.fao.org/.

58. Young, A. 1994. *Land degradation in South Asia: its severity, causes and effects upon the people.* FAO World Soil Resources Report 78. Rome: FAO.
Young, A. 1998. *Land resources: now and for the future.* Cambridge, UK: Cambridge University Press.

Young, A. 1999. Is there really spare land? A critique of estimates of available cultivable land in developing countries. *Environment, Development and Sustainability* 1:3–18.

59. FAO. 2000. *World soil resources report 90. Land resource potential and constraints at regional and country levels.* Based on the work of Bot, A., Nachtergaele, F., and Young, A. Rome: FAO.

60. Heldt, M. 2010. *Science and innovation can help farmers meet global challenges.* Presentation given at Investing in Science. London: Chatham House, 22 November.

61. FAO. 2009. *Statistical yearbook, 2009.* Rome: FAO.

62. FAO. 2010. *FAOSTAT.*

63. Oldeman, L., Hakkeling, R., and Sombroek, W. 1991. *World map of the status of human-induced soil degradation.* Wageningen: International Soil Reference and Information Centre (ISRIC) and UNEP; ISRIC—World Soil Information. 1987–1990. Global Assessment of Human-Induced Soil Degradation (GLASOD). Wageningen, The Netherlands: ISRIC. www.isric.org/projects/global-assessment-human-induced -soil-degradation-glasod.

64. Young. 1999. Is there really spare land?

65. BBC. 2011. *Hedge funds "grabbing land" in Africa.* http://www.bbc.co.uk/news/world -africa-13688683.

66. Oakland Institute. 2011. *Understanding land investment deals in Africa. Country report: Ethiopia.* http://media.oaklandinstitute.org/press-release-understanding-land-investment -deals-africa.

67. Oakland Institute. 2011. *Understanding land investment deals in Africa.*

68. Cotula, L., Vermeulen, S., Leonard, R., and Keeley, J. 2009. *Land grab or investment opportunity? Agricultural investment and international land deals in Africa.* Rome: FAO and London: IIED. ftp://ftp.fao.org/docrep/fao/011/ak241e/ak241e00.pdf.

69. Cotula, L. 2011. *Land deals in Africa: what is in the contracts?* London: IIED.

70. Cotula et al. 2009. *Land grab or investment opportunity?*

71. Comprehensive Assessment of Water Management in Agriculture. 2007. *Water for food, water for life: a comprehensive assessment of water management in agriculture.* London: Earthscan, and Colombo: International Water Management Institute. http://www .iwmi.cgiar.org/assessment/.

72. UN Development Program. 2006. *Human development report 2006. Beyond scarcity: power, poverty and the global water crisis.* New York: United Nations and Palgrave-McMillan.

73. World Bank. 2006. *Reengaging in agricultural water management: challenges and options.* Washington, DC: World Bank.

74. Rodell, M., Velicogna, I., and Famiglietti, J. 2009. Satellite-based estimates of groundwater depletion in India. *Nature* 460:999–1002.

75. Rodell et al. 2009. Satellite-based estimates of groundwater depletion.

76. FAO. 2010. *FAOSTAT.*

77. Trostle. 2008. *Global agricultural supply and demand.*

78. Karoly, D., Risbey, J., and Reynolds, A. 2003. Global warming contributes to Australia's worst drought. Sydney: WWF.

   Nicholls, N. 2004. The changing nature of Australian droughts. *Climate Change* 63:323–336.

   Schneider, K. 2009. *Warming takes center stage as Australian drought worsens.* Yale Environment 360, 2 April 2009. Yale School of Forestry and Environmental Studies. http://e360.yale.edu/content/feature.msp?id=2137.

79. *Scientific American.* 2010. Climate change, human failing behind Pakistan floods. http://www.scientificamerican.com/article.cfm?id=climate-change-human-fail.

   Marshall, M., and Hamzelou, J. 2010. Is climate change burning Russia? *New Scientist.* http://www.newscientist.com/article/dn19304-is-climate-change-burning-russia.html.

   Scientific American. 2010. Is the flooding in Pakistan a climate change disaster? http://www.scientificamerican.com/article.cfm?id=is-the-flooding-in-pakist.

2. What Is Hunger?

1. Homer. *The Odyssey* (12, 342). Translated in Ash, H. B. 1941. *Lucius Junius Moderatus Columella on Agriculture.* Vol. 1. Loeb Classical Library. Cambridge, MA: Harvard University Press and London: William Heinemann.

2. Food and Agriculture Organisation Statistics. 2010. *Food balance sheet, world.* http://faostat.fao.org.

3. UNICEF. 2009. *Tracking progress on child and maternal nutrition.* Geneva: UNICEF.

   UNICEF. 2009. *State of the world's children 2009, fast facts, developing countries.* Geneva: UNICEF.

4. World Food Programme (WFP). 2010. *Hunger glossary.* Rome: WFP. http://www.wfp.org/hunger/glossary.

5. WFP. 2010. *Hunger glossary.*

6. WFP. 2010. *Hunger glossary.*

7. UNICEF. 2007. *Progress for children: a world fit for children.* Statistical Review No. 6. New York: UNICEF.

8. FAO. 2006. *The state of food and agriculture 2006: food aid for food security?* Rome: FAO.

9. Waterlow, J., Armstrong, D., Fowden, L., and Riley, R., eds. 1998. *Feeding a world population of more than 8 billion people.* Oxford, UK: Oxford University Press.

   Hawkesworth S., Dangour, A., Johnston, D., Cock, K., Poole, N., Ruston, J., Uauy, R., and Waage, J. 2010. Feeding the world healthily: the challenge of measuring the effects of agriculture on health. *Philosophical Transactions of the Royal Society B* 365:3083–3097.

10. FAO/WHO. 1998. *Carbohydrates in human nutrition: Report of a joint FAO/WHO expert consultation.* Rome: WHO and FAO.

11. Lecerf, J-M. 2009. Fatty acids and cardiovascular disease. *Nutrition Review* 67:273–283.

   Srilakshmi, B. 2006. *Nutrition Science.* 2nd edition. New Delhi: New Age International Publishers.

   Tobin, A., and Dusheck, J. 2005. *Asking about life.* 3rd edition. Belmont, CA: Brooks/Cole-Thomson Learning.

   US Department of Health and Human Services and US Department of Agriculture. 2005. *Dietary guidelines for Americans.* http://www.healthierus.gov/dietarygiudelines.

   Northwestern University. 2002. *Nutrition fact sheet: lipids.* http://nuinfo-proto4.northwestern.edu/nutrition/factsheets/lipids.html.

   Nutrition Evidence Library. 2010. *What is the effect of saturated fat intake on increased risk of cardiovascular disease or type 2 diabetes?* Washington, DC: USDA. http://www.nutritionevidencelibrary.com/evidence.cfm?evidence_summary_id=250189.

12. WHO/FAO/UNU. 2007. *Protein and amino acid requirements in human nutrition.* Report of a joint WHO/FAO/UNU Expert Consultation. WHO Technical Report series 935. Geneva: WHO.

13. Millward, D., Layman, D., Tome, D., and Schaafsma, G. 2008. Protein quality assessment: impact of expanding understanding of protein and amino acid needs for optimal health. *American Journal of Clinical Nutrition* 87:S1576–S1581.

14. Pulses, broadly defined, are the edible seeds of legume crops such as beans, lentils and peas.

15. FAO/WHO/UNU. 2001. *Human energy requirements.* Report of a joint FAO/WHO/UNU Expert Consultation. Rome: FAO, WHO, UNU.

16. FAO/WHO/UNU. 2001. *Human energy requirements.*

17. FAO Statistics Division. 2008. *FAO methodology for the measurement of food deprivation: updating the minimum dietary energy requirements.* Rome: FAO. http://www.fao.org/fileadmin/templates/ess/documents/food_security_statistics/metadata/undernourishment_methodology.pdf.

18. FAO. 2010. *Food balance sheets.* Rome: FAO. http://faostat.fao.org/site/354/default.aspx.

19. Williams, C. 1935. Kwashiorkor: a nutritional disease of children associated with a maize diet. *Lancet* 226:1151–1152.

20. Jelliffe, D. 1963. The incidence of protein-calorie malnutrition of early childhood. *American Journal of Public Health Nations Health* 53:905–912.

21. Ashworth, A. 1985. Protein and "malnutrition," editorial. *Journal of Tropical Paediatrics* 31:288–289.

22. Jelliffe, D. 1959. Protein-calorie malnutrition in early childhood. *Journal of Paediatrics* 52:227.

   Jelliffe, D. 1963. The incidence of protein-calorie malnutrition of early childhood.

23. McLaren, D. 1974. The great protein fiasco. *Lancet* 2:93–96.

24. Joy, L., and Payne, P. 1975. *Nutrition and national development planning: three papers.* Brighton, UK: Institute of Development Studies, University of Sussex.

25. Victora, C., Adair, L., Fall, C., Hallal, P., Martorell, R., Richter, L., and Sachdev, H. 2008. Maternal and child undernutrition: consequences for adult health and human capital. *Lancet* 371:340–357.

    Ruel, M., Menon, P., Habicht, J-P., Loechl, C., Bergeron, G., Pelto, G., Arimond, M., Maluccio, J., Michaud, L., and Hankebo, B. 2008. Age-based preventative targeting of food assistance and behaviour change and communication for reduction of childhood undernutrition in Haiti: a cluster randomised trial. *Lancet* 371:588–595.

    Morris, P., Cogil, B., and Uauy, R. 2008. Effective international action against undernutrition: why has it proven so difficult and what can be done to accelerate the progress? *Lancet* 371:608–621.

    Black, R., Allen, L., Bhutta, Z., Caulfield, L., de Onis, M., Ezzati, M., Mathers, C., and Rivera, J. 2008. Maternal and child undernutrition: global and regional exposures and health consequences. *Lancet* 371:243–260.

26. United Nations Standing Committee on Nutrition. 2010. Maternal nutrition and the intergenerational cycle of growth failure. *Progress on nutrition: 6th report on the world nutrition situation.*

27. Muñoz, E. 2009. *New Hope for malnourished mothers and children.* Briefing Paper No. 7. Washington, DC: Bread for the World Institute.

28. Scaling Up Nutrition. 2010. Scaling up nutrition: a framework for action. March. http://www.unscn.org/files/Announcements/Scaling_Up_Nutrition-A_Framework _for_Action.pdf.

29. UNICEF. 2007. *Progress for children*; HarvestPlus. No date. *Nutrients.* http://www.harvest plus.org/content/nutrients.

30. Beaton, G., Martorell, R., Aronson, K., Edmonston, B., McCabe, G., Ross, A., and Harvey, B. 1993. *Effectiveness of vitamin A supplementation in the control of young child morbidity and mortality in developing countries.* Nutrition Policy Discussion Paper No. 13, State of the Art Series. Geneva: ACC/SCN.

31. WHO. 2010. *Worldwide prevalence of anaemia, 1993–2005.* http://www.who.int/vmnis/ anaemia/prevalence/summary/anaemia_status_summary/en/index.html.

32. Micronutrient Initiative. 2009. *Investing in the future. A united call to action on vitamin and mineral deficiencies.* Global Report Summary 2009. http://www.unitedcalltoaction .org.

33. Cohen, M., Tirado, C., Aberman, N-L., and Thompson, B. 2008. *Impact of climate change and bioenergy on nutrition.* Rome: IFPRI and FAO. http://www.fao.org/ag /agn/agns/files/HLC2_Food_Safety_Bioenergy_Climate_Change.pdf.

    Kuvibidila, S., and Vuvu, M. 2009. Unusual low plasma levels of zinc in non-pregnant Congolese women. *British Journal of Nutrition* 101:1783–1786.

34. Endocrine Today. 2007. *Worldwide iodine deficiency discussed at ATA 2007.* http://www .endocrinetoday.com/view.aspx?rid=25116.

35. Allen, L. 2003. Interventions for micronutrient deficiency control in developing countries: past, present and future. *Journal of Nutrition* 133:3875S–3878S.

36. Allen. 2003. Interventions for micronutrient deficiency control.

37. World Resources Institute. 1992. *World Resources 1992–1993*. Oxford, UK: Oxford University Press.

38. Black et al. 2008. Maternal and child undernutrition.

39. FAO. 2008. *FAO Methodology for the measurement of food deprivation; updating the minimum dietary energy requirements*. Rome: FAO Statistics Division.

40. FAO. 2008. *FAO Methodology for the measurement of food deprivation*.

41. FAO. 2009. *The state of food insecurity in the world 2009: economic crises—impacts and lessons learned*. Rome: FAO.

42. FAO. 2010. *The state of food insecurity in the world 2010: addressing food insecurity in protracted crises*. Rome: FAO.

43. FAO. 2010. *The state of food insecurity in the world 2010*.

44. Barrett, C. 2010. Measuring food insecurity. *Science* 327:825.

45. Shapouri, S., Rosen, S., Meade, B., and Gale, F. 2009. *Food security assessment, 2008–9 Outlook GFA-20*. Washington, DC: USDA Economic Research Service.

46. von Grebmer, K., Ruel, M., Menon, P., Nestorova, B., Olonfinbiyi, T., Fritschel, H., Yohannes, Y., von Oppeln, C., Towey, O., Golden, K., and Thompson, J. 2010. *Global hunger index. The challenge of hunger: focus on financial crisis and gender inequality*. Bonn: IFPRI, Concern Worldwide and Welthungerhilfe.

47. Conway, G., and Waage, J. 2010. *Science and innovation for development*. London: UK Collaborative on Development Sciences.

48. von Grebmer et al. 2010. *Global hunger index*.

49. von Grebmer et al. 2010. *Global hunger index*.

50. Barrett. 2010. Measuring food insecurity.

51. WFP. 2009. *Rwanda: comprehensive food security and vulnerability analysis and nutrition survey*. Rome: World Food Programme.

52. WFP. 2009. *Rwanda comprehensive food security and vulnerability analysis and nutrition survey*.

53. WFP. 2009. *Rwanda comprehensive food security and vulnerability analysis and nutrition survey*.

54. WFP. 2011. *Our work: comprehensive food security and vulnerability analysis*. http://www.wfp.org/food-security/reports/CFSVA.

55. Sen, A. 1981. *Poverty and famines: an essay on entitlement and deprivation*. Oxford, UK: Clarendon Press.

56. Sen. 1981. *Poverty and famines*.

57. Sen. 1981. *Poverty and famines*. Copyright © International Labour Organization 1981.

58. Sen, A. 1976. *Famines as failures of exchange entitlements*. Economic and Political Weekly. 11, Special Number, 1273–1280.

Sen, A. 1977. Starvation and exchange entitlements: a general approach and its application to the Great Bengal Famine. *Cambridge Journal of Economics* 1:33–60.

59. Maxwell, S. 1996. Food security: a post-modern perspective. *Food Policy* 21:155–170.

60. FAO. 1983. *World food security: a reappraisal of the concepts and approaches.* Director-General's Report. Rome: FAO.

61. Bickel, G., Nord, M., Price, C., Hamilton, W., and Cook, J. 2000. *Guide to measuring household food security.* Rev. ed. Alexandria, VA: USDA Food and Nutrition Service.
Smith, L., Alderman, H., and Aduayom, D. 2006. *Food insecurity in Sub-Saharan Africa: new estimates from household expenditure surveys.* Washington, DC: IFPRI.
Frongillo, E., and Nanama, S. 2006. Development and validation of an experience-based measure of household food insecurity within and across seasons in Northern Burkina Faso. *Journal of Nutrition* 136:1409S–1419S.

62. Maxwell, D. G. 1996. Measuring food insecurity. *Food Policy* 21:291–303.
Arimond, M., and Ruel, M. 2004. Dietary diversity is associated with child nutritional status: evidence from 11 demographic and health surveys. *Journal of Nutrition* 134:2579–2585.

63. Barrett. 2010. Measuring food insecurity.
Webb, P., Coates, J., Frongillo, E., Lorge Rogers, B., Swindale, A., and Bilinsky, P. 2006. Measuring household food insecurity: why it's so important and yet so difficult to do. *Journal of Nutrition* 136:1404S–1408S.

64. Barrett. 2010. Measuring food insecurity.

65. Chen, S., and Ravallion, M. 2007. Absolute poverty measures for the developing world, 1981–2004. *Proceedings of the National Academy of Sciences* 104:16757–16762.

66. Chen, S., and Ravallion, M., 2008. *The developing world is poorer than we thought.* Washington, DC: World Bank Development Research Group.

67. Von Braun, J. 2007. *The world food situation.* Presentation for the CGIAR Annual General Meeting, Beijing, December 3. http://www.slideshare.net/jvonbraun/the-world-food-situation.

68. Ravallion, M., Chen, S., and Sangraula, P. 2007. *New evidence on the urbanization of global poverty.* World Bank Policy Research Working Paper 4199. Washington, DC: World Bank.

69. International Fund for Agricultural Development. 2008. *Rural poverty portal.* http://www.ruralpovertyportal.org/web/guest/region.
Centre for International Earth Science Information Network. 2008. *Global distribution of poverty.* http://sedac.ciesin.columbia.edu/povmap/datasets/ds_nat_all.jsp.

70. World Bank. 2006. *India, inclusive growth and service delivery: building on India's success.* Development Policy Review Report No. 34580-IN. Washington, DC: World Bank.

71. Nagayets, O. 2005. *Current status and key trends.* Information brief prepared for the Future of Small Farms Research Workshop, Wye College, June 26–29.

72. FAO. 2010. *2000 World census of agriculture.* FAO Statistical Development Series 12. Rome: FAO. http://www.fao.org/docrep/013/i1595e/i1595e.pdf.

73. World Bank. 2008. *Poverty assessment for Bangladesh: creating opportunities and bridging the East-West divide.* Bangladesh Development Series No. 26. Dhaka: World Bank.

74. Jazairy, I., Alamgir, M., and Panuccio, T. 1992. *The state of the world rural poverty—an inquiry into its causes and consequences.* New York: New York University Press.
Lipton, M., and Longhurst, R. 1989. *New seeds and poor people.* London: Unwin Hyman.

75. World Bank. 2008. *Poverty assessment for Bangladesh.*

76. Gustafsson, B., and Sai, D. 2008. *Temporary and persistent poverty among ethnic minorities and the majority in rural China.* DP No. 3791. Bonn: Institute for the Study of Labor (IZA).

77. Hall, G., and Patrinos, HA. No date. *Indigenous peoples, poverty and human development in Latin America: 1994–2004.* http://www.crid.or.cr/digitalizacion/pdf/eng/doc/6327/doc16327.contenido.pdf.

78. World Bank. 2008. *Vietnam development report.* Hanoi: World Bank.

79. Chin, A., and Prakash, N. 2009. *The redistributive effects of political representation for minorities: evidence from India.* Centre for Research and analysis of Migration, Discussion paper series CDP No. 03/10.

80. Chin and Prakash. 2009. *The redistributive effects of political representation.*

81. Azad, N. 1999. *Engendered mobilization—the key to livelihood security: IFAD's experience in South Asia.* Rome: IFAD.
Osmani, S. R. 1998. *Food security, poverty and women: lessons from rural Asia,* Part I. Rome: IFAD/TAD.

82. Lapierre, D. 1986 *City of joy.* London: Arrow Books.

83. Sen. 1981. *Poverty and famines.* Page 1.

84. Sen, A. K. 1999. *Development as freedom.* Oxford, UK: Oxford University Press. Page 162.

3. The Green Revolution

1. Hanson, H., Borlaug, N., and Anderson, R. 1982. *Wheat in the third world.* Boulder, CO: Westview Press.

2. Thurow, R., and Kilman, S. 2009. *Enough: why the world's poorest starve in an age of plenty.* New York: Public Affairs Books. Page 4.

3. Culver, J., and Hyde, J. 2000. *American dreamer: a life of Henry A. Wallace.* New York: W.W. Norton.

4. Thurow and Kilman. 2009. *Enough.*

5. Stakman, E., Bradfield, R. and Mangelsdorf, P. 1967. *Campaigns against hunger.* Cambridge, MA: Harvard University Press.
de Alcantara, H. 1976. *Modernizing Mexican agriculture: socioeconomic implications of technological change, 1940–1970.* Report No. 76.5. Geneva: United Nations Research Institute for Social Development.

6. Maize is commonly referred to in the United States as corn.

7. Stakman et al. 1967. *Campaigns against hunger.*

8. Hanson et al. 1982. *Wheat in the third world.*
   Thurow and Kilman. 2009. *Enough.*

9. de Wit, C., van Laar, H., and van Keulen, H. 1979. Physiological potential of food production. In: Sneep, J., and Henrickson, A., eds. *Plant breeding perspectives.* Publ. No. 118, PUDOC. Wageningen, Netherlands: Centre for Agricultural Publishing and Documentation.
   Plucknett, D. 1993. *Science and agricultural transformation.* Lecture Series. Washington, DC: IFPRI.

10. Hanson et al. 1982. *Wheat in the third world.*

11. New World Encyclopedia. 2008. "Norman Borlaug." http://www.newworldencyclo pedia.org/entry/Norman_Borlaug?oldid=685200.
    Thurow and Kilman. 2009. *Enough.*

12. Wright, B. 1972. Critical requirements of new dwarf wheat for maximum production. In: *Proceedings of the second FAO/Rockefeller Foundation International Seminar on wheat improvement and production, March 1968.* Beirut: Ford Foundation.

13. Ganzel, B. 2007. *The Mexican agricultural program.* http://www.livinghistoryfarm.org /farminginthe50s/movies/borlaug_crops_16.html.

14. FAO. 2010. *FAOSTAT.* http://faostat.fao.org/site/567/default.aspx#ancor.

15. Brown, L. 1970. *Seeds of change.* New York: Praeger.

16. Thurow and Kilman. 2009. *Enough.*

17. Thurow and Kilman. 2009. *Enough.*

18. Barker, R., Herdt, R., and Rose, B. 1985. *The rice economy of Asia.* Washington, DC: Resources for the Future.

19. Baum, W. 1986. *Partners against hunger: the Consultative Group on International Agricultural Research.* Washington, DC: World Bank.

20. Chandler, R. 1982. *An adventure in applied science: a history of the International Rice Research Institute.* Los Baños, Philippines: IRRI.

21. Tso, T. No date. "Agriculture in 1949–1992." In: Tso, T., Tuan, F., and Faust, M., eds. *Agriculture in China 1949–2030.* Beltsville, MD: IDEALS.

22. Yang, D. 2008. China's agricultural crisis and famine 1959–1961: a survey and comparison to Soviet famines. *Comparative Economic Studies* 50:1–29.

23. Dikötter, F. 2010. *Mao's great famine: the history of China's most devastating catastrophe, 1958–62.* London: Bloomsbury.

24. Barker et al. 1985. *The rice economy of Asia.*

25. Brown. 1970. *Seeds of change.*

26. Brown. 1970. *Seeds of change.*

27. Dalrymple, D. 1986 *Development and spread of high yielding wheat varieties in developing countries.* Washington, DC: United States Agency for International Development.

28. Hanson et al. 1982. *Wheat in the third world.*

29. Pearse, A. 1980. *Seeds of plenty, seeds of want: social and economic implications of the Green Revolution.* Oxford, UK: Clarendon Press.

30. Pearse. 1980. *Seeds of plenty.*
    Castillo, G. 1975. *All in a grain of rice.* Laguna, Philippines: Southeast Asian Regional Center for Graduate Study and Research in Agriculture.

31. Action BioScience. 2002. *Biotechnology and the Green Revolution: Norman Borlaug—an original interview.* http://actionbioscience.org/biotech/borlaug.html.

32. Conway, G., and Pretty, J. 1991. *Unwelcome harvest: agriculture and pollution.* London: Earthscan.

33. Barker et al. 1985. *The rice economy of Asia.*
    Lipton, M., and Longhurst, R. 1989. *New seeds and poor people.* London: Unwin Hyman.
    Anderson, R., Brass, P., Levy, E., and Morris, B., eds. 1982. *Science, politics and the agricultural revolution in Asia.* Boulder, CO: Westview Press.
    Pearse. 1980. *Seeds of plenty.*
    Ruttan, V., and Binswanger, H. 1978. Induced innovation and the Green Revolution. In: Binswanger, H., and Ruttan, V., eds. *Induced innovation: technology, institutions and development.* Baltimore, MD: Johns Hopkins University Press.
    Hayami, Y., and Kikuchi. M. 1981. *Asian village economy at the crossroads.* Tokyo: Tokyo University Press.

34. Lipton and Longhurst. 1989. *New seeds and poor people.*

35. Lipton and Longhurst. 1989. *New seeds and poor people.*

36. Piesse, J., and Thirtle, C. 2009. Three bubbles and a panic: An explanatory review of recent food commodity price events. *Food Policy* 34:119–129. Reprinted with permission from Elsevier.

37. Dorward, A. 2011. *Getting real about food prices.* Development Viewpoints, No. 58. London: Centre for Development Policy and Research, School of Oriental and African Studies. http://eprints.soas.ac.uk/11094/.

38. Lipton and Longhurst. 1989. *New seeds and poor people.*

39. Freebain, D. 1995. Did the Green Revolution concentrate incomes? A quantitative study of research reports. *World Development.* 23:265–279.

40. FAO. 1992. *World food supplies and prevalence of chronic undernutrition in developing regions as assessed in 1992.* Document ESS/MISC/1992, Rome: FAO.
    FAO. 2010. *FAOSTAT (Prevalence of undernourishment in total population).*

41. Alexandratos, N., ed. 1995. *World agriculture: towards 2010, an FAO study.* Chichester, UK: John Wiley and Sons.

42. Frankel, F. 1971. *India's Green Revolution: economic gains and political costs.* Princeton, NJ: Princeton University Press.

43. Lockwood, B., Mukherjee, P., and Shand, R. 1971. *The HYV Program in India. Part I.* New Delhi: Planning Commission (India) with Australian National University.
    Schluter, M. 1971. *Differential rates of adoption of the new seed varieties in India: the problem of small farms.* Occasional Paper No. 47. Ithaca, NY: USAID/Department of Agricultural Economics, Cornell University.

44. Lipton and Longhurst. 1989. *New seeds and poor people.*

45. Burke, R. 1979. Green Revolution technology and farm class in Mexico. *Economic Development and Cultural Change* 28:135–154.

46. Hanson et al. 1982. *Wheat in the third world.*

47. Binswanger, H. 1978. *The economics of tractorization in South Asia.* Washington, DC: Agricultural Development Council.

48. Pingali, P., Bigot, Y., and Binswanger, H. P. 1987. *Agricultural mechanization and the evolution of farming systems in Sub-Saharan Africa.* Baltimore, MD: John Hopkins University Press.

49. Barker et al. 1985. *The rice economy of Asia.*

50. Cain, M., Rokeya Khanam, S., and Nahar, S. 1979. *Class, patriarchy and the structure of women's work in rural Bangladesh.* Working Paper 43. New York: Center for Policy Studies, Population Council.
    Greeley, M. 1987. *Post-harvest losses, technology and unemployment: the case of Bangladesh.* Boulder, CO: Westview.
    Scott, G., and Carr, M. 1985. *The impact of technology choice on rural women in Bangladesh: problems and opportunities.* Staff Working Paper 731. Washington, DC: World Bank.

51. Lipton and Longhurst. 1989. *New seeds and poor people.*

52. Hazell, P., and Ramasamy, C. 1991. *Green Revolution reconsidered: the impact of high yielding rice varieties in South India.* Baltimore, MD: John Hopkins Press and New Delhi: Oxford University Press.
    Hayami, Y., and Kikuchi, M. 2002. *A rice village saga: three decades of green revolution in the Philippines.* London: Macmillan Press.
    Jewitt, S., and Baker, K. 2007. The green revolution re-assessed: insider perspectives on agrarian change in Bulandshahr distract, Western Uttar Pradesh, India. *Geoforum* 38:73–89.

53. Hazell, P. 2009. *The Asian Green Revolution.* Discussion Paper 00911. Washington, DC: IFPRI. Page 10.

54. Hossain, M., Lewis, D., Bose, M., and Chowdhury, A. 2007. Rice research, technological progress, and poverty: the Bangladesh case. In: Adato, M., and Meinzen-Dick, R., eds. *Agricultural research livelihoods, and poverty: studies of economic and social impacts in six countries.* Baltimore, MD: Johns Hopkins University Press.

55. Hossain et al. 2007. Rice research.

56. Barker, R. 1978. Yield and fertiliser input. In: Evenson, R., ed. *Changes in rice farming in selected areas of Asia.* Los Baños, Philippines: IRRI.

57. Herdt, R., and Capule, C. 1983. *Adoption, spread and production impact of modern rice varieties in Asia.* Los Baños, Philippines: IRRI.

Pinstrup-Anderson, P., and Hazell, P. 1985. The impact of the Green Revolution and prospects for the future. *Food Reviews International* 1:1–25.

58. Rosegrant, M., and Evenson, R. 1995. *Total factor productivity and sources of long-term growth in Indian agriculture.* Environment and Production Technology Division Discussion Paper No. 7. Washington, DC: IFPRI.

59. Rosegrant and Evenson. 1995. *Total factor productivity.*

60. Rosegrant and Evenson. 1995. *Total factor productivity.* Page 26.

61. Chandler, R. 1982. *An adventure in applied science: a history of the International Rice Research Institute.* Los Baños, Philippines: IRRI.

4. The Political Economy of Food Security

1. Suetonius, *The Lives of the Caesars* Vol. 1: *Julius. Augustus. Tiberius. Gaius. Caligula.* Loeb Classical Library, No. 31 (J. C. Rolfe, trans.): 42.3.

2. Harlan, J. 1971. Agricultural origins: centers and noncenters. *Science* 174:468–473.

3. Wittfogel, K. 1957. *Oriental despotism: a comparative study of total power.* New Haven, CT: Yale University Press.

4. Butzer, K. 1976. *Early hydraulic civilisation in Egypt: a case study in cultural ecology.* Chicago: Chicago University Press.

5. Garnsey, P. 1988. *Famine and food supply in the Graeco-Roman world: responses to risks and causes.* Cambridge, UK: Cambridge University Press.

6. The National Bureau of Economic Research. No date. *The NBER political economy program.* http://www.nber.org/programs/pol/pol.html.

Wikipedia. 2011. "Political economy." http://en.wikipedia.org/wiki/Political_economy.

7. Drèze, J., and Sen, A. 1995. Introduction. In: Drèze, J., Sen, A., and Hussain, A. *The political economy of hunger: selected essays.* Oxford, UK: Oxford University Press.

8. Smith, M., Pointing, J., and Maxwell, S. 1992. *Household food security, concepts and definitions: an annotated bibliography.* Development Bibliography No. 8. Brighton, UK: Institute of Development Studies, University of Sussex.

FAO. 2003. Food security: concepts and measurement. In: FAO. *Trade reforms and food security: conceptualizing the linkages.* Rome: FAO. http://www.fao.org/docrep/005/y4671e/y4671e06.htm.

9. Tudge, C. 1977. *The famine business.* London: Faber and Faber.

10. United Nations. 1975. *Report of the world food conference.* Rome: UN.

11. Sen, A. 1976. Famines as failures of exchange entitlements. *Economic and Political Weekly* 11, Special No., 1273–1280.

Sen, A. 1977. Starvation and exchange entitlements: a general approach and its application to the Great Bengal Famine. *Cambridge Journal of Economics* 1:33–60.

12. World Bank. 1986. *Poverty and hunger: issues and options for food security in developing countries.* Washington, DC: World Bank. Page 1.

13. FAO. 2002. *The state of food insecurity in the world 2001.* Rome: FAO. http://www.fao.org/docrep/003/y1500e/y1500e06.htm#P0_2.

14. Sen, A. 1995. Food, economics and entitlements. In: Drèze et al. *The political economy of hunger.*

15. Anderson, K., ed. 2010. *The political economy of agricultural price distortions.* Cambridge, UK: Cambridge University Press.

16. Aker, J., Block, B., Ramachandran, V., and Timmer, C. 2011. *West African experience with the world rice crisis, 2007–2008.* CGD Working Paper 242. Washington, DC: Center for Global Development. http://www.cgdev.org/content/publications/detail/1424823.

17. Net News Publisher. 2010. *Senegal reaches over 55% self-sufficiency in rice production in 2009.* http://www.netnewspublisher.com/senegal-reaches-over-55-self-sufficiency-in-rice-production-in-2009/.
    Staatz, J., Dembélé, N., Kelly, V., and Adjao, R. 2008. Agricultural globalization in reverse: the impact of the food crisis in West Africa. Background paper for the Geneva Trade and Development Forum, Switzerland, September 17–20.

18. Dorward, A., and Chirwa, E. 2011. The Malawi agricultural input subsidy programme, 2005/06 to 2008/09. *International Journal of Agricultural Sustainability* 9:232–247.

19. Dorward and Chirwa. 2011. The Malawi agricultural input subsidy programme. Used with permission of Taylor & Francis Informa UK Ltd.—Journals; permission conveyed through Copyright Clearance Center, Inc.

20. Levy, S., ed. 2005. *Starter packs: a strategy to fight hunger in developing and transition countries? Lessons from the Malawi experience, 1998–2003.* Wallingford, UK: CABI.

21. The Economist 2008. *Malawi: can it feed itself?* http://www.economist.com/node/11294760.; IRIN News. 2011. *Malawi farm subsidy programme shrinks.* http://www.irinnews.org/report.aspx?ReportId=93954.

22. Staatz et al. 2008. Agricultural globalization in reverse.
    Hiensch, A. 2007. Surviving shocks in Ethiopia: the role of social protection for food security. Case study #4–2 of the program. *Food policy for developing countries: the role of the government in the global food system.* Cornell University.

23. Dorward and Chirwa. 2011. The Malawi agricultural input subsidy programme.

24. Lee, R. 2007. *Food security and food sovereignty.* Centre for Rural Economy Discussion Paper Series No. 11, March. Newcastle, UK: University of Newcastle upon Tyne.

25. Lee. 2007. *Food security and food sovereignty.* Page 6.

26. Hiensch. 2007. Surviving shocks in Ethiopia.

27. von Braun, J., Lin, J. and Torero, M. No date. Eliminating drastic food price spikes—a three pronged approach for reserves. Note for Discussion. Washington, DC: IFPRI. http://www.ifpri.org/sites/default/files/publications/reservenote20090302.pdf.

von Braun, J., and Torero, M. 2008. *Physical and virtual global food reserves to protect the poor and prevent market failure.* IFPRI Policy Brief 4. Washington, DC: IFPRI. http://www.ifpri.org/sites/default/files/pubs/pubs/bp/bp004.pdf.

28. Damary, R. 2011. *Kenya: strategic grain reserves stands at eight million bags.* http://allafrica.com/stories/201107120789.html.

    McKee, D. 2011. *Strategic grain reserves: Sub-Saharan Africa.* http://www.davidmckee.org/2011/06/03/strategic-grain-reserves/#toc-Sub-Saharan-Africa.

29. Grigg, D. 1993. *The world food problem.* 2nd edition. Oxford, UK: Blackwell Publishing.

30. FAO. 2011. *World food situation: cereal supply and demand brief.* Rome: FAO. http://www.fao.org/worldfoodsituation/wfs-home/csdb/en/.

31. Paarlberg, R. 2010. *Food politics: what everyone needs to know.* New York: Oxford University Press.

32. Beierle, T. 2002. *From Uruguay to Doha: agricultural trade negotiations at the World Trade Organization.* Discussion paper 02–13. Washington, DC: Resources for the Future.

    Elliot, K. 2007. *Agriculture and the Doha Round.* CGD/IIE Brief. Washington, DC: Center for Global Development and Peter G. Peterson Institute for International Economics.

33. Beierle. 2002. *From Uruguay to Doha.*

34. U.S. Department of Agriculture (USDA). 2001. *The road ahead: agricultural policy reform in the WTO, summary report.* Agriculture Economic Report No. 797, Washington, DC: Economic Research Service, USDA.

35. Beierle. 2002. *From Uruguay to Doha.*

36. USDA. 2001. *The road ahead.*

37. USDA. 2001. *The road ahead.*

38. USDA. 2001. *The road ahead.*

39. Beierle. 2002. *From Uruguay to Doha.*

40. Fairtrade Foundation. 2010. *Main findings of the Great Cotton Stitch-Up.* Fairtrade Foundation Press Release. http://www.fairtrade.org.uk/includes/documents/cm_docs/2010/m/2_main_findings_of_the_great_cotton_stitch_up.pdf.

    Alston, J., Sumner, D., and Brunke, H. 2007. *Impacts of reductions in US cotton subsidies on West African cotton producers.* Boston, MA: Oxfam America.

    Baffes, J. 2010. *Cotton subsidies, the WTO and the 'Cotton Problem.'* Policy Research Working Paper 5663. Washington, DC: World Bank.

41. International Fund for Agricultural Development. 2010. *Rural poverty report 2011—new realities, new challenges: new opportunities for tomorrow's generation.* Rome: IFAD.

42. Beierle. 2002. *From Uruguay to Doha.*

43. World Bank. 2007. *World development report 2008: agriculture for development.* Washington, DC: World Bank.

44. Beierle. 2002. *From Uruguay to Doha.*

45. Beierle. 2002. *From Uruguay to Doha.*

46. Beierle. 2002. *From Uruguay to Doha.*

47. Beierle. 2002. *From Uruguay to Doha.*
    Elliot. 2007. *Agriculture and the Doha Round.*

48. Alexandratos, N., ed. 1995. *World agriculture: towards 2010.* An FAO study. Chichester, UK: Wiley and Sons.
    FAO. 1995. *The state of food and agriculture, 1995. Agricultural trade: entering a new era.* Rome: FAO.

49. Beierle. 2002. *From Uruguay to Doha.*

50. Arbache, J., Dickerson, A., and Green, F. 2004. Trade liberalisation and wages in developing countries. *Economic Journal* 114: F73–F96.
    World Trade Organisation. No date. *Understanding the WTO.* http://www.wto.org/english/thewto_e/whatis_e/tif_e/tif_e.htm.
    Beierle. 2002. *From Uruguay to Doha.*

51. Beierle. 2002. *From Uruguay to Doha.*

52. Arbache et al. 2004. Trade liberalisation and wages.

53. Beattie, A. 2009. *False economy: a surprising history of the world.* New York: Riverhead Books.

54. Oxfam. No date. *Cultivating poverty: the impact of US cotton subsidies on Africa.* Oxfam Briefing Paper 30. http://www.oxfam.org.uk/resources/policy/trade/downloads/bp30_cotton.pdf.

55. Beattie. 2009. *False economy.*

56. Krueger, A., Schiff, M., and Valdés, A.1988. Agricultural incentives in developing countries: measuring the effect of sectoral and economy-wide policies. *World Bank Economic Review* 2:255–272.

57. Anderson, K. 2010. *Krueger/Schiff/Valdés revisited agricultural price and trade policy reform in developing countries since 1960.* Policy Research Working Paper 5165.Washington, DC: Development Research Group, World Bank.

58. Anderson. 2010. *Krueger/Schiff/Valdés revisited.*

59. Cook, J., Cylke, O., Larson, D., Nash, J., and Stedman-Edwards, P. 2010. *Vulnerable places, vulnerable people: trade liberalization, rural poverty and the environment.* Washington, DC: World Bank and Cheltenham, UK: Edward Elgar Publishing.

60. Cook et al. 2010. *Vulnerable places, vulnerable people.*

61. Cook et al. 2010. *Vulnerable places, vulnerable people.*

62. Cook et al. 2010. *Vulnerable places, vulnerable people.*

63. Singer, H., Wood, J., and Jennings, T. 1987. *Food aid: the challenge and the opportunity.* Oxford, UK: Clarendon Press.

64. Thomas, M., Sharp, K., Maxwell, S., Hay, R., Jones, S., Low, A., Clay, E., and Benson, C. 1989. *Food aid to Sub-Saharan Africa: a review of the literature.* WFP Occasional Paper 13. Rome: World Food Programme.

Boussard, J-M., Daviron, B., Gérard, F., and Voituriez, T. 2005. *Food security and agricultural development in Sub-Saharan Africa: building a case for more support.* CIRAD for FAO. Rome: FAO. http://www.fao.org/tc/tca/work05/CIRAD.pdf.

65. WFP. 2009. *Food Aid Information System.* http://www.wfp.org/fais/.

66. WFP. 2011. *Our work. P4P overview.* http://www.wfp.org/purchase-progress/overview.

67. WFP. 2011. *Our work.*

68. WFP. 2011. *Purchase for Progress. Mali: farmer's progress over two years of P4P.* http://www.wfp.org/purchase-progress/blog/mali-farmer%E2%80%99s-progress-over-two-years-p4p.

69. De Waal, A. 2004. *A tragedy in Darfur: on understanding and ending the horror.* Boston Review. http://bostonreview.net/BR 29.5/dewaal.html.
    De Waal, A. 2006. *Famine that kills: Darfur, Sudan, 1984–85.* Oxford, UK: Oxford University Press.
    Gettleman, J. 2008. *Darfur withers as Sudan sells food.* http://www.nytimes.com/2008/08/10/world/africa/10sudan.html.
    BBC. 2010. *Q and A: Sudan's Darfur conflict.* http://news.bbc.co.uk/1/hi/world/africa/3496731.stm.
    WFP. No date. *Sudan: overview.* http://www.wfp.org/countries/Sudan/Overview.
    Conway, G. 2009. *Science of climate change in Africa: impacts and adaptation.* Grantham Institute for Climate Change Discussion Paper No. 1. London: Grantham Institute, Imperial College London.

70. Gettleman. 2008. *Darfur withers.*
    BBC. 2010. *Q and A: Sudan's Darfur conflict.*
    WFP. No date. *Sudan: overview.*

71. Afrol News. 2010. *Niger food crisis deepens.* http://www.afrol.com/articles/16745.

72. Caritas. 2010. *Food crisis in Niger worse than in 2005 as millions face hunger.* http://www.caritas.org/newsroom/press_releases/PressRelease16_06_10.html.

73. Issa, O. 2010. *Niger facing growing food crisis.* http://ipsnews.net/news.asp?idnews=52513.

74. Afrol News. 2010. *Niger food crisis deepens.*

75. Aker, J. 2008. *How can we avoid another food crisis in Niger?* Berkeley, CA: Gianni Foundation of Agricultural Economics, University of California.

76. Afrol News. 2010. *Niger food crisis deepens.*
    Issa. 2010. *Niger facing growing food crisis.*

77. Afrol News. 2010. *Niger food crisis deepens.*

78. IRIN. 2011. *Niger: aid response to crisis in south.* Guardian development network. http://www.guardian.co.uk/world/2011/may/23/niger-food-aid-crop-fail.

79. AllAfrica.com. 2011. *Niger: food crisis threatens more than half of country's villages.* http://allafrica.com/stories/201112271363.html.

80. Ashdown, P. (chair). 2011. *Humanitarian Emergency Response Review (HERR).* London: DFID. www.dfid.gov.uk/documents/publications1/HERR.pdf.

5. A Doubly Green Revolution

1. Conway, G. 1997. *The doubly green revolution: food for all in the 21st century.* London: Penguin Books (Ithaca: Cornell University Press, 1999). Page 41.

2. Harr, J., and Johnson, P. 1988. *The Rockefeller century: three generations of America's greatest family.* New York: Charles Scribner's Sons.

3. Conway. 1997. *The doubly green revolution.*

4. Sen, A. 1999. *Development as freedom.* New York: Alfred Knopf.

5. Field, F. 2011. *The welfare state—never ending reform.* http://www.bbc.co.uk/history/british /modern/field_01.shtml.

6. Leke, A., Lund, S., Roxburgh, C., and van Wamelen, A. 2010. What's driving Africa's growth. *McKinsey Quarterly.* June.

7. Ghazvinian, J. 2007. *Untapped: the scramble for Africa's oil.* San Diego: Harcourt.

8. Brahmbhatt, M., Canuto, O., and Vostroknutova, E. 2010. Natural resources and development strategy after the crisis. In: Giugale, M. *The day after tomorrow: a handbook on the future economic policies in the developing world.* Washington, DC: World Bank.

9. Mellor, J. 1995. Introduction. In: Mellor, J., ed. *Agriculture on the road to industrialization.* Baltimore, MD: Johns Hopkins University Press.

10. Haggblade, S. 2007. *Returns to investment in agriculture.* Policy Synthesis: Food security research project No. 19. January. Lansing, MI: Michigan State University. Page 1.

11. World Bank. 2007. *World development report 2008: agriculture for development.* Washington, DC: World Bank.

12. Ahluwhalia, M. 2005. *Reducing poverty and hunger in India: the role of agriculture.* IFPRI 2004–5 Annual Report. Washington, DC: IFPRI.

13. India Planning Commission. 2007. *Agricultural strategy for eleventh plan, some critical issues.* http://planningcommission.gov.in/plans/planrel/53rdndc/AgricultureStrategy.pdf.

14. National Sample Survey Organisation. 2003. *Some aspects of operational land holdings in India 2002–3.* NSS Report No. 492. New Delhi: Ministry of Statistics and Programme Implementation, Government of India.

15. Conway, G., Lele, U., Peacock, J., and Pineiro, M. 1994. *Sustainable agriculture for a food secure world.* Washington, DC: Consultative Group on Agricultural Research and Stockholm: Swedish Agency for Research Co-operation with Developing Countries.

16. Nagayets, O. 2005. *Small farms: current status and key trends.* Information Brief, Research Workshop on the Future of Small Farms, Organised by IFPRI, Imperial College and ODI, Wye, June 2005.

17. Pingali, P. 2010. *Presentation: who is the smallholder farmer?* Des Moines, IA: Norman E. Borlaug International Symposium, The World Food Prize, October 13–15, 2010. http://ilriclippings.wordpress.com/2010/10/15/prabhu-pingali-of-gates-foundation -remarks-at-the-borlaug-dialogueworld-food-prize-ceremony-in-iowa/.

18. Wiggins, S. 2009. *Can the smallholder model deliver poverty reduction and food security for a rapidly growing population in Africa?* FAC Working Paper No. 08. London: Overseas Development Institute.

19. Cornia, G. 1985. Farm size, land yields and the agricultural production function: an analysis for fifteen developing countries. *World Development* 13:513–534.

    Eastwood, R., Lipton, M., and Newell, A. 2009. Farm size. In: Pingali, P., and Evenson, R., eds. *Handbook of Agricultural Economics,* vol. 4. Amsterdam: Elsevier.

    Collier, P., and Dercon, S. 2009. African agriculture in 50 years: smallholders in a rapidly changing world. In: *How to feed the world in 2050.* Proceedings of a technical meeting of experts, 24–26 June. Rome: FAO.

20. Hoogerbrugge, I., and Fresco, L. 1993. *Homegarden systems: agricultural characteristics and challenges.* Gatekeeper Series No. 39. London: International Institute for Environment and Development.

21. Soemarwoto, O., and Conway, G. 1991. The Javanese homegarden. *Journal for Farming Systems Research and Extension* 2:95–117.

22. FAO. 1989. Schematic representation of the structural composition of a Javanese homegarden. In: FAO. *Forestry and food security.* Rome: FAO.

23. Collier, P. 2009. Africa's organic peasantry; beyond romanticism. *Harvard International Review.* Page 31.

24. Wiggins. 2009. *Can the smallholder model deliver poverty reduction and food security?*

25. Wiggins. 2009. *Can the smallholder model deliver poverty reduction and food security?*

26. Wiggins. 2009. *Can the smallholder model deliver poverty reduction and food security?*

27. Wiggins. 2009. *Can the smallholder model deliver poverty reduction and food security?*

28. Collier and Dercon. 2009. African agriculture in 50 years.

    Byerlee, D., and Deininger, K. 2010. *The rise of large farms: drivers and development outcomes.* Helsinki: World Institute for Development Economic Research, United Nations University. http://www.wider.unu.edu/publications/newsletter/articles-2010 /en_GB/article-11–12–2010.

29. Simeh, A., and Ahmad, T. 2001. *The case study on the Malaysian Palm oil.* Regional workshop on commodity export diversification and poverty reduction in South and South-East Asia. UNCTAD, Bangkok, 3–5 April.

    Ghee, L., and Dorrall, R. 1992. Contract farming in Malaysia: with special reference to FELDA land schemes. In: Glover, D. and Ghee, L., eds. 1992. *Contract farming in South-East Asia: three country studies.* Kuala Lumpur, Malaysia: Institute for Advanced Studies, University of Malaya.

30. Pingali. 2010. *Presentation: who is the smallholder farmer?* Page 1.

31. FAO. 2009. *FAOSTAT.* http://faostat.fao.org/site/339/default.aspx.

32. Cassman, K. 1999. Ecological intensification of cereal production systems: yield potential, soil quality and precision agriculture. *Proceedings of the National Academy of Sciences* 96, 5952–5959.

33. Cassman. 1999. Ecological intensification of cereal production systems.

34. Cassman. 1999. Ecological intensification of cereal production systems.

35. Evans, L. 1993. *Crop evolution, adaptation and yield.* New York: Cambridge University Press.

36. Van Ittersum, M., and Rabbinge, R. 1997. Concepts in production ecology for analysis and quantification of agricultural input-output combinations. *Field Crops Research* 52:197–208.

37. Lobell, D., Cassman, K., and Field, C. 2009. Crop yield gaps: their importance, magnitudes and causes. *Annual Review of Environment and Resources* 34:179–204.

38. Ahmed, R. U. 1988. Rice price stabilization and food security in Bangladesh. *World Development* 16:1035–1050.

39. Rahman, S., and Rahman, M. 2008. Impact of land fragmentation and resource ownership on productivity and efficiency: the case of rice producers in Bangladesh. *Land Use Policy* 26:95–103.

40. Cumming, G., Barnes, G. Perz, S., Schmink, M., Sieving, K., Southworth, J., Binford, M., Holt, R., Stickler, C., and Van Holt, T. 2005. An exploratory framework for the empirical measurement of resilience. *Ecosystems* 8:975–987.

41. Walker, B., and Salt, D. 2006. *Resilience thinking: sustaining ecosystems and people in a changing world.* Washington, DC: Island Press.

42. Conway, G., and Waage, J. 2010. *Science and innovation for development.* London: UKCDS. Page 309.

43. Holling, C. 1973. Resilience and stability of ecological systems. *Annual Review of Ecology and Systematics* 4:1–23.

44. World Bank. 2011. *Measuring inequality.* Poverty Reduction and Equity. Washington, DC: World Bank. http://web.worldbank.org/WBSITE/EXTERNAL/TOPICS/EXTPOVERTY/EXTPA/0,,contentMDK:20238991~menuPK:492138~pagePK:148956~piPK:216618~theSitePK:430367,00.html.

45. Lipton, M. 1996. Personal communication.

46. World Hunger Education Service. 2011. *2011 World hunger and poverty facts and statistics.* http://www.worldhunger.org/articles/Learn/world%20hunger%20facts%202002.htm.

47. Hazell, P., and Ramasamy, C. 1991. *The Green Revolution reconsidered: the impact of high-yielding varieties in South India.* Published for IFPRI. Baltimore: John Hopkins University Press. http://www.ifpri.org/sites/default/files/publications/hazell91.pdf.

Harriss-White, B., and Janakarajan, S. 2004. *Rural India facing the 21st century: essays on long term village change and recent development policy.* London: Anthem Press.

48. Hazell and Ramasamy. 1991. *The Green Revolution reconsidered.*

49. Harriss-White, B., Janakarajan, S., and Colatei, D. 2004. Introduction: heavy agriculture and light industry in south Indian villages. In Harriss-White, B., and Janakarajan, S., eds. 2004. *Rural India facing the 21st century.* London: Anthem Press.

6. Sustainable Intensification

1. Hooper, W., and Ash, H. 1935. *Marcus Porcius Cato on agriculture. Marcus Terentius Varro on agriculture.* Cambridge, MA: Loeb Classical Library, Harvard University Press and London: William Heinemann (my translation).

2. Conway, G., and Barbier, E. 1990. *After the Green Revolution: sustainable agriculture for development.* London: Earthscan.

3. World Commission on Environment and Development. 1987. *Our common future.* Oxford, UK: Oxford University Press.

4. Darwin, C. 1890. *A naturalist's voyage: journal of researches into the natural history and geology of the countries visited during the voyage of H.M.S. 'Beagle' round the world, under the command of Capt. Fitz Roy, R.N.* (new edition, 2001) London: John Murray.

5. May, R., and McLean, A., eds. 2007. *Theoretical ecology: principles and applications.* 3rd edition. Oxford, UK: Oxford University Press.
   Begon, M., Harper, J., and Townsend, C. 1996. *Ecology: individuals, populations and communities.* 2nd edition. Oxford, UK: Blackwell Scientific Publications.

6. Altieri, M. 1995. *Agroecology: the science of sustainable agriculture.* 2nd edition. Boulder, CO: Westview Press.
   Lowrance, R., Stinner, B., and House, G., eds. 1984. *Agricultural ecosystems: unifying concepts.* Chichester, UK: John Wiley and Sons.
   National Research Council. 1989. Alternative Agriculture Committee on the role of alternative farming methods in modern agriculture. Washington, DC: National Academy Press.
   Committee on Sustainable Agriculture and the Environment in the Humid Tropics. 1993. *Sustainable agriculture and the environment in the humid tropics.* Washington, DC: National Academy Press.
   Paoletti, M., Stinner, B., and Lorenzoni, G., eds. 1989. *Agricultural ecology and environment.* Proceedings of an International Symposium on Agricultural Ecology and Environment, Padova, Italy, 5–7 April. Amsterdam: Elsevier.
   Tivy, J. 1990. *Agricultural ecology.* New York: Longman Scientific and Technical.

7. Gypmantasiri, P., Wiboonpongse, A., Rerkasem, B., Craig, I., Rerkasem, K., Ganjanapan, L., Titayawan, M., Seetisarn, M., Thani, P., Jaisaard, R., Ongprasert, S., Radnachaless, T., and Conway, G. 1980. *An interdisciplinary perspective of cropping systems in the Chiang*

Mai valley: key questions for research. Chiang Mai, Thailand: Faculty of Agriculture, University of Chiang Mai.

8. Walker, B., Norton, G., Barlow, N., Conway, G., Birley, M., and Comins, H. 1978. A procedure for multidisciplinary ecosystem research with reference to South African Savanna Ecosystem Project. *Journal of Applied Ecology* 15:408–502.

9. Lowrance et al., eds. 1984. *Agricultural ecosystems: unifying concepts.*
    Spedding, C. 1975. *The biology of agricultural systems.* London: Academic Press.

10. Conway, G. 1987. The properties of agroecosystems. *Agricultural Systems* 24:95–117.

11. Conway. 1987. The properties of agroecosystems.

12. Conway, G. 1985. Agroecosystem analysis. *Agricultural Administration* 20:31–55.
    Conway. 1987. The properties of agroecosystems.

13. Conway, G. 1987. *Helping poor farmers—a review of Foundation activities in farming systems and agroecosystems research and development.* New York: Ford Foundation.

14. Trenbath, B., Conway, G. and Craig, I. 1990. Threats to sustainability in intensified agricultural systems: analysis and implications for management. In: Gliessman, S., ed. *Agroecology: researching the ecological basis for sustainable agriculture.* New York: Springer-Verlag.

15. Rahman, S. 2010. Six decades of agricultural land use change in Bangladesh: effects on crop diversity, productivity, food availability and the environment, 1948–2006. *Singapore Journal of Tropical Geography* 31:254–269.
    Lunkad, S., and Sharma, A. 2008. Combating negative impact of green revolution on groundwater, soil and land in Haryana, India. In: Paliwal, B., ed. 2008. *Global groundwater resources and management: selected papers from the 33rd International Geological Congress (33rd IGC).* Oslo, Norway, August. http://www.cprm.gov.br/33IGC /1286697.html.
    FAO. 1998. *Rice crisis looms in Asia.* Rome: FAO Spotlight September. http://www.fao .org/ag/magazine/9809/spot1.htm.

16. Black, M., and King, J. 2009. *The atlas of water.* London: Earthscan Publications.
    Rodell, M., Velicogna, I., and Famiglietti, J. 2009. Satellite-based estimates of groundwater depletion in India. *Nature* 460:999–1002.

17. Conway, G., and Pretty, J. 1991. *Unwelcome harvest: agriculture and pollution.* London: Earthscan.
    Pingali, P. 2001. Environmental consequences of agricultural commercialization in Asia. *Environment and Development Economics* 6:483–502.

18. Anand, G. 2010. *Green Revolution in India wilts as subsidies backfire.* http://online.wsj .com/article/SB10001424052748703615904575052921612723844.html.

19. Lee, D., and Barrett, C., eds. 2001. *Tradeoffs or synergies? Agricultural intensification, economic development and the environment.* Wallingford, UK: CABI Publishing.

20. Conway, G., and Waage, J. 2010. *Science and innovation for development.* London: UKCDS. Page 131.

21. Begon et al. 1996. *Ecology*.

    Ives, A. 2005. Community diversity and stability: changing perspectives and changing definitions. In: Cuddington, K., and Beisner, B., eds. *Ecological paradigms lost: routes to theory change*. Amsterdam: Academic Press.

    May and McLean. 2007. *Theoretical ecology*.

22. Ives. 2005. Community diversity and stability.

23. Millennium Ecosystem Assessment. 2005. *Ecosystems and human well-being: synthesis*. Washington, DC: Island Press. http://www.millenniumassessment.org/documents /document.356.aspx.pdf.

24. Conway and Barbier. 1990. *After the Green Revolution*.

25. Vandermeer, J. 1989. *The ecology of intercropping*. Cambridge, UK: Cambridge University Press.

26. Woomer, P., Lan'gat, M., and Tungani, J. 2004. Innovative maize-legume intercropping results in above- and below-ground competitive advantages for understorey legumes. *West African Journal of Applied Ecology* 6:85–94.

27. Brooks, T., Mittermeier, R., da Fonseca, G., Gerlach, J., Hoffmann, M., Lamoreux, J., Mittermeier, C., Pilgrim, J., and Rodrigues, A. 2006. Global biodiversity conservation priorities. *Science* 313:58–61.

    Millennium Ecosystem Assessment. 2005. *Ecosystems and human well-being: synthesis*.

28. World Bank. 2006. *Strengthening forest law enforcement and governance*. Washington, DC: World Bank.

    Costanza R., d'Arge, R., de Groot, R., Farber, S., Grasso, M., Hannon, B., Limburg, K., Naeem, S., O'Neill, R., Paruelo, J., Raskin, R., Sutton, P., and van den Belt, M. 1997. The value of the world's ecosystem services and natural capital. *Nature* 387:253–260.

29. FAO. 1996. *Seeds of Life*. World Food Summit. Rome: FAO.

    Chang, T. 1992. Availability of plant germplasm for use in crop improvement. In: Stalker, H., and Murphy, J. *Plant breeding in the 1990s*. Wallingford, UK: CABI Publishing.

30. Global Crop Diversity Trust. No date. *Crop diversity*. http://www.croptrust.org.

31. FAO Commission on genetic resources for Food and Agriculture. 2009. *Draft Second Report on the State of the World's Plant Genetic Resources for Food and Agriculture. Intergovernmental Technical Working Group on Plant Genetic Resources for Food and Agriculture*. Rome: FAO.

32. The International Treaty on Plant Genetic Resources for Food and Agriculture. No date. *Overview*. http://www.planttreaty.org/.

33. UN. 2009. *Millennium Development Goals Report 2009*. New York: United Nations.

34. Quisumbing, A., Brown, L., Feldstein, H., Haddad, L., and Pen-a, C. 1995. *Women: the key to food security*. Food Policy Report. Washington, DC: IFPRI.

35. FAO. 1985. *Women and developing agriculture*. Women in Agriculture Series No. 4. Rome: FAO.

36. Tripath, R., MacRae, M., and Kent, R. 2009. *Unheard voices, women marginal farmers speak out: a Zambian case study.* Dublin: Concern Worldwide. http://www.concern.net/sites/concern.net/files/documents/unheard-voices/zambiancasestudy.pdf.

37. von Braun, J., de Haen, H., and Blanken, J. 1991. *Commercialization of agriculture under population pressure: effects on production, consumption and nutrition in Rwanda.* Research Report 85. Washington, DC: IFPRI.

38. Mehra, R., and Rojas, M. 2009. *Women, food security and agriculture in a global marketplace, a significant shift.* Washington, DC: International Center for Research on Women.

    IFAD, FAO and World Bank. 2009. *Gender in agriculture sourcebook.* Washington, DC: World Bank.

39. Ehiri, J., ed. 2009. *Maternal and child health, global challenges, programs and policies.* New York: Springer.

    Smith, L. C., Ramakrishnan, U., Ndiaye, A., Haddad, L., and Martorell, R. 2003. *The importance of women's status for child nutrition in developing countries.* IFPRI Research Report 131. Washington, DC: IFPRI.

40. Brown, L., and Haddad, L. 1994. *Time allocation patterns and time burdens: a gendered analysis of seven countries.* Washington, DC: IFPRI.

41. Leonard, H. 1989. Overview: environment and the poor. In: Leonard, H., ed. *Environment and the poor: development strategies for a common agenda.* U.S.—Third World Policy Perspectives, No. 11, Overseas Development Council, referring to Sadik, N. 1988. Women as resource managers. In: UNPF. 1985. *State of the World Population, 1985.* New York: United Nations Population Fund.

42. Chambers, R., and Conway, G. 1992. *Sustainable rural livelihoods: practical concepts for the 21st century.* IDS discussion paper, No. 296. Brighton, UK: Institute of Development Studies.

    Conway, G. 2011. Exploring sustainable livelihoods. In: Cornwall, A., and Scoones, I., eds. *Revoutionising development: reflections on the work of Robert Chambers.* London and Washington, DC: Earthscan.

43. Chambers and Conway. 1992. *Sustainable rural livelihoods.* Page 7.

44. Chambers and Conway. 1992. *Sustainable rural livelihoods.*

45. World Food Programme. 2009. *Rwanda: comprehensive food security and vulnerability analysis and nutrition survey.* Rome: World Food Programme.

46. WFP. 2009. *Rwanda: comprehensive food security and vulnerability analysis and nutrition survey.* Page 12.

47. WFP. 2009. *Rwanda: comprehensive food security and vulnerability analysis and nutrition survey.*

48. Krishna, A. 2004. Escaping poverty and becoming poor: who gains, who loses, and why? *World Development* 32:121–136.

49. Krishna, A., Kristjanson, P., Radeny, M., and Nindo, W. 2004. Escaping poverty and becoming poor in 20 Kenyan villages. *Journal of Human Development* 5:211–226.

50. Krishna, A. 2010. *One illness away: why people become poor and how they escape poverty.* Oxford, UK: Oxford University Press.

51. Krishna, et al. 2004. Escaping poverty and becoming poor in 20 Kenyan villages.

52. Maxwell, S., ed. 1989. *To cure all hunger: food policy and food security in Sudan.* London: Intermediate Technology.

53. Davies, S. 1996. *Adaptable livelihoods: coping with food insecurity in the Mahelian Sahel.* London: Macmillan Press and New York: St. Martin's Press. Reproduced with permission of Palgrave Macmillan.

54. Davies. 1996. *Adaptable livelihoods.*

55. Davies. 1996. *Adaptable livelihoods.*

56. Scoones, I. 1996. New challenges for range management in the 21st century. *Outlook on Agriculture* 25:253–256.

57. World Bank. 2005. *Agricultural growth for the poor: an agenda for development.* Directions in Development 2005. Washington, DC: World Bank.

58. Alden Wily, L. 2000. *Land tenure reform and the balance of power in Eastern and Southern Africa.* Natural resource perspectives No. 58, June. London: Overseas Development Institute.

59. World Bank. 2005. *Agricultural growth for the poor.*

60. Binswanger, H., and Deininger, K. 1997. Explaining agricultural and agrarian policies in developing countries. *Journal of Economic Literature* 35:1958–2005.

7. Appropriate Technology

1. Juma, C. 2011. *The new harvest: agricultural innovation in Africa.* New York: Oxford University Press. Page xviii.

2. Coultate, T. 1996. *Food—the chemistry of its components.* 3rd edition. London: Royal Society of Chemistry.
    Evens, M. *Vitamin A.* Bristol, UK: University of Bristol. http://www.chm.bris.ac.uk/motm/carotene/beta-carotene_vita.html.

3. Conway, G., and Waage, J. 2010. *Science and innovation for development.* London: UKCDS.

4. Conway and Waage. 2010. *Science and innovation for development.*

5. Conway and Waage. 2010. *Science and innovation for development.*

6. Blaylock, A. 2007. *The future of controlled-release fertilisers.* Agrium Advanced Technologies presentation at the International Nitrogen Conference. Costa do Sauipe, Brazil. October 1–5.
    Fairchild, D., and Malzer, G. 2008. *The future of spatial N management.* The Mosaic Company, Plymouth, MN. Presentation. January 1.

7. ICRISAT. 2001. Things grow better with Coke®: micro-fertilizer system sparks 50–100 percent millet yield increases in the Sahel. *SATrends.* Issue 2. Patancheru, India: ICRISAT. http://www.icrisat.org/what-we-do/satrends/01jan/1.htm#Things Grow Better with Coke®.

8. Paul, A. 2010. *Micro-dosing could end famine.* New Internationalist. http://findarticles .com/p/articles/mi_m0JQP/is_436/ai_n56379197/.

9. ICRISAT. 2010. *What ICRISAT thinks . . . about Niger's hunger crisis.* www.icrisat.org /what-we-do/wit/wit_1/wit_1.htm.
    ICRISAT. 2010. *Hunger in Niger could have been prevented.* http://www.icrisat.org/news room/latest-news/one-pager/africa-hunger/africa-hunger-crisis.htm.
    Paul. 2010. *Micro-dosing could end famine.*

10. AATF. 2008. *Striga control in maize.* http://www.aatf-africa.org/striga/.

11. AATF. No date. *Project 1: Striga control in maize—managing a cereal killer.* http://aatf.africa .org/userfiles/Striga-Project-Brief.pdf.

12. Robert, E., Craufurd, R., and Le Cochet, F. 1961. Estimation of percentage of natural cross-pollination: experiment on rice. *Nature* 190:1084–1085.

13. Rerkasem, B., and Rerkasem, K. 2005. *On-farm conservation of rice biodiversity.* Paper presented at FAO international workshop: In-situ conservation of native landraces and their wild relatives. Bangkok, August 29–September 2.

14. Rerkasem and Rerkasem. 2005. *On-farm conservation of rice biodiversity.*

15. Oupkaew, P., Pusadee, T., Sirabanchongkran, A., Rerkasem, K., Jamjod, S., and Rerkasem, B. 2011. Complexity and adaptability of a traditional agricultural system: case study of a gall midge resistant rice landrace from northern Thailand. *Genetic Resources and Crop Evolution* 58:361–372.

16. Rerkasem and Rerkasem. 2005. *On-farm conservation of rice biodiversity.*

17. Oupkaew et al. 2011. Complexity and adaptability of a traditional agricultural system.

18. Oupkaew et al. 2011. Complexity and adaptability of a traditional agricultural system.

19. IDE. 2006. *Bangladesh.* Denver, CO: International Development Enterprises. http:// www.ideorg.org/OurStory/History.aspx.

20. IDE. 2006. *Bangladesh.*

21. KickStart. 2011. *Our impact.* San Francisco: KickStart. http://www.kickstart.org/what -we-do/impact/.

22. IDE. 2006. *I am not worried about my children's future any more—an IDE success story from Bangladesh.* Denver, CO: International Development Enterprises.

23. Conway and Waage. 2010. *Science and innovation for development.*

24. Buter, S. 2010. *Ambridge rules the waves.* Financial Times. December 30. http://www.ft .com/cms/s/2/88b91582–0ee5–11e0–9ec3–00144feabdc0.html#axzz1TVWbmscl.

25. Brockes, E. 2001. *A long way from Ambridge.* http://www.guardian.co.uk/world/2001/oct /23/afghanistan.terrorism3.

Howard, R., Rolf, F., van de Veen, H., and Verhoeven, J. 2003. The power of the media—a handbook for peacebuilders. The Hague: European Centre for Conflict Prevention, European Centre for Common Ground, Institute for Media, Policy and Civil Society (IMPACS). http://www.gppac.net/documents/Media_book_nieuw /p2_2_afghanistan.htm.

26. Smith, D. 2009. *Africa calling: mobile phone usage sees record rise after huge investment.* http:// www.guardian.co.uk/technology/2009/oct/22/africa-mobile-phones-usage-rise.

27. Conway and Waage. 2010. *Science and innovation for development.*

28. Alliance for a Green Revolution in Africa. 2011. *AGRA's market access program.* http:// www.agra-alliance.org/section/work/markets1.

29. Kasozi, E. 2009. *Uganda: using a mobile phone to reduce farming cost.* The Monitor. http:// www.allafrica.com/stories/200907081084.html.
   Ron, L., and Katragadda, L. 2009. *Taking Africa's data to the next level.* Google Africa Blog. http://www.google-africa.blogspot.com/2009/06/google-sms-to-serve-needs-of -poor-in.html.
   Grameen Foundation. 2009. *Uganda projects: AppLab Uganda.* http://www.grameen foundation.applab.org/section/applab-initiatives.

30. Shore, K. 2005. *Work in progress—rural Pondicherry's wireless internet.* Ottowa: The International Development Research Centre. http://www.idrc.ca/en/ev-47023-201-1 -DO_TOPIC.html.

31. Internet World Stats. 2009. *World internet usage and population statistics.* http://www .internetworldstats.com/stats.htm.

32. BBC News. 2009. *East Africa gets high-speed web.* http://news.bbc.co.uk/1/hi/world /africa/8165077.stm.

33. Meridian Institute. 2005. *Nanotechnology and the poor—opportunities and risks.* Dillon, CO: Meridian Institute.

34. UN. 1992. *The Convention on Biological Diversity, Article 2—Use of Terms.* United Nations— Treaty Series.

35. Conway, G. 2003. Biotechnology and the war on poverty. In: Serageldin, I., and Persley, G. *Biotechnology and sustainable development: voices of the south and north.* Proceedings of a Conference at the Bibliotheca Alexandrina, Alexandria, Egypt, March 16–20. Wallingford, UK: CABI Publishing.

36. Bjarnason, M., and Vasal, S. 1992. Breeding of quality protein maize (QPM). *Plant Breeding Reviews* 9:181–216.

37. CGIAR. 2007. *Double agent.* CGIAR News. Washington, DC: CGIAR.
   Wu, K-M., Lu, Y-H., Feng, H-Q., Jiang, Y-Y., and Zhao, J-Z. 2008. Suppression of cotton bollworm in multiple crops in China in areas with *BT* toxin containing cotton. *Science* 321:1676–1678.

38. James, C. 2009. Global Status of Commercialized Biotech/GM Crops: 2009. *ISAAA Brief* No. 41. Ithaca, NY: ISAAA.

39. International Service for the Acquisition of Agri-biotech Applications. 2009. *Burkina Faso Farmers Gaining from BT Cotton.* http://www.isaaa.org/kc/cropbiotechupdate /article/default.asp?ID=5068.

ISAAA. 2009. *Brief 41–2009: Global status of commercialized biotech/GM crops: 2009. The first fourteen years, 1996 to 2009.* http://www.isaaa.org/resources/publications/briefs /41/executivesummary/default.asp.

40. World Bank. 2007. *World development report 2008.* Washington, DC: Agriculture for Development, World Bank.

41. World Bank. 2007. *World development report 2008.* Page 178.

42. Dorward, A., Kydd, J., Morrison, J., and Urey, I. 2004. A policy agenda for pro-poor agricultural growth. *World Development* 32:73–89.

43. Dorward, et al. 2004. A policy agenda for pro-poor agricultural growth.

44. Alston, J., Mara, M., Pardey, P. and Wyatt, T. 2000. Research returns redux: a meta analysis of the returns to agricultural R & D. *Australian Journal of Agricultural and Resource Economics.* 44:185–215.

World Bank. 2005. *Agricultural growth for the poor: an agenda for development. Directions in Development 2005.* Washington, DC: World Bank.

45. von Braun, J., and Díaz-Bonilla, E. 2008. Globalization of agriculture and food: causes, consequences, and policy implications. In: von Braun, J., and Díaz-Bonilla, E., eds. *Globalization of food and agriculture and the poor.* New Delhi, India: Oxford University Press.

46. UNECA. 2009. *Economic report on Africa 2009. Challenges to agricultural development in Africa.* Addis Ababa, Ethiopia: United Nations Economic Commission for Africa. African Union.

47. UNECA. 2009. *Economic report on Africa 2009.*

48. Conway, G. 1972. Ecological aspects of pest control in Malaysia. In: Farvar, J., and Milton, J., eds. *The Careless Technology: Ecological aspects of international development.* Garden City, NY: Natural History Press, Doubleday and Co.

49. Alene, A., and Coulibaly, O. 2009. The impact of research on productivity and poverty in Sub-Saharan Africa. *Food Policy* 34:198–209.

50. Conway and Waage. 2010. *Science and innovation for development.*

51. Conway and Waage. 2010. *Science and innovation for development.* Page 6.

52. Wagner, C. 2008. *The new invisible college, science for development.* Washington, DC: Brookings Institution Press.

53. Conway and Waage. 2010. *Science and innovation for development.*

54. Conway and Waage, 2010. *Science and innovation for development.*

55. Wagner. 2008. *The new invisible college, science for development.*

56. Raitzer, D. 2003. *Benefit-cost meta-analysis of investment in the international agricultural research centres of the CGIAR.* Consultative Group on International Agricultural

Research Science Council. Prepared on Behalf of the CGIAR Standing Panel on Impact Assessment. Rome: FAO.

Raitzer, D., and Kelley, T. 2008. Benefit-cost meta-analysis of investment in the international agricultural research centres of the CGIAR. *Agricultural Systems* 96:108–123.

57. Evenson, R., and Gollin, D., eds. *Crop variety improvement and its effect on productivity: the impact of international agricultural research*. Wallingford, UK: CABI Publishing.

58. Quilligan, E. 2010. *New maize brings hope to Kenya's drylands*. El Batan, Mexico: CIMMYT. http://blog.cimmyt.org/?p=6853.

59. Renkow, M., and Byerlee, D. 2010. The impacts of CGIAR research: a review of recent evidence. *Food Policy* 35:391–402.

60. CGIAR. 2009. *Change Management*. http://www.cgiar.org/changemanagement/index .html, http://www.sciencecouncil.cgiar.org/home/en/; http://www.cgiarfund.org /cgiarfund/.

8. Creating Markets

1. World Food Programme. 2011. *Ghana: farmers come to trust markets*. Rome: WFP. http://www.wfp.org/purchase-progress/blog/ghana-farmers-come-trust-markets.

2. Tripp, R. 2003a. *The enabling environment for agricultural technology in Sub-Saharan agriculture and the potential role of donors*. Natural Resource Perspectives, No. 84, London: Overseas Development Institute.

    Tripp, R. 2003b. *Strengthening the enabling environment for agricultural technology development in Sub-Saharan Africa*. Working Paper 212. London: Overseas Development Institute.

    Brinkerhoff, D. 2004. *The enabling environment for implementing the Millennium Development Goals: government actions to support NGOs*. Paper presented at: George Washington University Conference "The Role of NGOs in Implementing the Millennium Development Goals." Washington, DC, May 12–13.

3. World Bank. 1992. *The political economy of agricultural pricing policy*. Baltimore, MD: Johns Hopkins University Press.

4. Binswanger, H., Deininger, K., and Feder, G. 1995. Power, distortions, revolt and reform in agricultural land relations. In: Behrman, J. and Srinivasan, T. *Handbook of development economics vol IIIB*. Amsterdam: Elsevier.

5. Lipton, M., and Lipton, M. 1993. Creating rural livelihoods: some lessons for South Africa from experience elsewhere. *World Development* 21:1515–1548.

    Hazell, P. and Garrett, J. 1996. *Reducing poverty and protecting the environment: the overlooked potential of less-favoured lands*. 2020 Brief 39. Washington, DC: IFPRI.

6. International Fund for Agricultural Development. 2010. *Rural poverty report 2011—new realities, new challenges: new opportunities for tomorrow's generation*. Rome: IFAD.

7. Onumah, G., Davis, J., Kleih, U., and Proctor, F. 2007. *Empowering smallholder farmers in markets: changing agricultural marketing systems and innovative responses by producer*

organizations. Working Paper 2, Empowering smallholder farmers in markets (ES-FIM), Wageningen, The Netherlands: Wageningen University and Research Centre. IFAD. 2010. *Rural poverty report 2011.*

8. Parikh, A. 2007. *Trade liberalisation: impact on growth and trade in developing countries.* Singapore: World Scientific Publishing; Ganesh-Kumar, A., Roy, D. and Gulati, A. 2010. *Liberalizing food grains markets: experiences, impact and lessons from South Asia.* IFPRI Issue Brief 64. Washington, DC: IFPRI.

9. Babu, S., and Sanyal, P. 2009. Case study #7–2: persistent food insecurity from policy failures in Malawi. In: Per Pinstrup-Andersen and Fuzhi Cheng, eds., *Case studies in food policy for developing countries.* Ithaca, NY: Cornell University Press.

10. Von Braun, J. 2009. Addressing the food crisis: governance, market functioning and investment in public goods. *Food Security* 1:9–15.

11. Brown, O., and Gibson, J. 2006. *Boom or bust: developing countries' rough ride on the commodity price rollercoaster.* Manitoba, Canada: International Institute for Sustainable Development.

12. IFAD. 2010. *Rural poverty report 2011.*

13. Berdegué, J., Biénabe, E., and Peppelenbos, L. 2008. *Keys to inclusion of small-scale producers in dynamic markets: innovative practices in connecting small-scale producers with dynamic markets.* Regoverning Markets Innovative Practice Series. London: International Institute for Environment and Development.

14. FAO. 2008. *The state of food and agriculture in Asia and the Pacific region.* Rome: FAO. http://www.fao.org/docrep/010/ai411e/AI411E02.htm.

15. IFAD. 2010. *Rural poverty report 2011.*

16. ECLAC, FAO, and IICA. 2009. *The outlook for agricultural and rural development in the Americas: a perspective on Latin America and the Caribbean.* San Jose, Costa Rica: IICA.

17. Henson, S. 2006. *New markets and their supporting institutions: opportunities and constraints for demand growth.* Background paper for the World Bank, *World Development Report 2008: Agriculture for development.* http://go.worldbank.org/GLF6HRYFI0.

18. IFAD. 2010. *Rural poverty report 2011.* Adapted from Henson. 2006. *New markets and their supporting institutions.* Page 120.

19. GKPnet.projects. No date. SokoniSMS: Empowering farmers through SMS market in Kenya. http://www.gkpnet.org/projects/public/ict4dinitiatives/view.do?gkpprojectid=32600.

20. Ethiopia Commodity Exchange. 2011. *Company profile.* http://www.ecx.com.et/CompanyProfile.aspx.

21. Gabre-Madhin, E. 2011. *1000 days of ECX!* http://www.ecx.com.et/CEOmessages.aspx.

22. Economic Report on Africa 2009. *Challenges to agricultural development in Africa.* Addis Ababa, Ethiopia: United Nations Economic Commission for Africa and African Union.

23. Pingali, P., and Rosegrant, M. 1995. Agricultural commercialization and diversification: processes and policies. *Food Policy* 20:171–185.

24. Von Braun. 2009. Addressing the food crisis.

25. Pingali and Rosegrant. 1995. Agricultural commercialization and diversification.

26. Jongwanich, J. 2009. The impact of food safety standards on processes food exports from developing countries. *Food Policy* 34:447–457.

27. Jongwanich. 2009. The impact of food safety standards.

28. Kirtsen, J., and Sartorius, K. 2002. Linking agri-business and small-scale farmers in developing countries: is there a new role for contract farming? *Development in Southern Africa* 19:503–529.

   Narrod, C., Roy, D., Okello, J., Avendaño, B., Rich, K., and Thorat, A. 2009. Public-private partnerships and collective action in high value fruit and vegetable supply chains. *Food Policy* 34:8–15.

29. Jongwanich. 2009. The impact of food safety standards.

   Narrod et al. 2009. Public-private partnerships and collective action.

30. Jongwanich. 2009. The impact of food safety standards.
   Narrod et al. 2009. Public-private partnerships and collective action.

31. Narrod et al. 2009. Public-private partnerships and collective action.

32. Okello, J., and Swinton, S. 2005. Compliance with international food safety standards in Kenya's green bean industry: comparison of a small- and a large-scale farm producing for export. *Applied Economic Perspectives and Policy* 29:269–285.

33. Narrod et al. 2009. Public-private partnerships and collective action.
   Kirtsen and Sartorius. 2002. Linking agri-business and small-scale farmers.

34. Narrod et al. 2009. Public-private partnerships and collective action.

35. Canadian International Development Agency. 2008. *Faso Jigi: a people's hope.* http://www.acdi-cida.gc.ca/acdi-cida/acdi-cida.nsf/eng/FRA-42715145-QBN.

36. IFAD. 2010. *Rural poverty report 2011.*

37. Berdegué, J. 2008. Rural producer organisations in Chile: cooperating to compete—easier said than done. *Capacity.org.* Issue 34, August. http://www.capacity.org/en/journal/archives/producer_organisations_and_value_chains.

38. Losch, B., Fréguin-Gresh, S., and White, E. 2010. *Structural dimensions of liberalization on agriculture and rural development: a cross-regional analysis on rural change.* Synthesis report of the RuralStruc Program. Washington, DC: World Bank.

39. Kirtsen and Sartorius. 2002. Linking agri-business and small-scale farmers.

40. Kirtsen and Sartorius. 2002. Linking agri-business and small-scale farmers.
   Investopedia. No date. *Spot markets.* http://www.investopedia.com/terms/s/spotmarket.asp.

41. Birthal, P., and Joshi, P. Case study #6–4: Smallholder farmers access to markets for high-value agricultural commodities in India. In: Per Pinstrup-Andersen and Fuzhi Cheng, eds, *Case studies in food policy for developing countries.* Ithaca, NY: Cornell University Press.

Birthal, P., Joshi, P., and Gulati, A. 2005. *Vertical coordination in high-value food commodities: Implications for smallholders. Markets, Trade, and Institutions Division.* Discussion Paper No. 85. Washington, DC: IFPRI. http://www.ifpri.org/divs/mtid/dp/mtidp85.htm.

42. Berdegué et al. 2008. *Keys to inclusion of small-scale producers in dynamic markets.*

43. Aliguma, L., Magala, D., and Lwasa, S. 2007. *Uganda: connecting small-scale producers to markets: the case of the Nyabyumba United Farmers Group in Kabale district.* Regoverning Markets Innovative Practice Series. London: International Institute for Environment and Development.

44. Datamonitor Research Store. 2010. *Organic food: global industry guide.* http://www.datamonitor.com/store/Product/organic_food_global_industry_guide_2010?productid=C9A72F75–510A-4CC7-AA28–9DE6FE209E1A.

45. World Fair Trade Organization. 2011. *About Fair Trade.* http://www.wfto.com/index.php?option=com_frontpageandItemid=1.

46. World Fair Trade Organization. 2011. *About Fair Trade.*

47. Fairtrade Foundation. 2009. *Global Fairtrade sales increase by 22%.* http://www.fairtrade.org.uk/press_office/press_releases_and_statements/jun_2009/global_fairtrade_sales_increase_by_22.aspx.

48. Fairtrade Foundation. 2009. *Toledo Cacao Growers' Association: a profile of a cocoa farmers' cooperative in Southern Belize.* http://www.fairtrade.org.uk/producers/cacao/toledo_cacao_growers_association_belize/default.aspx.

49. Alpert, E. 2011. *Agro-dealers in Malawi help make sure seeds grow.* http://www.one.org/blog/2011/03/30/agro-dealers-in-malawi-help-make-sure-seeds-grow/.

50. Alliance for a Green Revolution in Africa. 2011. *Early Accomplishments.* http://www.agra-alliance.org/section/about/earlyaccomplishments.

51. NYASA Times. 2011. *Malawi's rural agro-dealers want fertilizer subsidy.* http://www.nyasatimes.com/national/malawi%E2%80%99s-rural-agro-dealers-want-fertilizer-subsidy.html.
   Alpert. 2011. *Agro-dealers in Malawi help make sure seeds grow.*

52. CNFA. 2008. *Our work—Kenya: Agrodealer Strengthening Program.* http://www.cnfa.org/kasp.

53. Skees, J., Hazell, P., and Miranda, M. 1999. *New approaches to crop yield insurance in developing countries.* Environment and Production Technology Division Discussion Paper No. 55. Washington, DC: IFPRI.

54. Collins, D. Morduch, J., Rutherford, S., and Ruthven, O. 2009. *Portfolios of the poor: how the world's poor live on $2 a day.* Princeton, NJ: Princeton University Press.

55. Chaia, A., Dalal, A., Goland, T., Gonzalez, M., Morduch, J., and Schiff, R. 2009. *Half the world is unbanked.* Framing Note. Financial Access Initiative. http://www.larevista agraria.info/sites/default/files/revista/r-agra119/LRA-119.pdf.
   Skees et al. 1999. *New approaches to crop yield insurance.*

56. Skees et al. 1999. *New approaches to crop yield insurance.*

57. Skees et al. 1999. *New approaches to crop yield insurance.*

58. Levin, T. and Reinhard, D. 2007. *Microinsurance aspects in agriculture.* Discussion Paper. Munich, Germany: Munich Re Foundation.

59. IFAD. 2010. *Rural poverty report 2011.*

60. Mude, A., Barrett, C., Carter, M., Chantarat, S., Ikegami, M., and McPeak, J. 2010. *Project summary—Index based livestock insurance for Northern Kenya's arid and semi-arid lands: the Marsabit pilot.* ILRI. http://mahider.ilri.org/bitstream/10568/494/1/IBLI _PROJECT_SUMMARY_0110.pdf.

61. ILRI. No date. Personal communication between Brenda Wandera and Robert Ouma of ILRI and Sara Delaney.

62. IFAD. 2010. *Rural poverty report 2011.*

63. Skees et al. 1999. *New approaches to crop yield insurance.*

64. Pretty, J. 1995. *Regenerating agriculture: policies and practice for sustainability and self-reliance.* London: Earthscan.

65. Fernandez, A. 1992. *The MYRADA Experience: alternative management systems for savings and credit of the rural poor.* Bangalore: MYRADA.
    Ramaprasad, V., and Ramachandran, V. 1989. *Celebrating awareness.* Bangalore: MY-RADA and New Delhi: Foster Parents Plan International.

66. Hossain, M. 1988. *Credit alleviation of rural poverty: the Grameen Bank in Bangladesh.* IF-PRI Research Report 65. Washington, DC: IFPRI; Grameen Bank. 2009. *Historical data series in USD.* http://www.grameen-info.org/index.php?option=com_con tentandtask=viewandid=177andItemid=144.
    Jain, P. S. 1996. Managing credit for the rural poor: lessons from the Grameen Bank. *World Development* 24:79–89.

67. Yunus, M., ed. 1984. *Jorimon of Beltoil Village and others: in search of a future.* Dhaka, Bangladesh: Grameen Bank.

68. Yunus. 1984. *Jorimon of Beltoil Village and others.*

69. IFAD. 2010. *Rural poverty report 2011.*

70. ICRISAT. 2010. *What ICRISAT thinks . . . about Niger's hunger crisis.*http://www.icrisat .org/what-we-do/wit/wit_1/wit-september-2005.pdf.
    ICRISAT. 2010. *Hunger in Niger could have been prevented.*http://www.icrisat.org/news room/latest-news/one-pager/africa-hunger/africa-hunger-crisis.htm.

71. IFAD. 2008. *Empowering farmers in Tanzania through the warehouse receipt system.* http:// www.ruralpovertyportal.org/c/document_library/get_file?p_I_id=60350and folerId=100461andname=DLFE-1441.pdf.

72. Grameen Bank. 2011. *About us.* Dhaka, Bangladesh: Grameen Bank. www.grameen -info.org.

73. The Economist. 2010. *A better mattress: microfinance focuses on lending. Now the industry is turning to deposits.* http://www.economist.com/node/15663834.

74. IFAD. 2010. *Rural poverty report 2011.*

75. Fan, S. 2010. *Halving hunger: meeting the first Millennium Development Goal through "business as unusual."* Washington, DC: IFPRI.

76. Birthal and Joshi. 2007. Smallholder farmers access to markets.

77. USAID. 2002. *FY 2002 annual report for Tanzania.* Dar es Salaam, Tanzania: USAID.

78. IFAD. 2010. *Rural poverty report 2011.*
    Dorward, A., Kydd, J., Morrison, J., and Urey, I. 2004. A policy agenda for pro-poor agricultural growth. *World Development* 32:73–89.

79. FAO. 2009. *FAOSTAT.* Rome: FAO. http://faostat.fao.org/default.aspx; CIA. 2010. *World factbook.* Washington, DC: CIA. https://www.cia.gov/library/publications/the-world-factbook/.

80. African Development Bank and UN Economic Commission for Africa. 2003. *Review of the implementation status of the Trans African Highways and the missing links—Volume 2: description of corridors.* Stockholm: SWECO International AB, Nordic Consulting Group AB.

81. Yara International ASA. 2011. *The Beira agricultural corridor.* http://www.yara.com/sustainability/africa_program/partnerships/beira_agricultural_corridor/index.aspx. Beira Corridor. 2011. "About." http://www.beiracorridor.com/index.php.

82. Santos, R. 2010. *The Maasai tribe goes mobile.* http://2010globalmarketing.wordpress.com/2010/07/24/695/.

83. Graham, F. 2010. *M-Pesa: Kenya's mobile wallet revolution.* http://www.bbc.co.uk/news/business-11793290.

84. IT News Africa. 2010. *Vodacom and Nedbank launch M-PESA in South Africa.* http://www.itnewsafrica.com/2010/03/vodacom-and-nedbank-to-launch-m-pesa-in-south-africa/.

85. Jack, W., and Suri, T. 2010. *The economics of M-PESA.* http://www.mit.edu/~tavneet/M-PESA.pdf.

86. Jack and Suri. 2010. *The economics of M-PESA.*
    Graham. 2010. *M-Pesa: Kenya's mobile wallet revolution.*

87. Jack and Suri. 2010. *The economics of M-PESA.*

88. IFAD. 2010. *Rural poverty report 2011.*

9. Designer Crops

1. James, C. 2010. Global status of commercialized biotech/GM crops: 2010. *ISAAA Brief* No.42. Ithaca, NY: ISAAA. http://www.isaaa.org/resources/publications/briefs/42/executivesummary/default.asp.

2. Acquaah, G. 2007. *Principles of plant genetics and breeding.* Oxford, UK: Blackwell Publishing.

3. Gale, M., and Youssefian, S. 1985. Dwarfing genes in wheat. In: Russell, G., ed. *Progress in plant breeding.* London: Butterworths.

4. Hedden, P. 2003. The genes of the Green Revolution. *Trends in Genetics* 19:5–9.

5. Kingsbury, N. 2009. *Hybrid: The history and science of plant breeding.* Chicago: University of Chicago Press.

6. Duvick, D. 1997. Heterosis: feeding people and protecting natural resources: In: Coors, J. G., and Pandey, S., eds. *International Symposium on the Genetics and Exploitation of Heterosis in Crops.* Mexico City, Mexico.

7. Kingsbury. 2009. *Hybrid;* Duvick, D. 2001. Biotechnology in the 1930s: the development of hybrid maize. *Nature Reviews Genetics* 2:69–74.

8. IRRI. 2009. *Cereal knowledge bank maize.* Los Baños, Philippines: IRRI. http://www.knowledgebank.irri.org/ckb/index.php/quality-seeds/what-is-hybrid-maize/4-what-is-an-opv.

9. Agricultural commodity prices and futures. 2011. *Mexican corn production and imports.* http://www.agricommodityprices.com/futures_prices.php?id=483.

10. Pingali, P., and Pandey, S. 2000. *Part 1—meeting world maize needs: technological opportunities and priorities for the public sector.* CIMMYT world maize facts and trends. http://apps.cimmyt.org/Research/Economics/map/facts_trends/maizeft9900/pdfs/maizeft9900_Part1a.pdf.

11. Longping, Y. 2004. *Hybrid rice for food security in the world.* FAO Rice Conference 12–13 February. Rome: FAO. http://www.fao.org/rice2004/en/pdf/longping.pdf.

12. Li, J., Xin, Y., and Yuan, L. 2009. *Hybrid rice technology development: ensuring China's food security.* IFPRI Discussion Paper 00918. Washington, DC: International Food Policy Research Institute.

13. Rice Knowledge Bank. 2009. *Hybrid seed rice production: restorer line.* Los Baños, Philippines: IRRI. http://www.knowledgebank.irri.org/extension/hybridricehybrid-rice-parental-lines/hybridricerestorer-line.

14. Rice Knowledge Bank. 2009. *Hybrid seed rice production.*

15. Li, et al. 2009. *Hybrid rice technology development.* Page 5.

16. Li et al. 2009. *Hybrid rice technology development.* Figure 1. Reproduced with permission from the International Food Policy Research Institute, www.ifpri.org. The original version of the publication in which this figure appears can be found online at http://www.ifpri.org/publication/ifpridp00918.

17. Gibson J., and Smith, C. 1989. The incorporation of biotechnologies into animal breeding strategies. In: Babiuk L., Phillips J., and Moo-Young, M., eds. *Animal biotechnology. Comprehensive biotechnology first supplement.* Oxford, UK: Pergamon Press.

18. Van Vleck L. 1981. Potential genetic impact of artificial insemination, sex selection, embryo transfer, cloning and selfing in dairy cattle. In: Brackett B., Seidel Jr G., and Seidel S., eds. *New technologies in animal breeding.* New York: Academic Press.

19. FAO. 2010. *Current status and options for livestock biotechnologies in developing countries.* FAO International Technical Conference. Guadelajara, Mexico, March 1–4. http://www.fao.org/docrep/meeting/019/k6695e.pdf.

20. Cunningham, E., and Syrstad, O. 1987. *Crossbreeding Bos indicus and Bos taurus for milk production in the tropics.* FAO Animal Production and Health Paper 68. Rome: FAO.

21. Syrstad, O. 1989. The role and mechanisms of genetic improvement in production systems constrained by nutritional and environmental factors. http://www.fao.org/docrep/003/t0413e/T0413E05.htm. In: Speedy, A., and Sansoucy, R. *Feeding dairy cows in the tropics.* Proceedings of the FAO expert Consultation. Bangkok, Thailand. July.

22. Syrstad, O. 1996. Dairy cattle crossbreeding in the tropics. The choice of crossbreeding strategy. *Tropical Animal Health and Production* 28:223–229.

23. Thorpe, W., Morris, C., and Kangethe, P. 1994. Crossbreeding of Ayrshire, Brown Swiss and Sahiwal cattle for annual and lifetime milk yield in the lowland tropics of Kenya. *Journal of Dairy Science* 77:415–427.

24. Garnett, T. 2011. Where are the best opportunities for reducing greenhouse gas emissions in the food system (including the food chain). *Food Policy* 36:S23–S32.

25. Stalker, H., and Murphy, J., eds. 1992. *Plant breeding in the 1990s: proceedings of the Symposium on Plant Breeding in the 1990s.* Wallingford, UK: CAB International.

26. Ruane, J., and Sonnino, A. 2007. Marker-assisted selection as a tool for genetic improvement of crops, livestock, forestry and fish in developing countries: an overview of the issues. In: Guimarães, E., Raine, J., Scherf, B., Sonnino, A., and Dargie, J., eds. 2007. *Marker-assisted selection: current status and future perspectives in crops, livestock, forestry and fish.* Rome: FAO.

27. Koebner, R. 2003. MAS in cereals: green for maize, amber for rice, still red for wheat and barley. In: *Proceedings of the International Workshop on Marker-Assisted Selection: A Fast Track to Increase Genetic Gain in Plant and Animal Breeding?* http://www.fao.org/biotech/docs/Koebner.pdf.

28. Jeffers, D. 2001. *Maize pathology research: increasing maize productivity and sustainability in biologically stressed environments.* El Batan, Mexico: CIMMYT.

29. Voesenek, L., and Bailey-Serres, J. 2009. Plant biology: genetics of high-rise rice. *Nature* 460:959–960.

30. Hattori, Y., Nagai, K., Furukawa, S., Song, X-J., Kawano, R., Sakakibara, H., Wu, J., Matsumoto, T., Yoshimura, A., Kitano, H., Matsuoka, M., Mori, H., and Ashikari, M. 2009. The ethylene response factors SNORKEL1 and SNORKEL2 allow rice to adapt to deep water. *Nature* 460:1026–1030.

31. Xu, K., Xu, X., Fukao, T., Canlas, P., Maghirang-Rodriguez, R., Heuer, S., Ismail, A. M., Bailey-Serres, J., Ronald, P. C., and Mackill, D. J. 2006. Sub1A is an ethylene-response-factor-like gene that confers submergence tolerance to rice. *Nature* 442:705–708.

32. Voesenek and Bailey-Serres. 2009. Plant biology.

33. Dolgin, E. 2009. The resistant rice of the future: cross breeding could create rice varieties that can survive flooding and fungi. *Nature online.* doi:10.1038/news.2009.841.

34. Siangliw, M., Toojinda, T., Tragoonrung, S., and Vanavichit, A. 2003. Thai jasmine rice carrying QTLch9 (SubQTL) is submergence tolerant. *Annals of Botony* 91:255–261.

35. Department for International Development. 2010. *Sowing the seeds of scuba rice.* http://www.dfid.gov.uk/stories/case-studies/2010/sowing-the-seeds-of-scuba-rice/.

36. Blair, M., Fregene, M., Beebe, S., and Ceballos, H. 2007. Marker-assisted selection in common beans and cassava. In Guimarães E., Ruane J., Scherf B., Sonnino A., and Dargie J., eds. 2007. *Marker-assisted selection: current status and future perspectives in crops, livestock, forestry and fish.* Rome: FAO. http://www.fao.org/docrep/010/a1120e/a1120e00.htm.

37. William, H., Morris, M., Warburton, M., and Hoisington, D. 2007. Technical, economic and policy considerations on marker-assisted selection in crops: lessons from the experience at an international agricultural research centre. In Guimarães et al. 2007. *Marker-assisted selection.*

38. Acquaah. 2007. *Principles of plant genetics and breeding.*

39. Conway, G., and Waage, J. 2010. *Science and innovation for development.* London: UK Collaborative on Development Sciences.
    Bhandari N. 2008. *Going bananas: fighting hunger with Africa Harvest.* http://www.peopleandplanet.net/doc.php?id=3434.
    Riungu, C. 2008. Banana research earns Dr. Wambugu world fame. *The East African.* Nairobi. 29 September–5 October.

40. Mbogo, S. 2008. Kenya: farmers turn to new banana variety. *Business daily.* Nairobi. 12 May.
    Africa Harvest. 2009. *Chura Community Project.* http://africaharvest.org/chura.php.

41. Somado, E. Guei, R., and Keya, S., eds. 2008. *NERICA, the New Rice for Africa, a Compendium.* Cotonou, Benin: WARDA.

42. Conway and Waage. 2010. *Science and innovation for development.*
    Somado et al. 2008. *NERICA, the New Rice for Africa.*

43. Kijima, Y., Otsuka, K., and Ssenkuuma, D. 2007. *Assessing the impact of a NERICA on income and poverty in Central and Western Uganda.* FSAID Discussion Paper Series on International Development Strategies. No. 2007–10–001. http://www.fasid.or.jp/daigakuin/fa_gr/kyojyu/pdf/discussion/2007–10–001.pdf.

44. Adekambi, S., Diagne, A., Simtowe, F., and Blaou, G. 2009. *The impact of agricultural technology adoption on poverty: the case of NERICA rice varieties in Benin.* Paper prepared for presentation at the International Association of Agricultural Economists' Conference 2009. Beijing, China, August 16–22. http://ageconsearch.umn.edu/bitstream/51645/2/473.pdf.

45. Acquaah. 2007. *Principles of plant genetics and breeding.*

46. Larkin, P.1994. *Genes at work: biotechnology*. Collingwood, Australia: CSIRO Publishing.

47. Acquaah. 2007. *Principles of plant genetics and breeding*.

48. Peel, M. 2001. *A basic primer on biotechnology*. NDSU Extension Service. http://www .ag.ndsu.edu/pubs/plantsci/crops/a1219w.htm.

49. The World Food Prize. 2011. *2000: Vasal and Villegas*. http://www.worldfoodprize.org /en/laureates/20002009_laureates/2000_vasal_and_villegas/.
    Villegas, E. 1994. Factors limiting quality protein maize (QPM) development and utilization, pp. 79–88. In: Larkins, B., and Mertz, E., eds. *Quality protein maize: 1964–1994*. Proceedings of the international symposium on quality protein maize. December 1–3, EMBRAPA/CNPMS. Sete Lagoas, Minas Gerais, Brazil.

50. Bjarnason, M., and Vasal, S. 1992. Breeding of quality protein maize (QPM). *Plant Breeding Reviews* 9:181–216.

51. Schoof, R. 2010. *Sweet potato a vanguard in global fight to curb severe malnutrition*. The Modesto Bee, November 13. http://www.modbee.com/2010/11/13/1426973/ vegetable-a-vanguard-in-global.html.
    HarvestPlus. 2009. *Sweet potato*. http://www.harvestplus.org/content/sweet-potato.

52. HarvestPlus. 2009. *Sweet potato*.

53. Golden Rice Humanitarian Board. 2007. *The science behind Golden Rice*. http://www .goldenrice.org/Content2-How/how1_sci.html.
    Potrykus, I., (Pers. Comm. 27 Oct 2009).

54. Golden Rice Project. 2010. *FAQ*. http://www.goldenrice.org/Content3-Why/why3 _FAQ.html.

55. The New Nation. 2010. *Golden rice by 2012, hopes Bangladesh Rice Research Institute*. http://greenbio.checkbiotech.org/news/golden_rice_2012_hopes_bangladesh_ rice_research_institute.

56. Anderson, K., Jackson, L., and Nielsen, C. 2005. Genetically modified rice adoption: implications for welfare and poverty alleviation. *Journal of Economic Integration* 20:771–788.

57. James. 2010. Global status of commercialized biotech/GM crops.

58. James. 2010. Global status of commercialized biotech/GM crops.

59. James. 2010. Global status of commercialized biotech/GM crops.

60. James. 2010. Global status of commercialized biotech/GM crops.

61. James. 2010. Global status of commercialized biotech/GM crops.

62. Mooney, H., and Bernardi, G., eds. 1990. *Introduction of genetically modified organisms into the environment*. Scientific Committee on Problems of the Environment. Chichester, UK: John Wiley and Sons.
    Casper, R., and Landsmann, J., eds. 1992. *The biosafety results of field tests of genetically modified plants and microorganisms*. Braunschweig, Germany: Biologische Bundensansalt für Land-und Fortwirtschaft.

63. Lewis P. 1992. Mutant foods create risks we can't yet guess. *The New York Times,* June 16.

64. Lemaux, P. 2008. Genetically engineered plants and foods: a scientist's analysis of the issues (part I). *Annual Review of Plant Biology* 59:771–812.

65. McHughen A. 2006. *Plant genetic engineering and regulation in the U.S.* University of California Agriculture and Natural Resources. Agricultural Biotechnology in California: Safety of Genetically Engineered Food. Publ. 8179.

66. Flachowsky G. 2007. Feeds from genetically engineering plants—results and future challenges. *ISB News Rep.* March 4–7.

67. Lemaux. 2008. Genetically engineered plants and foods.

68. Lemaux. 2008. Genetically engineered plants and foods.

69. Lemaux. 2008. Genetically engineered plants and foods.

70. Lemaux. 2008. Genetically engineered plants and foods. Page 797.

71. Rissler, J. and Melloon, M. 1996. *The ecological risks of genetically engineered crops.* Cambridge, MA: MIT Press.
    Lemaux, P. 2009. Genetically engineered plants and foods: a scientist's analysis of the issues (part II). *Annual Review of Plant Biology* 60:511–519.

72. Warwick S., Láeg'ere A., Simard M-J., and James T. 2008. Do escaped transgenes persist in nature? The case of an herbicide resistance transgene in a weedy *Brassica rapa* population. *Molecular Ecology* 17:1387–1395.

73. Clegg, M., Giddings, L., Lewis, C., and Barton, J., eds. 1993. *Report of the International Consultation on Rice Biosafety in Southeast Asia.* September 1–2. Technical Paper, Biotechnology Series No. 1. Washington, DC: World Bank.

74. Paarlberg, R. 2009. *Starved for science: how biotechnology is being kept out of Africa.* Cambridge, MA and London, UK: Harvard University Press.

75. Vidal, J. 1999. How Monsanto's mind was changed. *Guardian Weekly.* October 14–20.

76. Meldosi, A. 2011. GM bananas. *Nature Biotechnology* 29:472.

77. The Royal Society. 2009. *Reaping the benefits: science and the sustainable intensification of global agriculture.* London: Royal Society.

10. The Livestock Revolution

1. Delgado, C., Rosegrant, M., Steinfeld, H., Ehui, S., and Courbois, C. 1999. Livestock to 2020. *The next food revolution.* Food, Agriculture and the Environment Discussion Paper No. 28. Washington, DC: IFPRI, Rome: FAO and Nairobi: ILRI.

2. de Haan C., Steinfeld H., and H. Blackburn. 1997. *Livestock and the environment: finding a balance.* Fressingfield, UK: WRENmedia.
   Blench R. 2000. '*You can't go home again'—extensive pastoral livestock systems: issues and options for the future.* ODI/FAO report. London: Overseas Development Institute.

3. Thornton P., Kruska, R., Henninger, N., Kristjansen, P., Reid, R., Atieno, F., Odero, A., and Ndegwa, T. 2002. *Mapping poverty and livestock in the developing world*. Nairobi: ILRI. http://www.ilri.org/InfoServ/Webpub/fulldocs/Mappoverty/index.htm.

4. McSweeney C., Dalrimple B., Gobius K., Kennedy P., Krause D., Merckie R., and Xue G. 1999. The application of rumen biotechnology to improve the nutritive value of fibrous feedstuffs: pre- and post ingestion. *Livestock Production Science* 59:265–283.

5. Delgado et al. 1999. Livestock to 2020. *The next food revolution*. Page 2. Reproduced with permission from the International Food Policy Research Institute, www.ifpri .org. This discussion paper can be found at this web address: http://www.ifpri.org /sites/default/files/pubs/2020/dp/dp28.pdf.

6. FAO. 2006. *World agriculture: towards 2030/2050. Interim report. Prospects for food, nutrition, agriculture and major commodity groups.* Global Perspective Studies Unit. Rome: FAO.

7. FAO. 2009. *The state of food and agriculture: livestock in the balance.* Rome: FAO.

8. Demment, M., and Allen, L., eds. 2003. *Animal source foods to Improve micronutrient nutrition and human function in developing countries,* Supplement to the *Journal of Nutrition* 133:3875S–4061S.

9. Randolph, T., Schelling, E., Grace, D., Nicholson, C., Leroy, J., Cole, D., Demment, M., Omore, A., Zinsstag, J., and Ruel, M. 2007. Role of livestock in human nutrition and health for poverty reduction in developing countries. *Journal of Animal Science* 85:2788–2800.

10. Office of Dietary Supplements. National Institutes of Health. 2007. *Iron.* http://ods .od.nih.gov/factsheets/iron/.

11. Allen, L. 2003. Interventions for Micronutrient Deficiency Control in Developing Countries: Past, Present and Future. In: Demment and Allen, eds. *Animal source foods.*
Allen, L. 2000. *Ending hidden hunger: the history of micronutrient deficiency control.* Background analysis for the World Bank–UNICEF Nutrition Assessment Project. Washington, DC: World Bank.

12. Allen. 2003. Interventions for micronutrient deficiency control.

13. Allen. 2003. Interventions for micronutrient deficiency control.

14. Neumann, C., Bwibo, N., Murphy, S., Sigman, M., Whaley, S., Allen, L., Guthrie, D., Weiss, R., and Demment, M. 2003. Animal source foods improve dietary quality, micronutrient status, growth and cognitive function in Kenyan school children: background, study design and baseline findings. In: Demment and Allen. *Animal source foods.*
Grillenberger, M., Neumann, C., Muprhy, S. Bwibo, N., Van't Veer, P., Hautvast, P., and West C. 2003. Food supplements have a positive impact on weight gain and the addition of animal source foods increases lean body mass of Kenyan schoolchildren. In: Demment and Allen. *Animal source foods.*
Whaley, S., Sigman, M., Neumann, C., Bwibo, N., Guthrie, D., Weiss, R., Alber, S., and Muprhy, S. 2003.The impact of dietary intervention on the cognitive development of Kenyan school children. In: Demment and Allen. *Animal source foods.*

Left, S. 2005. *Raising children as vegans 'unethical', says professor.* Guardian.co.uk, Monday 21 February. http://www.guardian.co.uk/society/2005/feb/21/health.food.

15. Wachs, T., and McCabe, G. 2001. Relation of maternal intelligence and schooling to offspring nutritional intake. *International Journal of Behavioural Development* 25: 444–449.

16. Whaley et al. 2003.The impact of dietary intervention on the cognitive development. Page 3971S.

17. Left. 2005. *Raising children as vegans 'unethical'.*

18. Pimentel D., and Pimentel, M. 2003. Sustainability of meat-based and plant-based diets and the environment. *American Journal of Clinical Nutrition* 78:660S–3S.

19. FAO. 2009. *The state of food and agriculture;* Livestock in Development. 1999. *Livestock in poverty-focused development.* Crewkerne, UK: Livestock in Development.

20. FAO. 2011. *Rural income generating activities.* Rome: FAO. http://www.fao.org/es/ESA /riga/english/index_en.htm.
    FAO. 2009. *The state of food and agriculture.*

21. Heffernan, C. 2004. Livestock and the poor: issues in poverty-focused livestock development. Chapter 15. In: Owen, E., Smith, T., Steele, M., Anderson, S., Duncan, A., Herrero, M., Leaver, J., Reynolds, C., Richards, J., and Ku-Vera, J. Responding to the livestock revolution: the role of globalisation and implications for poverty alleviation. *British Society of Animal Science,* publication 33.

22. FAO. 2009. *The state of food and agriculture.*

23. Steinfeld, H., Gerber, P., Wassenaar, T., and Castel, V. 2006. *Livestock's long shadow. Environmental issues and options.* Rome: FAO.

24. FAO. 2009. *The state of food and agriculture.* Page vi.

25. Steinfeld, H., Wassenaar, T. and Jutzi, S. 2006. Livestock production systems in developing countries: status, drivers, trends. *Scientific and technical review—International Office of Epizootics* 25: 505–516.

26. Stoddart, L., Smith, A., and Box, T. 1975. *Range management.* New York: McGraw Hill Book Co.

27. Behnke, R., Scoones, I., and Kerven, C. 1993. Rangeland ecology at disequilibrium: new models of natural variability and pastoral adaptation in African savannas. London: Overseas Development Institute. Adapted from Caughley, G. 1979. What is this thing called carrying capacity? In: Boyce, M. and Hayden-Wing, L., eds. *North American elk: ecology, behaviour and management.* Laramie, Wyoming: University of Wyoming Press.
    Bell, R. 1985. Carrying capacity and offtake quotas. In: Bell, R. and McShane Caluzi, E., eds. *Conservation and wildlife management in Africa.* Washington, DC: US Peace Corps.

28. FAO. 2009. *The state of food and agriculture.*

29. Scoones, I. 1996. New challenges for range management in the 21st century. *Outlook on Agriculture* 25:253–256.

30. Earthwitness. 2009. *Maasai Mara—wildlife on the brink.* http://www.earthwitness.net/2009/04/maasai-mara-wildlife-on-brink.html.

31. Ogutu, O., Piepho, P., Dublin, T., Bhola, N., and Reid, R. 2009. Dynamics of Mara–Serengeti ungulates in relation to land use changes. *Journal of Zoology* 278:1–14.

32. Earthwitness. 2009. *Maasai Mara;* Cattlefencing. 2010. *Participatory land-use planning empowers the pastoral community of Kenya's Kitengela Maasailand.* http://www.ilri.org/ilrinews/index.php/archives/2314.

33. Earthwitness. 2009. *Maasai Mara.*

34. Cattlefencing. 2010. *Participatory land-use planning.*

35. Earthwitness. 2009. *Maasai Mara.*

36. Cattlefencing. 2010. *Participatory land-use planning.*

37. Behnke, R., and Scoones, I. 1993. Rethinking range ecology: implications for rangeland management in Africa. In: Behnke et al. Range ecology at disequilibrium; Toulmin, C. 1994. Tracking through drought: options for destocking and restocking. In: Scoones, I. 1991. Wetlands in drylands: key resources for agricultural and pastoral production in Africa. *Ambio* 20:366–371.

38. FAO. 2001. *Mixed crop-livestock farming. A review of traditional technologies based on literature and field experience.* FAO Animal Production and Health Papers No. 152. http://www.fao.org/docrep/004/y0501e/y0501e00.htm#toc.
Van Keulen, H., and Schiere, H. 2004. Crop-livestock systems: old wine in new bottles? In: *New directions for a diverse planet. Proceedings of the 4th International Crop Science Congress,* Brisbane, 26 September-October.

39. FAO. 2001. *Mixed crop-livestock farming.*

40. Van Keulen and Schiere. 2004. Crop-livestock systems.

41. Delgado, C. 2003. Rising consumption of meat and milk in developing countries has created a new food revolution. *American Society for Nutritional Sciences* 133:3907S–3910S.

42. Tetrapak. 2011. *Tetra Pak Dairy Index,* Issue 4. Lausanne, Switzerland: Tetra Pak. http://www.tetrapak.com/Document%20Bank/Food_categories/Dairyindex4_2011.pdf.

43. Sharma, V., Staal, S., Delgado, C., and Singh, R. 2003. *Policy, technical and environmental determinants and implications of the scaling-up of milk production in India.* Annex III, Research Report of IFPRI-FAO Livestock Industrialization Project: Phase II. Washington, DC: IFPRI.

44. Staal, S., Pratt, A., and Jabbar, M. 2008. *Dairy development for the resource poor, part 3: Pakistan and India.* Dairy development case studies. Pro-Poor Livestock Policy Initiative. Rome: FAO.

45. Staal et al. 2008. *Dairy development for the resource poor.*

46. Staal et al. 2008. *Dairy development for the resource poor.*

47. Public Relations Society of India. 2006. *National cooperative PR.* http://www.prsi.co.in/amul.htm.

48. Peacock, C. 1996. *Improving goat production in the Tropics. A manual for development workers.* London: Farm-Africa and Oxfam.

49. APHCA. 2008. *Goats—undervalued assets in Asia.* Proceedings of the APHCA-ILRI regional workshop on goat production systems and markets. Luang Prabang, Lao PDR, 24–25 October 2006. Bangkok: Animal Production and Health Commission for Asia and the Pacific and International Livestock Research Institute.

50. Ayele, Z. and Peacock, C. 2003. Improving access to and consumption of animal source foods in rural households: the experiences of a women focused goat development program in the highlands of Ethiopia. *Journal of Nutrition* 133:3981S–3986S.

    FARM-Africa. 2007. *The goat model: a proven approach to reducing poverty among small-holder farmers in Africa by developing profitable goat enterprises and sustainable support services.* Working Paper No. 9, June. Addis Ababa: FARM-Africa.

51. Ayele and Peacock. 2003. Improving access to and consumption of animal source foods. Page 3985S.

52. FAO. 2009. *The State of Food and Agriculture.*

53. Bingsheng, K., and Yijun, H. 2008. Poultry sector in China: structural changes during the past decade and future trends. In: *Poultry in the 21st century: avian influenza and beyond.* Proceedings of the International Poultry Conference, Bangkok, November 2007. FAO Animal Production and Health Proceedings No. 9. Rome: FAO.

54. Odhiambo, Z. 2007. *Commercialising biotech crops—the global picture.* New Agriculturalist. http://www.new-ag.info/focus/focusItem.php?a=341.

55. New Agriculturalist. 2008. *Improved cowpea for crops and livestock in West Africa.* http://www.new-ag.info/focus/focusItem.php?a=351.

56. New Agriculturalist. 2008. *Sweet sorghum for food, feed and fuel.* http://www.new-ag.info/focus/focusItem.php?a=352.

57. Steinfeld, H., de Haan, C., and Blackburn, H. No date. *Livestock environment interactions: issues and options.* http://www.fao.org/docrep/x5305e/x5305e00.htm.

58. FAO. 2009. *The state of food and agriculture.*

59. FAO. 2009. *The state of food and agriculture.*

60. Mc Sweeney, C., Dalrymple, B., Gobius, K., Kennedy, P., Krause, D., Mackie, R., and Xue, G. 1999. The application of rumen biotechnology to improve the nutritive value of fibrous feedstuffs: pre- and post ingestion. *Livestock Production Science* 59:265–283.

    Cook, B., Pengelly, B., Brown, S., Donnelly, J., Eagles, D., Franco, M., Hanson, J., Mullen, B., Partridge, I., Peters, M., and Schultze-Kraft, R. 2005. *Tropical forages: an interactive selection tool.* Brisbane, Australia: CSIRO, DPI&F, CIAT and ILRI. http://www.tropicalforages.info/index.htm.

61. CIAT. No date. *Tropical forages: a multipurpose genetic resource.* CIAT in Focus. Crop Commitments. Cali, Colombia: Centro Internacional de Agricultura Tropical (CIAT). http://webapp.ciat.cgiar.org/ciatinfocus/pdf/foragesfocus.pdf.

62. CIAT. No date. *Tropical forages.*

63. CIAT. No date. *Tropical forages.*

Miles, J., Maass, B., and do Valle, C., eds. 1996. *Brachiaria: biology, agronomy and improvement.* Cali, Colombia: CIAT and Campo Grande, Brazil: Empresa Brasilieira de Pesquisa Agropecuária (EMBRAPA).

64. McSweeney et al. 1999. The application of rumen biotechnology.

65. Tien, M., and Kirk, K. 1983. Lignin degrading enzyme from the Hymenomycete *Phanerochaete chrysosporium* Burds. *Science* 221:661–663.

Tien, M., and Tu, C. 1987. Cloning and sequencing of a cDNA for a ligninase from *Phanerochaete chrysosporium. Nature* 326:520–523.

66. Conway, G., and Waage, J. 2010. *Science and innovation for development.* London: UK Collaborative on Development Sciences.

67. Roeder, P., and Karl, R. 2009. *The global effort to eradicate rinderpest.* IFPRI Discussion Paper 923. Washington, DC: IFPRI.

11. Farmers as Innovators

1. Chambers, R. 2005. *Ideas for development.* London and Sterling, VA: Earthscan.

2. World Bank.1977. *Agricultural extension: the training and visit system.* Washington, DC: World Bank.

3. Page, S., Elahi Baksh, M., Duveiller, E., and Waddington, S. 2009. Putting the poorest farmers in control of disseminating improved wheat seed: a strategy to accelerate technology adoption and alleviate poverty in Bangladesh. *Food Security* 1:99–109.

4. Page et al. 2009. Putting the poorest farmers in control.

5. Page et al. 2009. Putting the poorest farmers in control.

6. Chambers, R. 1983. *Rural development: putting the last first.* Harlow, UK: Longman.

7. Conway, G. 1985. Agroecosystem analysis. *Agricultural Administration* 20:31–55.

Gypmantasiri, P., Wiboonpongse, A., Rerkasem, B., Craig, I., Rerkasem, K., Ganjanapan, L., Titayawan, M., Seetisarn, M., Thani, P., Jaisaard, R., Ongprasert, S., Radnachaless, T., and Conway, G. 1980. *An interdisciplinary perspective of cropping systems in the Chiang Mai valley: key questions for research.* Chiang Mai, Thailand: Faculty of Agriculture, University of Chiang Mai.

8. Kelompok Penelitian Agro-ekosistem (KEPAS). 1985. *The critical uplands of Eastern Java: an agroecosystem analysis.* Jakarta: Agency for Agricultural Research and Development-(KEPAS).

KEPAS. 1985. *Swampland agroecosystems of Southern Kalimantan.* Jakarta: KEPAS.

KEPAS. 1986 *Agro-ekosistem daerah kering di nusa tenggara timur.* Jakarta: KEPAS.

KKU-Ford Cropping Systems Project. 1982. *An agroecosystem analysis of Northeast Thailand.* Khon Kaen, Thailand: Faculty of Agriculture, Khon Kaen University.

KKU-Ford Cropping Systems Project. 1982 *Tambon and Village Agricultural Systems in Northeast Thailand.* Khon Kaen, Thailand: Faculty of Agriculture, Khon Kaen University.

Conway, G. 1986. *Agroecosystem Analysis for Research and Development.* Bangkok: Winrock International.

9. Conway, G., Alam, Z., Husain, T., and Mian, M. 1985. *An agroecosystem analysis for the Northern Areas of Pakistan.* Gilgit, Pakistan: Aga Khan Rural Support Programme.

10. Carruthers, I., and Chambers, R. 1981. *Rapid Rural Appraisal: rationale and repertoire.* IDS Discussion Paper 155. Brighton, UK: Institute for Development Studies, University of Sussex.

Conway, G., and McCracken, J. 1990. Rapid Rural Appraisal and Agroecosystem Analysis. In: Altieri, M., and Hecht, S., eds. *Agroecology and small farm development.* Florida: CRC Press.

Khon Kaen University. 1987. *Proceedings of the International Conference on Rapid Rural Appraisal.* 2–5 September, Khon Kaen University, Khon Kaen, Thailand.

11. Ethiopian Red Cross. 1988. *Rapid Rural Appraisal: a closer look at rural life in Wollo.* Addis Ababa: Ethiopian Red Cross Society and London: International Institute for Environment and Development.

12. Ethiopian Red Cross. 1988. *Rapid Rural Appraisal.*

13. Chambers, R. 1997. *Whose reality counts? Putting the last first.* London: Intermediate Technology Publications.

14. Chambers, R. 1996. *Behaviour and attitudes: a missing link in agricultural science?* Paper presented at the Second International Crop Science Congress, 17–24 November, New Delhi.

15. Drinkwater, M. 1993. Sorting fact from opinion: the use of direct matrix to evaluate finger millet varieties. *RRA Notes* 17:24–28.

Manoharan, M., Velayudham, K., and Shunmugavalli, N. 1993. PRA: an approach to find felt needs of crop varieties. *RRA Notes* 18:66–68.

The Women of Sangams Pastapur, Medak, Andhra Pradesh and Pimbert, M. 1991. Farmer participation in on-farm varietal trials: multilocational testing under resource-poor conditions. *RRA Notes* 10:3–8.

16. Manoharan et al. 1993. PRA: an approach to find felt needs.

Drinkwater, M. 1993. Sorting fact from opinion: the use of a direct matrix to evaluate finger millet varieties. *RRA Notes* 17:24–28.

17. Chambers. 1997. *Whose reality counts?*

Cornwall, A., Gujit, I., and Welbourn, A. 1994. Acknowledging process: methodological challenges for agricultural research and extension. In: Scoones, I., and Thompson, J. 1994. *Beyond Farmer First: rural people's knowledge, agricultural research and extension practice.* London: Intermediate Technology Publications.

18. Cornwall, A., and Pratt, G. 2010. The use and abuse of participatory rural appraisal: reflections from practice. *Agriculture and Human Values* 28:263–272.

Mueller, J., Assanou, I., Guimbo, I., and Almedon, A. 2010. Evaluating rapid participatory rural appraisal as an assessment of ethnoecological knowledge and local biodiversity patterns. *Conservation Biology* 24:140–150.

Gona, J., Xiong, T., Mushit, M., Newton, C., and Hartley, S. 2010. Identification of people with disabilities using participatory rural appraisal with key informants: a pragmatic approach with action potential promoting validity and low cost. *Disability and Rehabilitation* 32:79–85.

Bolajoko, M., Moses, G., Gambori-Bolajoko, K., Ifende, V., Emenna, P., and Bala, A. 2011. Participatory rural appraisal of livestock diseases among the Fulani community of the Barkin Ladi local government area, Plateau State, Nigeria. *Journal of Veterinary Medicine and Animal Health* 3:11–13.

19. CAAEP. 2007. *Commune agro-ecosystem analysis.* CAAEP Users Manual. Phnom Penh: Cambodia Australia Agricultural Extension Project.

20. Department of Agricultural Extension, Cambodia. 2009. *An agro-ecosystem analysis of Sna Ansar Commune.* Phnom Penh: Department of Agricultural Extension.

21. Simpson, B., and Owens, M. 2002. *Farmer field schools and the future of agricultural extension in Africa.* Rome: SD Dimensions. Sustainable Development Department (SD), FAO.

22. Simpson and Owens. 2002. *Farmer field schools.*

23. Davis, K., Nknonya, E., Kato, E., Mekonnen, D., Odendo, M., Miiro, R., and Nkuba, J. 2010. *Impact of farmer field schools on productivity and poverty in East Africa.* IFPRI Discussion Paper 00992. June. Washington, DC: IFPRI.

24. Simpson and Owens. 2002. *Farmer field schools.*

25. Conway, G. 1997. Practical innovation: partnerships between scientists and farmers. In: Waterlow, J., Armstrong, D., Fowden, L., and Riley, R., eds. *Feeding a world population of more than eight million people: a challenge to science.* Oxford, UK: Oxford University Press.

26. Hooper, W., and Ash, H. 1935. *Marcus Porcius Cato on agriculture. Marcus Terentius Varro on agriculture.* Loeb Classical Library, Cambridge, MA: Harvard University Press and London: William Heinemann.

27. Pretty, J. 1995. *Regenerating agriculture: policies and practice for sustainability and self-reliance.* London: Earthscan Publications Ltd.

28. Pretty, J. 1995. *Regenerating agriculture.*

29. Reij, C., Scoones, I., and Toulmin, C., eds. 1996. *Sustaining the soil: indigenous soil and water conservation in Africa.* London: Earthscan Publications Ltd.
Richards, P. 1985. *Indigenous agricultural revolution.* London: Hutchinson.
Scoones and Thompson. 1994. *Beyond Farmer First.*

30. Goodell, G., Kenmore, P., Litsinger, J., Bandong, J., De La Cruz, C., and Lumaden, M. 1982. Rice insect pest management technology and its transfer to small-scale farmers in the Philippines. In: IRRI. *Report of an exploratory workshop on the role of anthropologists and other social scientists in interdisciplinary teams developing improved food production technology.* Los Baños, Philippines: IRRI.

31. Conway, S., and Conway, G. 1997. *Lanna: a million ricefields.* Unpublished manuscript.

32. Ashby, J., Quiros, C., and Rivers, Y. 1989. Farmer participation in technology development: work with crop varieties. In: Chambers, R., Pacey, A. and Thrupp, L., eds. 1989. *Farmer First: farmer innovation and agricultural research.* London: Intermediate Technology Publications.

Ashby, J., Quiros, C., and Rivers, Y. 1987. *Farmer participation in on-farm trials.* Agricultural Administration (Research and Extension) Network, Discussion Paper 22. London: Overseas Development Institute.

33. Ashby, J., Quiros, C., and Rivers, Y. 1989. Experience with group techniques in Colombia. In: Chambers et al., eds. 1989. *Farmer First.* Page 129.

34. Sperling, L., and Scheidegger, U. 1995. Participatory selection of beans in Rwanda: results, methods and institutional issues. *Gatekeeper Series,* No. 51. London: International Institute for Environment and Development.

35. Nakkazi, E. 2010. *Rwanda releases 15 new beans varieties.* New Science Journalism. http://www.newsciencejournalism.net/index.php?/news_articles/view/rwanda _releases_15_new_beans_varieties.

Rwanda Agricultural Research Institute. 2010. *15 new bean varieties in Rwanda released by ISAR.* http://www.isar.rw/spip.php?article176.

36. CIMMYT. 2005. *Taking the same path: teaming up with universities in southern Africa.* El Batan, Mexico: CIMMYT. http://www.cimmyt.org/en/newsletter/86–2005/273 -taking-the-same-path-teaming-up-with-universities-in-southern-africa.

37. Betuco. No date. *Farmer's voices are heard here.* http://betuco.be/voorlichting/Mother per cent20and per cent20baby per cent20trails per cent20.pdf.

38. CIMMYT. 2005. *A world tour: programme director profiles.* http://www.cimmyt.org/en /newsletter/86–2005/297-a-world-tour-program-director-profiles.

39. IDS. 1996. *The power of participation: PRA and policy.* IDS Policy Briefing, Issue 7. Brighton, UK: Institute of Development Studies, University of Sussex.

40. Bunch, R. 1983. *Two ears of corn: a guide to people-centred agricultural improvement.* Oklahoma City: World Neighbours.

Bunch, R. 1989. Encouraging farmer's experiments. In: Chambers et al., eds. 1989. *Farmer First.*

Pretty. 1995. *Regenerating agriculture.*

41. Conway, G. 1988. Rapid Rural Appraisal for sustainable development: experiences from the northern areas of Pakistan. In: Conroy, C. and Litvinoff, M., eds. *The greening of aid.* London: Earthscan.

Conway et al. 1985. *An Agroecosystem Analysis.*

42. Aga Khan Foundation. 2007. *Rural development in Pakistan.* www.akdn.org/rural_devel opment/pakistan.asp.

43. Pretty. 1995. *Regenerating agriculture.*

Shah, P. 1994. Participatory watershed management in India: the experience of the Aga Khan Rural Support Programme. In: Scoones and Thompson. *Beyond Farmer First.*

44. Rhoades, R. 1987. Farmers and experimentation. *Agricultural Administration (R and E) Network Paper 21.* London: Overseas Development Institute.

    Rhoades, R., and Booth, R. 1982. Farmer-back-to-farmer: a model for generating acceptable agricultural technology. *Agricultural Administration* 11:127–137.

45. Rhoades, R. 1989. The role of farmers in the creation of agricultural technology. In: Chambers et al., eds. 1989. *Farmer First.*

46. Rodale Institute. 1989. Diffuse Light Storage of potato seed. *International Ag-Sieve.* 2: No. 7. http://www.fadr.msu.ru/rodale/agsieve/txt/vol2/7/art4.html.

47. Ashby et al. 1987. Farmer participation in on-farm trials.

48. Millar, D. 1994. Experimenting farmers in northern Ghana. In: Scoones and Thompson. *Beyond Farmer First.*

49. SRI International Network and Resources Center. 2011. *About SRI.* Ithaca, NY: Cornell International Institute for Food Agriculture and Development (CIIFAD). http://sri .ciifad.cornell.edu/index.html; http://www.future-agricultures.org/farmerfirst/files /T1c_Uphoff.pdf.

50. Trlica, M. 2010. *Grass growth and response to grazing.* Fact sheet 6.108. Colorado State University. http://www.ext.colostate.edu/pubs/natres/06108.html.

51. Sheehy, J. E., Peng, S., Dobermann, A., Mitchell, P., Ferrer, A., Jianchang, Y., Zou, Y., Zhong, X., and Huang, J. 2004. Fantastic yields in the system of rice intensification: fact or fallacy? *Field Crops Research* 88:1–8.

52. Sato, S., and Uphoff, N. 2007. *Raising factor productivity in irrigated rice production: opportunities with the system of rice intensification.* Wallingford, UK: CAB International. http://www.iai.ga.a.u-tokyo.ac.jp/j-sri/meeting/sri-uphoff-sato.pdf.

53. Kassam, A., Stoop, W., and Uphoff, N. 2011. Review of SRI modifications in rice crop and water management and research issues for making further improvements in agricultural and water productivity. In: Uphoff, N and Kassam, A., eds. Paddy and the water environment. *Journal of the International Society of Paddy and Water Environment Engineering* 9:163–180.

54. Gujja, B., and Thiyagarajan, T. 2009. New hope for Indian food security? The System of Rice Intensification. *The Gatekeeper Series,* 143, November. London: IIED.

55. Uphoff, N., Kassam, A., and Stoop, W. 2008. A critical assessment of a desk study comparing crop production systems: the example of the 'system of rice intensification' versus 'best management practice'. *Field Crops Research* 108:109–114.

56. Kassam et al. 2011. Review of SRI modifications. Page 166.

57. Thakur, A., Rath, S., Patil, D., and Kumar, A. 2011. Effects on rice plant morphology and physiology of water and associated management practices of the System of Rice Intensification and their implications for crop performance. *Paddy and Water Environment* 9:13–24.

    Lin, X., Zhu, D. and Lin, X. 2011. Effects of water management and organic fertilization with SRI crop practices on hybrid rice performance and rhizosphere dynamics. *Paddy and Water Environment.* 9:33–39.

Mishra, A. and Salokhe, V. 2011. Rice root growth and physiological responses to SRI water management and implications for crop productivity. *Paddy and Water Environment*. 9:41–52.

Barison, J. and Uphoff, N. 2011. Rice yield and its relation to root growth and nutrient-use efficiency under SRI and conventional cultivation: an evaluation in Madagascar. *Paddy and Water Environment*. 9:65–78.

Anas, I., Rupela, O., Thiyagarajan, T. and Uphoff, N. 2011. A review of studies on SRI effects on beneficial organisms in rice soil rhizospheres. *Paddy and Water Environment*. 9:53–64.

58. Glover, D. 2011. The System of Rice Intensification: time for an empirical turn. *NJAS-Wageningen Journal of Life Sciences*. Doi: 10.1016/j.nas.2010.11.006.

Krupník, T. 2010. *Exploring the spread of the System of Rice Intensification (SRI) in Madagascar*. Trip Report. Bill and Melinda Gates Foundation, May 1–13.

59. Uphoff, N. 2007. Farmer innovations improving the system of rice intensification (SRI). *Jumal Tanah dan Lingkungan* 9:45–56.

60. Uphoff. 2007. Farmer innovations improving the system of rice intensification. Pages 4–18.

61. Bunch, R. 1989. Encouraging farmers' experiments. In: Chambers et al., eds. *Farmer First*.

12. Controlling Pests

1. Matteson, P., Gallagher, K., and Kenmore, P. 1992. Extension of integrated pest management for planthoppers in Asian irrigated rice: empowering the user. In: Denno, R., and Perfect, T., eds. *Ecology and management of planthoppers*. London: Chapman and Hall.

2. 'Pests' include insects, mites, nematodes and vertebrate pests such as rats and Quelea birds. 'Pathogens' cause diseases and include fungi, bacteria, viruses and, in the case of livestock, various protozoa and worms. 'Weeds' are any plants that adversely compete with crop plants.

3. Alexandratos, N. 1995. *World agriculture: towards 2010: An FAO study*. Rome: FAO.

4. Foundation for Advancements in Science and Education (FASE). 1996. *FASE Research Report*. Pasadena, CA: FASE.

5. Stephenson, G. 2003. Pesticide use and world food production: risks and benefits. In: Coats, J. and Yamamoto, H., eds. *Environmental fate and effects of pesticides*. Washington, DC: American Chemical Society.

6. EJF. 2007. *The deadly chemicals in cotton*. London, UK: Environmental Justice Foundation in collaboration with Pesticide Action Network UK.

7. WHO. 2006. *National Workshop on the Prevention of Pesticide Poisoning in China*. http://www.wpro.who.int/china/meetings/meeting_20061024e.htm.

Jørs, E., Morant, R., Aguilar, G., Huici, O., Lander, F., Bælum, J., and Konradsen, F. 2006. Occupational pesticide intoxications among farmers in Bolivia: a cross-sectional study. *Environmental Health:A GlobalAccess Science Source* 5:10. doi:10.1186/1476–069X-5–10.

8. Jørs et al. 2006. Occupational pesticide intoxications.

9. USDA. Personal communication with Sara Delaney.

10. Thundiyil, J., Stober, J., Besbilli, N., and Pronczuk, J. 2008. Acute pesticide poisoning: a proposed classification tool. *Bulletin of the WHO* 86:161–240.

   Konradsen, F. 2007. Acute pesticide poisoning—a global public health problem. *Danish Medical Bulletin* 54:58–59. http://www.danmedbul.dk/DMB_2007/0107/0107-artikler/DMB3886.pdf.

   WHO. 2004. *The impact of pesticides on health: preventing intentional and unintentional deaths from pesticide poisoning.* http://www.who.int/mental_health/prevention/suicide/en/PesticidesHealth2.pdf.

11. Jeyaratnam, J. 1990. Acute pesticide poisoning: a major global health problem. *World Health Statistics Quarterly* 43:139–144.

12. Jeyaratnam. 1990. Acute pesticide poisoning.

13. Phillips, M., Yang, G., Zhang, Y., Wang, L., Ji, H., and Zhou, M. 2002. Risk factors for suicide in China: a national case-control psychological autopsy study. *The Lancet* 360:1728–1736.

14. Jeyaratnam. 1990. Acute pesticide poisoning.

15. Litchfield, M. H. 2005. Estimates of acute pesticide poisoning in agricultural workers in less developed countries. *Toxicological Reviews* 244:271–278.

16. Orozco, F. Cole, D., Forbes, G., Kroschel, J., Wanigaratne, S., and Arica, D. 2009. Monitoring adherence to the International Code of Conduct: highly hazardous pesticides in Central Andean agriculture and farmers' rights to health. *International Journal of Occupational and Environmental Health* 15:255–268.

17. Loevinsohn, M. 1987. *Insecticide use and increased mortality due to unintentional pesticide poisonings.* Geneva, Switzerland: Working paper presented to the Consultation in Planning Strategy for the Prevention of Pesticide Poisoning. Doc No. WHO/VCB/86.926.

18. Rola, A. 1989. *Pesticides, health risks and farm productivity: a Philippine experience.* Agricultural Policy Research Program Monograph, 89–01. Los Baños, Philippines: University of the Philippines.

19. Loevinsohn. 1987. *Insecticide use and increased mortality.*

20. Rola, A., and Pingali, P. 1993. *Pesticides, productivity and farmers' health: an economic assessment.* Los Baños, Philippines: IRRI.

21. Conway, G., and Pretty, J. 1991. *Unwelcome harvest: agriculture and pollution.* London: Earthscan.

22. Fishel, F. 2009. *Pesticide toxicity profile: Neonicitinoid pesticides.* Gainesville, FL: University of Florida, IFAS Extension. http://edis.ifas.ufl.edu/pi117.

23. Conway, G. 1971. Better methods of pest control. In: Murdoch, W., ed. *Environment: resources, pollution and society.* Stanford: Sinauer Associates Inc.

    Dent, D. 1991. *Insect pest management.* Wallingford, UK: CAB International.

24. Conway, G. 1972. Ecological aspects of pest control in Malaysia. In: Farvar, J. and Milton, J., eds. *The careless technology: ecological aspects of international development.* Garden City, NY: Natural History Press and Doubleday.

    Conway, G. 1976. Man versus pests. In: May, R., ed. *Theoretical ecology: principles and applications.* 2nd edition. Oxford, UK: Blackwell Scientific Publications.

25. Conway. 1972. Ecological aspects of pest control in Malaysia.

    Conway. 1976. Man versus pests.

26. Conway. 1972. Ecological aspects of pest control in Malaysia.

    Conway and Pretty. 1991. *Unwelcome harvest.*

    Pretty, J. N., ed. 2005. *The pesticide detox: towards a more sustainable agriculture.* London and Washington, DC: Earthscan.

27. Crosland, N. 1989. Laboratory to experiment. *Proceedings of the 5th International Congress of Toxicology,* July. Brighton, UK.

28. Pretty. 2005. *The pesticide detox.*

29. Saxena, R. 1987. Antifeedants in tropical pest management. *Insect Science and its Applications* 8:731–736.

30. Soil Association. 2007. *Pesticides and organic farming—a last resort.* http://92.52.112.178/web/sa/saweb.nsf/librarytitles/24E22.HTMl/$file/Pesticides per cent20and per cent20organic per cent20farming per cent20-per cent20A per cent20last per cent20resort.pdf.

31. Bahlai, C., Xue, Y., McCreary, C., Schaafsma, A., and Hallett, R. 2010. Choosing organic pesticides over synthetic pesticides may not effectively mitigate environmental risk in soybeans. *PLoS ONE.* http://www.plosone.org/article/info:doi/10.1371/journal.pone.0011250.

    University of Guelph. 2010. *Organic pesticides not always 'greener' choice, study finds.* Ontario: University of Guelph. http://www.uoguelph.ca/news/2010/06/organic_pestici_1.html.

32. Gold, L., Slone, T., Ames, B., and Manley, N. 2001. Pesticide residues in food and cancer risk: a critical analysis. In: Krieger, R., ed. *Handbook of pesticide toxicology.* 2nd edition. San Diego, CA: Academic Press. http://potency.berkeley.edu/text/handbook.pesticide.toxicology.pdf.

33. Ames, B., Profet, M., and Gold, L. 1990. Dietary pesticides (99.99 per cent all natural). *Proceedings of the National Academy of Sciences USA* 87:7777–7781.

34. Altieri, M. 1995. *Agroecology: the science of sustainable agriculture.* 2nd edition. Boulder, CO: Westview Press.

35. Hassanali, A., Herren, H., Khan, Z., Pickett, J., and Woodcock, C. 2008. Integrated pest management: the push-pull approach for controlling insect pests and weeds of cereals, and its potential for other agricultural systems including animal husbandry. *Philosophical Transactions of the Royal Society London* 363:611–621.

Rothamsted Research Chemical Ecology Group. 2010. *Push-pull habitat manipulation for control of maize stemborers and the witchweed Striga.* http://www.rothamsted.ac.uk/bch /CEGroup/ChemEcolGroupArea6.html.

icipe. 2010. *Push-Pull, a novel conservation agriculture technology.* http://www.push-pull.net /works.shtml.

36. Agricultural Research Council. 2010. *Crop protection.* http://www.arc.agric.za/home .asp?pid=637.

37. Hassanali et al. 2008. Integrated pest management.

38. Neuenschwander, P., and Herren, H. 1988. Biological control of the cassava mealybug *Phenacoccus manihoti,* by the exotic parasitoid *Epidinocarsis lopezi* in Africa. *Philosophical Transactions of the Royal Society of London B* 318:319–333.

Kiss, A., and Meerman, F. 1991 *Integrated pest management in African agriculture.* Technical Paper 142, African Technical Department Series, Washington, DC: World Bank.

Norgaard, R. 1988. The biological control of cassava mealybug in Africa. *American Journal of Agricultural Economics* 70:366–371.

39. Campbell, R. 1989. *Biological control of microbial plant pathogens.* Cambridge, UK: Cambridge University Press.

40. AATF. 2008. *Annual Report 2008: addressing farmers' constraints through scientific interventions.* Nairobi: African Agricultural Technology Foundation.

41. AATF. 2004. *Mycotoxin control in food gains.* Proceedings of a small group meeting. 22–14 June, Nairobi. Nairobi: AATF.

42. AATF. 2010. *AATF project 6 (PI006). Aflatoxin biological control project.* Technical Progress report: January–June. Nairobi: AATF. http://www.aatf-africa.org/userfiles/Technical -Report-Jan-June10.pdf.

43. Conway. 1971. Better methods of pest control.

44. Smith, R. 1972. The impact of the Green Revolution on plant protection in tropical and subtropical areas. *Bulletin of the Entomological Society of America* 18:7–14.

45. Loevinsohn, M., Litsinger, J., and Heinrichs, E. 1988. Rice insect pests and agricultural change. In: Harris, M., and Rogers, C., eds. *The entomology of indigenous and naturalized systems in agriculture.* Boulder, CO: Westview Press.

46. Dover, M., and Croft, B. 1984. *Getting tough: public policy and the management of pesticide resistance.* Washington, DC: World Resources Institute.

Georghiou, G. 1986. The magnitude of the problem. In: National Research Council. *Pesticide resistance: strategies and tactics for management.* Committee on Strategies for the Management of Pesticide Resistant Pest Populations, Board on Agriculture, National research Council, Washington, DC: National Academy Press.

Whalon, M., Mota-Sanchez, D., and Hollingworth, R. 2008. *Global pesticide resistance in arthropods*. Oxford, UK: Oxford University Press.

47. Whalon et al. 2008. *Global pesticide resistance in arthropods*.

48. Brent, K., and Holloman, D. 2007. *Fungicide resistance in crop pathogens; how can it be managed?* 2nd edition. Brussels: Fungicide Resistance Action Committee, Crop Life International.

49. Weed Science. 2011. *International survey of herbicide resistant weeds*. http://www.weedscience.org/In.asp.

50. Sudderuddin, K. 1979. Insecticide resistance in agricultural pests with special reference to Malaysia. In: *Proceedings of MAPPS Seminar*. 1–2 March, Kuala Lumpur, Malaysia.

51. Conway, G., and Waage, J. 2010. *Science and innovation for development*. London: UK Collaborative on Development Sciences.

52. Hanson, H., Borlaug, N., and Anderson, R. 1982. *Wheat in the third world*. Boulder, CO: Westview Press.

53. Cornell University. No date. *Durable rust resistance in wheat project*. Ithaca, NY: Cornell University. http://wheatrust.cornell.edu/.

54. Antony, N. 2011. '*Super wheat' resists devastating rust*. http://www.scidev.net/en/news/-super-wheat-resists-devastating-rust.html.

55. USDA-CSREES Coordinate Agricultural Projects. No date. *MAS Wheat: disease resistance. Stem rust resistance*. maswheat.ucdavis.edu/protocols/StemRust/index.htm.
Cornell University. No date. *Durable rust resistance*.

56. IRRI. 1996. *Bt rice: research and policy issues*. IRRI Information Series No. 5, Los Baños, Philippines: IRRI.

57. Conway and Waage. 2010. *Science and innovation for development*.
Cornell University. 2011. *Collaboration on Insect Management for Brassicas in Asia and Africa*. Ithaca, NY: Cornell University. http://vivo.cornell.edu/display/grant50950.

58. Gonsalves, D. 1998. Control of papaya ringspot virus in papaya: a case study. *Annual Review of Phytopathology* 36:415–437.

59. Tabashnik, B., Gassmann, A., Crowder, D., and Carriére, Y. 2008. Insect resistance to *Bt* crops: evidence versus theory. *Nature Biotechnology* 26:199–202.
Science Daily. 2008. *First documented case of pest resistance to biotech cotton*. http://www.sciencedaily.com/releases/2008/02/080207140803.htm.

60. James, C. 2006. Global status of commercialized biotech/GM crops: 2006. *ISAAA Briefs* 35:1–9.

61. Monsanto. 2010. *Cotton in India*. http://www.monsanto.com/newsviews/Pages/india-pink-bollworm.aspx.

62. Tabashnik, B., Carriére, Y., Dennehy, T., Morin, S., Sisterson, M., Roush, R., Shelton, A., and Zhao, J. 2003. Insect resistance to transgenic *Bt* crops: lessons from the laboratory and field. *Journal of Economic Entomology* 96:1031–1038.
Tabashnik et al. 2008. Insect resistance to *Bt* crops.

63. Conway. 1971. Better methods of pest control.

64. Flint, M., and van den Bosch, R. 1981. *Introduction to integrated pest management*. New York: Plenum Press.

    Cate, J. and Hinkle, M. 1994. *Integrated pest management: the path of a paradigm*. Washington, DC: National Audubon Society.

65. Wood, B. 2002. Pest control in Malaysia's perennial crops: a half century perspective tracking the pathway to integrated pest management. *Integrated Pest Management Reviews* 7:173–190.

66. Kenmore, P. 1980. *Ecology and outbreaks of a tropical pest of the green revolution, the brown planthopper, Nilaparvata lugens (Stal)*. PhD thesis. Berkeley, CA: University of California.

67. Kenmore, P. 1986. *Status report on integrated pest control in rice in Indonesia with special reference to conservation of natural enemies and the rice brown planthopper (Nilaparvata lugens)*. Jakarta: FAO.

68. Barbier, E. 1987. Natural resources policy and economic framework. In: Tarrant, J., Barbier, E., Greenberg, R., Higgins, M. and Lintner, F. *Natural resources and environmental management in Indonesia. Annex 1*. Jakarta: United States Agency for International Development (USAID).

69. Herdt, R., and Capule, C. 1983. *Adoption, spread, and production impact of modern rice varieties in Asia*. Los Baños, Philippines: IRRI.

70. Matteson et al. 1992. Extension of integrated pest management for planthoppers.

71. Kenmore, P. 1991. Indonesia's integrated pest management-a model for Asia. Manila, Philippines: FAO Intercountry IPC Rice Programme.

72. Heong, K. 2009. Are planthopper problems caused by a breakdown in ecosystem services? In: Heong, K., and Hardy, B., eds. *Planthoppers: new threats to the sustainability of intensive rice production systems in Asia*. Los Baños, Philippines: IRRI.Page 228.

73. Thrupp, L., ed. 1996. *New partnerships for sustainable agriculture*. Washington, DC: World Resources Institute.

74. Pretty. 2005. *The pesticide detox*.

75. Bentley, J., Rodrígues, G., and González, A. 1993. Science and the people: Honduran campesinos and natural pest control inventions. In: Buckles, D., ed. *Gorras y sombreros: caminos hacia la colaboración entre técnicos y campesionosia*. El Zamarano, Honduras: Department of Crop Protection.

    Bentley, J. 1994. Stimulating farmer experiments in non-chemical pest control in Central America. In: Scoones, I. and Thompson, J., eds. *Beyond Farmer First: rural people's knowledge, agricultural research and extension practice*. London: Intermediate Technology Publications.

76. Pretty. 2005. *The pesticide detox*.

    Kenmore. 1991. Indonesia's integrated pest management.

    Matteson et al. 1992. Extension of integrated pest management for planthoppers.

404

Winarto, Y. 1994. Encouraging knowledge exchange: integrated pest management in Indonesia. In: Scoones and Thompson. *Beyond Farmer First.*

77. van der Fliert, E. 1993. *Integrated pest management: farmer field schools generate sustainable practices.* WAU Paper 93-3. Wageningen, The Netherlands: Wageningen Agricultural University.

78. Lu, Y., Wu, K., Jiang, Y., Xia, B., Li, P., Feng, H., Wyckhuys, K., and Gio, Y. 2010. Mirid bug outbreaks in multiple crops correlated with wide-scale adoption of *Bt* cotton in China. *Science* 328:1151–1154.

79. Orr, A. 2003. Integrated pest management for resource-poor African farmers: is the emperor naked? *World Development* 31:831–845.

80. Heong, K. 2011 Personal communication.
    Kazushige, S., Liu, G., and Qiang, Q. 2009. Prevalence of whitebacked planthoppers in Chinese hybrid rice and whitebacked planthopper resistance in Chinese japonica rice. In: Heong, K., and Hardy, B., eds. *Planthoppers: new threats to the sustainability of intensive rice production systems in Asia.* Los Baños, Philippines: IRRI.

13. Rooted in the Soil

1. Ash, H. 1941. *Lucius Junius Moderatus Columella on agriculture.* Vols. 1–3. Loeb Classical Library, Cambridge, MA: Harvard University Press and London: William Heinemann. (II, I. 6–13).

2. National Soil Resources Institute. 2001. *A guide to better soil structure.* Silsoe, UK: Cranfield University. http://www.soil-net.com/legacy/downloads/resources/structure_brochure.pdf.
   Brady, N., and Weil, R. 2008. *The nature and properties of soil.* 14th edition. New Jersey: Prentice Hall.

3. Conway, G., and Pretty, J. 1991. *Unwelcome harvest: agriculture and pollution.* London: Earthscan. Taken from: Brady, N. 1984. *The nature and property of soils.* New York: Macmillan. Page 165.

4. Mitchell, D., and Ingco, M. 1993. *The world food outlook.* Washington, DC: World Bank.

5. Steen, I. 1998. Phosphorus availability in the 21st century: management of a non-renewable resource. *Phosphorous and Potassium* 217:25–31. http://www.nhm.ac.uk/research-curation/research/projects/phosphate-recovery/pandk217/steen.htm.

6. FAO/IFA/IFDC. 1992. *Fertilizer use by crop.* Document ESS/Misc./1992/3. Rome: FAO.

7. Yang, H. 2006. Resource management, soil fertility and sustainable crop production: experiences of China. *Agriculture, Ecosystems and Environment* 116:27033. In: Bellarby, J., Foereid, B., Hastings, A. and Smith, P. 2008. *Cool farming: climate impacts of agriculture and mitigation potential.* Amsterdam: Greenpeace International.

8. World Bank. 2009. *World Bank development indicators 2009—agricultural inputs.* Washington, DC: World Bank.

9. Erisman, J., Sutton, M., Galloway, J., Klimont, Z., and Winiwarter, W. 2008. How a century of ammonia synthesis changed the world. *Nature Geoscience* 1:636–639.

10. Powlson, D. 2011. *Greenhouse gas emissions associated with nitrogen fertiliser.* Presentation at Reducing Greenhouse Gases. February 28–March 1. London: Royal Society.

11. Handa, B. 1983. *Effect of fertiliser use on groundwater quality in India, Groundwater in Resources Planning* II:1105–1109.

12. Idaho Southwest District Health. 2010. *Nitrate in drinking water.* http://www.public healthidaho.com/pdf/Nitrate-Nitrite-in-Drinking-Water.pdf.

13. Pobel, D., Riboli, E., Cornée, J., Hémon, B., and Guyader, M. 1995. Nitrosamine, nitrate and nitrite in relation to gastric cancer: A case-control study in Marseille, France. *European Journal of Epidemiology* 11:67–73.
    Forman, D. 2004. Commentary: nitrites, nitrates and nitrosation as causes of brain cancer in children: epidemiological challenges. *International Journal of Epidemiology* 33:1216–1218.
    Ward, M., Kilfoy, B., Weyer, P., Anderson, K., Folsom, A., and Cerhan, J. 2010. Nitrate intake and the risk of thyroid cancer and thyroid disease. *Epidemiology* 21:389–395.
    Chiu, H-F., Tsai, S-S., and Yang, C-Y. 2007. Nitrate in drinking water and risk of death from bladder cancer: an ecological case-control study in Taiwan. *Journal of Toxicology and Environmental Health* Part A, 70:1000–1004.

14. Super, M., de V. Hesse, H., MacKenzie, D., Dempster, W., Du Plessis, J., and Ferreira, J. 1981. An epidemiological study of well water nitrates in a group of SW African/ Namibian infants. *Water Research* 15:1265–1270.

15. Ju, XT., Xing, G-X., Chen, X-P., Zhang, S-L., Zhang, L-J., Liu, X-J., Cui, Z-L., Yin, B., Christie, P., Zhu, Z-L., and Zhang, F-S. 2009. Reducing environmental risk by improving N management in intensive Chinese agricultural systems. *Proceedings of the National Academy of Sciences USA* 106:3041–3046.

16. Prasad, R., and De Datta, S. 1979. Increasing fertilizer nitrogen efficiency in wetland rice. In: IRRI. *Nitrogen and rice.* Los Baños, Philippines: IRRI.

17. Sanchez, P. 1976. *Properties and management of soils in the Tropics.* New York: John Wiley and Sons.

18. Viets, F., Humbert, R., and Nelson, C. 1967. Fertilizers in relation to irrigation. In: Hagan, R., Haise, H., and Edminster, R. *Irrigation of agricultural lands.* Madison, WI: American Society of Agronomy.

19. Roy, A. 2008. *Managing access to farm inputs.* Presented at the World Bank Symposium, Cultivating Innovation: A Response to the Food Price Crisis. Washington, DC: IFDC.

20. Mikkelson, D., De Datta, S., and Obcemea, W. 1978. Ammonia losses from flooded rice soils. *Soil Science Society of America Journal* 42:725–730.
    Lindau, C., Bollich, R., DeLaune, A., Mosier, A., and Bronson, K. 1993. Methane mitigation in flooded Louisiana rice fields. *Biology and Fertility of Soils* 15:174–178.

21. Ladha, J., Kirk, G., Bennett, J., Peng, S., Reddy, C., Reddy, P., and Singh, U. 1998. Opportunities for increased nitrogen-use efficiency from improved lowland rice germplasm. *Field Crops Research* 56:41–71.

22. Postgate, J. 1990. Fixing the nitrogen fixers. *New Scientist:* February 3, 1990.
    Markmann, K., and Parniske, M. 2009. Evolution of root endosymbiosis with bacteria: how novel are nodules? *Trends in Plant Science* 14:77–86.

23. Royal Society. 2009. *Reaping the benefits: science and the sustainable intensification of global agriculture.* London: The Royal Society.

24. Charpentier, M., and Oldroyd, G. 2010. How close are we to nitrogen-fixing cereals? *Current Opinion in Plant Biology* 13:556–564.

25. Eswaran, H., Lal, R., and Reich, P. 2001. Land degradation: an overview. In: Bridges, E., Hannam, I., Oldeman, L., Pening de Vries, F., Scherr, S., and Sompatpanit, S., eds. *Responses to land degradation.* Proceedings of the 2nd International Conference on Land Degradation and Desertification, Khon Kaen, Thailand. New Delhi: Oxford Press.

26. Oldeman, L., Hakkeling, R., and Sombroek, W. 1990. *World map of the status of human-induced soil degradation.* Wageningen: International Soil Reference and Information Centre (ISRIC) and UNEP.
    Scherr, S. and Yadav, S. 1996. *Land degradation in the developing world: implications for food, agriculture and environment to 2020.* Food, Agriculture and Environment Discussion Paper 14, Washington, DC: IFPRI.

27. Bai, Z., Dent, D., Olsson, L., and Schaepman, M. 2008. *Proxy global assessment of land degradation and improvement. 1. Identification by remote sensing.* Report 2008/01. ISRIC. Wageningen, Netherlands: World Soil Information; Bai, Z. and Dent, D. 2007. *Land degradation and improvement in Argentina. 1. Identification by remote sensing.* Wageningen, The Netherlands: International Soil Reference Information Center—World Soil Information.

28. Nachtergaele, F. And Petri, M. 2008. *Mapping land use systems at global and regional scales for land degradation assessment analysis.* Version 1.0. Technical report n.8 of the LADA FAO/UNEP Project; Nkonya, E., Gerber, N., Baumgartner, P., von Braun, J., De Pinto, A., Graw, V., Kato, E., Kloos, J. and Walter, T. 2011. *The economics of land degradation. Toward an integrated global assessment.* Development Economics and Policy series, volume 66. Frankfurt: Peter Lang GmbH; Nkonya, E., Gerber, N., von Braun, J. and De Pinto, A. 2011. *Economics of land degradation. The costs of action versus inaction.* Washington, DC: International Food Policy Research Institute.

29. Scoones, I., Reij, C., and Toulmin, C. 1996. Sustaining the soil: indigenous soil and water conservation in Africa. In: Reij, C., Scoones, I., and Toulmin, C., eds. *Sustaining the soil: indigenous soil and water conservation in Africa.* London: Earthscan.
    Stocking, M. 1993. *Soil erosion in developing countries: where geomorphology fears to tread!* Discussion Paper 241. Norwich, UK: School of Development Studies, University of East Anglia.

30. Bojo, J., and Cassells, D. 1995. *Land degradation and rehabilitation in Ethiopia; a reassessment*. AFTES Working Paper 17. Washington, DC: World Bank.

31. UNEP. 1984. *General assessment of progress in the implementation of the plan of action to combat desertification 1978–1984: report of the Executive Director*. Governing Council, Twelfth Session, UNEP/GC. 12/9. Nairobi: United Nations Environment Programme.

32. Swift, J. 1996. Desertification: narratives, winners and losers. In: Leach and Mearns. *The lie of the land*.

33. Lal, R. 2001. Soil degradation by erosion. *Land Degradation and Development* 12:519–539.

34. Lal. 2001. Soil degradation by erosion; Lal, R. 2002. Soil erosion and the global carbon budget. *Environment International*. 29:437–450.

35. Pretty, J. 1995. *Regenerating agriculture: policies and practice for sustainability and self-reliance*. London: Earthscan.
    Scoones et al. 1996. Sustaining the soil.

36. Conway and Pretty. 1991. *Unwelcome harvest*.

37. Pretty. 1995. *Regenerating agriculture*.

38. Kerr, J., and Sanghi, N. 1992. *Soil and water conservation in India's semi arid tropics*. Gatekeeper Series SA34. London: Sustainable Agriculture Programme, International Institute for Environment and Development.

39. Abujamin, S., Abdurachman, A., and Suwardjo, H. 1985. *Contour grass. Strips as a low-cost conservation practice*. Extension Bulletin No. 225. Taiwan: Food and Fertilizer Technology Center.

40. Peet, M. 1995. *Sustainable practices for vegetable production in the south, conservation tillage*. http://ncsu.edu/sustainable/tillage/tillage.html.
    Hobbs, P., Sayre, K., and Gupta, R. 2007. The role of conservation agriculture in sustainable agriculture. *Philosophical Transactions of the Royal Society B* 363:543–555.
    Fawcett, R., and Towery, D. 2002. *Conservation tillage and plant biotechnology. How new technologies can improve the environment by reducing the need to plow*. West Lafeyette, IN: Conservation Technology Information Center.

41. Hobbs, P., Sayre, K., and Gupta, R. 2008. The role of conservation agriculture in sustainable agriculture. *Philosophical Transactions of the Royal Society B*. 363:543–555.

42. Derpsch, R. 2008. *Historical review of no-tillage cultivation of crops*. Proceedings of the 1st JIRCAS Seminar on Soybean Research. No-tillage Cultivation and Future Research Needs, March 5–6, Brazil. JIRCAS Working Report No. 13.
    Blevins, R., and Frye, W. 1993. Conservation tillage: an ecological approach to soil management. In: Sparks, D. 1993. *Advances in agronomy*, Vol. 51. San Diego, CA: Academic Press.
    Brock B., Canterberry J., and Naderman G. 2000. Ten milestones in conservation tillage: history and role in the North Carolina conservation program. In: Sutherland J., ed. *Proceedings of the 43rd annual meeting of the Soil Science Society of North Carolina*.

January 18–19. Raleigh, NC: SSSNC. http://www.soil.ncsu.edu/about/century/tenmilestones.html.

43. Hobbs, P., Gupta, R., Malik, R., and Dhillon, S. No date. *The adoption of conservation agriculture for rice-wheat systems in South Asia: a case study from India.* Rice-Wheat Consortium. http://www.css.cornell.edu/faculty/hobbs/File per cent20posters/poster1.pdf.

44. Lal, R. 2010. Enhancing efficiency in agro-ecosystems through soil carbon sequestration. *Crop Science* 50:S-120–131.

45. Lal, 2010. Enhancing efficiency. Redrawn from: Petchawee, S., and Chaitep, W. 1995. Organic matter management for sustainable agriculture. In: Lefroy, R., Blair, G., and Crasswell, E., eds. *Organic matter management in upland systems in Thailand.* Canberra: ACIAR. Page S-126.

46. Lal, R. 2010. Enhancing efficiency. Redrawn from: Ganzhara, N. 1998. Humus, soil properties and yield. *Eurasian Soil Science* 31:738–745. Page S-125.

47. Soil Association. 2007. *Pesticides and organic farming—a last resort.* http://www.whyorganic.org/web/sa/saweb.nsf/librarytitles/24E22.HTMl/$file/Pesticides per cent20and per cent20organic per cent20farming per cent20-per cent20A per cent20last per cent20resort.pdf.

48. Research Institute of Organic Agriculture. 2007. *Statistics on global organic farming 2007.* http://www.organic-world.net/statistics-world.html.

49. Hine, R., and Pretty, J. 2006. Promoting production and trading opportunities for organic agricultural products in East Africa. Capacity building study 3 in: *Organic agriculture and food security in East Africa.* University of Essex.

50. Badgley, C., Moghtader, J., Quintero, E., Zakem, E., Chappell, M., Avilés-Vàzquez, K., Samulon, A., and Perfecto, I. 2007. Organic agriculture and the global food supply. *Renewable Agriculture and Food Systems* 22:86–108.

51. Goulding, K., Jarvis, S., and Whitmore, A. 2008. Optimising nutrient management for farm systems. *Philosophical Transactions of the Royal Society B* 363:667–680. With data from: Rasmussen, P., Goulding, K., Brown, J., Grace, P., Janzen, H., and Korschens, M. 1998. Long-term agroecosystem experiments: assessing agricultural sustainability and global change. *Science* 282:893–896. Reprinted with permission from the American Association for the Advancement of Science.

52. Goulding, K., Trewavas, A., and Giller, K. 2011. Feeding the world: a contribution to the debate. *World Agriculture* 2:32–38.

53. Pretty, J., Noble, A., Bossio, D., Dixon, J., Hine, R., Penning de Vries, F., and Morison, J. 2006. Resource-conserving agriculture increases yields in developing countries. *Environmental Science and Technology* 40:1114–1119.

54. Ronald, P., and Adamchak, R. 2008. *Tomorrow's table: Organic farming, genetics and the future of food.* New York: Oxford University Press. http://pamelaronald.blogspot.com/.

55. Conway and Pretty. 1991. *Unwelcome harvest.*
Pretty. 1995. *Regenerating Agriculture.*

56. Pretty et al. 2006. Resource-conserving agriculture.
    Kohle, S., and Mitra, B. 1987. Effects of Azolla as an organic source of nitrogen in rice-wheat cropping systems. *Journal of Agronomy and Crop Science* 159:212–215.

57. Agarwal, P., and Garrity, D. 1987. *Intercropping of legumes to contribute nitrogen in low-input upland rice-based cropping systems.* International Symposium on Nutrient Management for Food Crop Production in Tropical Farming Systems, Malang, Indonesia.

58. Bares, D., Heichel, G., and Sheaffer, C. 1986. *Nitro alfalfa may foster new cropping system.* News. November 20. St. Paul, MN: Minnesota Extension Service.

59. Yamoah, C., Agboola, A., and Wilson, G. 1986. Nutrient contribution and maize performance in alley cropping systems. *Agroforestry Systems* 4:247–254.

60. Hooper, W., and Ash, H. 1935. *Marcus Porcius Cato on agriculture. Marcus Terentius Varro on agriculture.* Cambridge, MA: Loeb Classical Library, Harvard University Press and London: William Heinemann (my translation).

61. Augstburger, F. 1983. Agronomic and environmental potential of manure in the Bolivian valleys and highlands. *Agricultural Ecosystems and Environment* 10:335–346.

62. Kotschi, J., Water-Bayer, A., Adelheim, R., and Hoesle, U. 1988. *Ecofarming in agricultural development.* Eschborn, Germany: GTZ.

63. Agarwal and Garrity. 1987. *Intercropping of legumes to contribute nitrogen.*

64. Johansen, C. 1993. Two legumes unbind phosphate. *International Ag-Sieve* 5:1–3.

65. IITA. 1992. *Sustainable food production in Sub-Saharan Africa. 1. IITA's contribution.* Ibadan, Nigeria: International Institute of Tropical Agriculture.

66. Palmer, J. 1992. The sloping agricultural land technology. In: Hiemstra, W., Reijntjes, C., and Van Der Werf, E., eds. *Let farmers judge.* London: Intermediate Technology Publications.

67. Greenland, D. 1995. *Contributions to agricultural productivity and sustainability from research on shifting cultivation, 1960 to present.* Unpublished mimeo.

68. IITA. 1992. *Sustainable food production in Sub-Saharan Africa.*

69. Zhaohua, Z. 1988. A new farming system-crop / *Paulownia* intercropping. In: *Multipurpose tree species for small farm development.* Little Rock, AR: IDRC/ Winrock.

70. Gypmantasiri, P., Wiboonpongse, A., Rerkasem, B., Craig, I., Rerkasem, K., Ganjanapan, L., Titayawan, M., Seetisarn, M., Thani, P., Jaisaard, R., Ongprasert, S., Radnachaless, T., and Conway, G. 1980. *An interdisciplinary perspective of cropping systems in the Chiang Mai Valley: key questions for research.* Chiang Mai, Thailand: Faculty of Agriculture, University of Chiang Mai.

71. Conway and Pretty. 1991. *Unwelcome harvest.*
    Pretty et al. 2005. *Resource-conserving agriculture.*

72. Scoones, I., and Toulmin, C. 1993. *Socio-economic dimensions of nutrient cycling in agropastoral systems in dryland Africa.* Paper for ILCA Nutrient Cycling Conference. Addis Ababa: ILRI.

McCown, R., Haaland, G., and de Haan, C. 1979. The interaction between cultivation and livestock production. *Ecological Studies* 34:297–332.

73. Rerkasem, M., and Rerkasem, M. 1984. *Organic manures in intensive cropping systems.* Chiang Mai, Thailand: Multiple Cropping Project, Faculty of Agriculture, University of Chiang Mai.

74. Mbegu, A. 1996. Making the most of compost: a look at *wafipa* mounds in Tanzania. In: Reij et al. 1996. *Sustaining the soil.*

75. Kisian'gani quoted in Pretty et al. 2005. *Resource-conserving agriculture.*

76. ERCS/IIED. 1988. *Wollo: a closer look at rural life.* Addis Ababa: Ethiopian Red Cross Society and London: IIED.
    Chambers, R. 1990. *Microenvironments unobserved.* Gatekeeper Series SA 22, Sustainable Agriculture Programme. London: IIED.

77. Pretty et al. 2005. *Resource-conserving agriculture.*

78. Shah, P., Bharadwaj, G., and Ambastha, R. 1991. Participatory impact monitoring of a soil and water conservation programme by farmers, extension volunteers, and AKRSP. *RRA Notes* 13:127–131. London: IIED.

79. Kerr and Sanghi. 1992. *Soil and water conservation in India's semi arid tropics.*

80. FAO. 1991. *Issues and perspectives in sustainable agriculture and rural development.* Main Document. FAO Newsletter, Conference on Agriculture and the Environment, s-Hertogenbosch, Netherlands, April 15–19.

14. Sustained by Water

1. Kofi Annan Foundation. 2010. *Africa can feed itself: green revolution takes root.* http://kofiannanfoundation.org/newsroom/news/2010/03/africa-can-dfeed-itself-green-revolution-takes-root.

2. Spedding, C. 1996. *Agriculture and the citizen.* London: Chapman and Hall.

3. Al-Kaisi, M., and Broner. I. 2009. *Crop water use and growth stages.* Colorado State University. Extension Fact Sheet No. 4.715. http://www.ext.colostate.edu/pubs/crops/04715.html.
    Smith, A., Coupland, G., Dolan, L., Harberd, N., Jones, J., Martin, C., Sablowski, R. and Amey, A. 2009. *Plant biology.* New York: Garland Science.

4. Winter, T., Harvey, J., Franke, O., and Alley, W. 1998. *Ground water and surface water, a single resource.* U.S. Geological Survey Circular 1139. http://www.connectedwaters.unsw.edu.au/resources/fact/water_resources.html.

5. FAO. 2009. *AQUASTAT database.* Rome: FAO. http://www.fao.org/nr/aquastat.

6. International Water management Institute (IWMI). 2007. *Water for food, water for life: a comprehensive assessment of water management in agriculture.* London and Colombo: Earthscan and International Water management Institute (IWMI).

7. UNEP. 2006. *Challenges to international waters—regional assessments in a global perspective.* Nairobi: United Nations Environment Programme.

8. Millennium Ecosystem Assessment. 2005. *Current state and trends assessment.* Washington, DC: Island Press.
    UNDP. 2006. *Human development report 2006. Beyond scarcity: power poverty and the global water crisis.* New York: United Nations, Palgrave-McMillan.

9. World Bank. 2008. *World development report 2008: agriculture for development.* Washington, DC: World Bank. Data from Smakhtin, Revenga, and Döll. Adapted from: United Nations Development Program. 2006. *Human development report 2006. Beyond scarcity: power, poverty, and the global water crisis.* New York: United Nations, Palgrave-Macmillan.

10. IWMI. 2007. *Water for food, water for life.*

11. UN Water. 2007. *Coping with water scarcity: challenge of the twenty-first century.* Prepared for World Water Day 2007. http://www.unwater.org/wwd07.

12. Kundzewicz, Z., Mata. L., Arnell, N., Döll, P., Jiménez, B., Miller, K., Oki, T., Şen, Z., and Shiklomanov, I. 2007. Freshwater resources and their management. In: Parry, M., Canziani, O., Palutikof, J., van der Linden, P., and Hanson, C., eds. *Climate change 2007: impacts, adaptation and vulnerability.* Working Group II Contribution to the Fourth Assessment Report of the Intergovernmental Panel on Climate Change. Cambridge, UK Cambridge University Press. Page 197.

13. Kameri-Mbote, P. 2007. Water, conflict and cooperation: lessons from the Nile River Basin. *Navigating Peace.* No. 4. January. Washington, DC: Woodrow Wilson International Center for Scholars.

14. Kameri-Mbote. 2007. Water, conflict and cooperation.
    http://www.nilebasin.org/newsite/index.php?option = com_contentandview = sectionandlayout=blogandid=5andItemid=68andlang=en.

16. Nile Basin Initiative. 2011. *Egypt and Ethiopia agree to reset relations.* http://www.nilebasin.org/newsite/index.php?option=com_content&view=category&layout=blog&id=40&Itemid=50&lang=en; Mekonnen, D. Z. 2010. The Nile Basin Cooperative Framework Agreement negotiations and the adoption of a 'water security' paradigm: flight into obscurity or a logical cul-de-sac? *The European Journal of International Law.* 21:421–440.

17. Barker, R., Herdt, R., and Rose, B. 1985. *The rice economy of Asia.* Washington, DC: Resources for the Future.

18. FAO. 2010. *FAOSTAT.* Rome: FAO. http://faostat.fao.org/.

19. Farvar, J., and Milton, J., eds. 1972. *The careless technology: ecological aspects of international development.* Garden City, NY: Natural History Press, Doubleday.
    Amte, B. 1989. *Cry, the beloved Narmada.* Chandrapur, Maharashtra, India: Maharogi Sewa Samiti.
    Scudder, T. 1989. Conservation vs. development: river basin projects in Africa. *Environment* 31:4–9, 27–32.

Goldsmith, E., and Hilyard, N. 1984 and 1986. *Social and environmental effects of large dams, Vols. 1 and 2*. Camelford, UK: Wadebridge Ecological Centre.

20. Alexandratos, N. 1995. *World agriculture: towards 2010: An FAO study*. Chichester, UK: Wiley & Sons.

21. Alexandratos. 1995. *World agriculture: towards 2010*.
Government of India. 2006. *Report of the minor irrigation census 2000–2001*. New Delhi: Ministry of Water Resources.

22. IRRI. 1995. *IRRI 1994–1995: water a looming crisis*. Los Baños, Philippines: International Rice Research Institute.

23. Black, M., and King, J. 2009. *The atlas of water*. London: Earthscan.

24. World Bank. 2006. *Reengaging in agricultural water management: challenges and options*. Washington, DC: World Bank.

25. Rodell, M., Velicogna, I., and Famiglietti, J. 2009. Satellite-based estimates of groundwater depletion in India. *Nature* 460:999–1002.

26. World Bank. 2003. *India: revitalizing Punjab's agriculture*. New Delhi: World Bank.

27. UNEP. 2006. *Challenges to international waters*.

28. Howe, C. 2002. Policy issues and institutional impediments in the management of groundwater: lessons from case studies. *Environment and Development Economics* 7:625–641.

29. Postel, S. 1999. *Pillar of sand: can the irrigation miracle last?* New York: W. W. Norton.

30. Millennium Ecosystem Assessment. 2005. *Current state and trends assessment*.

31. Smedema, L., and Shiati, K. 2002. Irrigation and salinity: a perspective review of the salinity hazards of irrigation development in the arid zone. *Irrigation and Drainage Systems* 16:161–174.

32. Gypmantasiri, P., Wiboonpongse, A., Rerkasem, B., Craig, I., Rerkasem, K., Ganjanapan, L., Titayawan, M., Seetisarn, M., Thani, P., Jaisaard, R., Ongprasert, S., Radnachaless, T., and Conway, G. 1980. *An interdisciplinary perspective of cropping systems in the Chiang Mai Valley: key questions for research*. Chiang Mai, Thailand: Faculty of Agriculture, University of Chiang Mai.

33. Wickramasekera, P. 1981. *Water management under channel irrigation: a study of the Minipe Settlement in Sri Lanka*. Sri Lanka: Mimeo, Department of Economics, University of Peredeniya.

34. Ramamurthy quoted in Chambers, R. 1990. *Microenvironments unobserved*. Gatekeeper Series SA 22. London: Sustainable Agriculture Programme, IIED.

35. Wade, R. 1982. The system of administrative and political corruption: canal irrigation in South India. *Journal of Development Studies* 18:287–328.

36. Transparency International. 2008. *Global corruption report 2008: corruption in the water sector*. Cambridge, UK: Cambridge University Press.

37. Pant, N. 1981. *Some aspects of irrigation administration (a case study of Kosi Project).* Naya Prokash, Calcutta, 6.

38. Ford Foundation. 1994. Saving the village tank. *Bulletin, New Delhi Office* 1:3–5. New Delhi: Ford Foundation.

39. Asian Development Bank. 2006. *Rehabilitation and management of tanks in India: a study of select states.* Philippines: Asian Development Bank.

40. FAO. 2009. *FAOSTAT.*
    You, L. 2008. *Africa infrastructure country diagnostic. Irrigation investment needs in Sub-Saharan Africa.* Background Paper 9. Washington, DC: World Bank.

41. Pretty. 1995. *Regenerating agriculture.*
    Reij, C. 1991. *Indigenous soil and water conservation in Africa.* Gatekeeper Series SA 27. London: Sustainable Agriculture Programme, IIED; Reij, C., Scoones, I., and Toulmin, C., eds. 1996. *Sustaining the soil: indigenous soil and water conservation in Africa.* London: Earthscan.

42. Goldsmith, E., and Hildyard, N. 1984. *The social and environmental impacts of large dams.* Camelford, UK: Wadebridge Ecological Centre.

    World Commission on Dams. 2000. *Dams and development: a new framework for decision making.* London: Earthscan.

43. China Three Gorges Project. 2002. *Biggest flood control benefit in the world.* http://www.ctgpc.com.cn/en/benefifs/benefifs_a.php.

44. Conway and Waage. 2010. *Science and innovation for development.*
    Risbud, A. 2006. *Cheap drinking water from the ocean.* Technology Review, June, 12. http://www.technologyreview.com/energy/16977/.

45. Pretty, J. 1995. *Regenerating agriculture: policies and practice for sustainability and self-reliance.* London: Earthscan.
    Gubbels, P. 1994. Farmer-driven research and the Project Agro-Forestier in Burkina Faso. In: Scoones, I., and Thompson, J., eds. *Beyond farmer first.* London: Intermediate Technology.
    Wedum, J., Doumba, Y., Sanogo, B., Dicko, G. and Cissé, O. 1996. Rehabilitating degraded land: *Zaï* in the Djenné Circle of Mali. In: Reij, C., Scoones, I., and Toulmin, C., eds. 1996. *Sustaining the soil: indigenous soil and water conservation in Africa.* London: Earthscan.

46. Cofie, O., Barry, B., and Bossio, D. 2004. *Human resources as a driver of bright spots: the case of rainwater harvesting in West Africa.* Conference paper 19, NEPAD/IGAD Regional Conference: Agricultural Successes in the Greater horn of Africa, Nairobi. Ghana and Sri Lanka: IWMI.
    Dreschel, P., Olaleye, A., Adeoti, A., Thiombiano, L., Barry, B., and Vohland, K. No date. *Adoption driver and constraints of resource conservation technologies in Sub-Saharan Africa.* http://westafrica.iwmi.org/Data/Sites/17/Documents/PDFs/Adoption Constraints-Overview.pdf.

Doumbia, M., Berthe, A. and Aune, J. 2005. *Integrated plant nutrition management in Mali summary report 1998–2004*. Oslo: Drylands Coordination Group Report No. 36.

47. Bastian, E., and Gräfe, W. 1989. Afforestation with multipurpose trees in *media lunas*: a case study from the Tarija basin, Bolivia. *Agroforestry Systems* 9:93–126.

48. Scoones, I. 1991. Wetlands in drylands: key resources for agricultural and pastoral production in Africa. *Ambio* 20:366–371.

49. Pretty. 1995. *Regenerating agriculture.*

50. Long, S., and Ort, D. 2010. More than taking the heat: crops and global change. *Current Opinion in Plant Biology* 13:241–248.

51. Castiglioni, P., Warner, D., Benson, R., Anstrom, D., Harrison, J., Stoecker, M., Abad, M., Kumar, G., Salvador, S., D'Ordine, R., Navarro, S., Back, S., Fernandes, M., Targolli, J., Dasgupta, S., Bonin, C., Luethy, M., and Heard, J. 2008. Bacterial RNA chaperones confer abiotic stress tolerance in plants and improved grain yield in maize under water-limited conditions. *Plant Physiology* 147:446–455.

52. AATF. 2011. *Water efficient maize for Africa.* http://www.aatf-africa.org/wema/en/.

53. Alexandratos. 1995. *World agriculture: towards 2010*; Williams, M. 1996. *The transition in the contribution of living aquatic resources to food security.* Food, Agriculture, and Environment Discussion Paper 13. Washington, DC: IFPRI.

54. ScienceDaily. 2009. *Aquaculture's growth seen as continuing.* http://www.sciencedaily.com/releases/2009/01/090102082248.htm.

55. FAO. 2010. *The state of the world fisheries and aquaculture.* Rome: FAO.

56. FAO. 2010. *The state of the world fisheries.*

57. Alexandratos. 1995. *World agriculture: towards 2010.*
    FAO. 2010. *The state of the world fisheries.*

58. Bimbao, M., Cruz, A., and Smith, I. 1992. An economic assessment of rice-fish culture in the Philippines. In: Hiemstra, W., Reijntjes, C., and Van Der Werf, E., eds. *Let farmers judge.* London: Intermediate Technology Publications.

59. Kamp, K., Gregory, R., and Chowhan, G. 1993. Fish cutting pesticide use. *ILEIA Newsletter* 2/93:22–23.

60. Holmes, B. 1996. Blue revolutionaries. *New Scientist*, December 7.

61. FAO. 1992. *Review of the state of world fishery resources. Part 2: inland fisheries and aquaculture.* Fisheries Circular 710 (Rev. 8). Rome: FAO.

62. Brummett, R., and Noble, R. 1995. *Aquaculture for African smallholders.* Penang, Malaysia: International Centre for Living Aquatic Resources Management (ICLARM).

63. Noble, R., and Rashidi, B. 1990. *Aquaculture technology transfer to smallholder farmers in Malawi, Southern Africa.* Naga: The ICLARM Quarterly, October.

64. UN Commission for Sustainable Development. 2008. *Farmer-scientist research partnerships for integrated aquaculture.* CSD-8: Sustainable Development Success Stories. http://www.un.org/esa/dsd/dsd_aofw_mg/mg_success_stories/csd8/SARD-6.htm.

65. UN CSD. 2008. *Farmer-scientist research partnerships*.

66. Conway and Waage. 2010. *Science and innovation for development*.
    World Bank. 1993. *Water resources management: a World Bank policy paper*. Washington, DC: International Bank for Reconstruction and Development, 10.

67. Moench, M., and Stapleton, S. 2007. *Water, climate, risk and adaptation*. Working Paper. Boulder, CO: Institute for Social and Environmental Transition.

68. USAID Water Team. 2002. *Integrated water resources management: a framework for action in freshwater and coastal systems*. Washington, DC: USAID. http://waterwiki.net/images /8/80/IWRM_Framework_Freshwaterandcoastal.pdf.

15. Adapting to Climate Change

1. Canuto, O. 2011. *An inconvenient truth for Latin America*. Growth and Crisis Blog. Washington, DC: World Bank. http://blogs.worldbank.org/growth/inconvenient-truth -latin-america.

2. Toulmin, C. 2007. *Africa's development prospects up in smoke?* Colin Trapnell Memorial Lecture, Green College, Oxford.

3. Mertz, O., Halsnæs, K., Oleson, J., and Rasmussen, K. 2009. Adaptation to climate change in developing countries. *Environmental Management* 43:743–752.

4. Brooks, N., Adger, W. N. and Kelly, P. M. 2005. The determinants of vulnerability and adaptive capacity at the national level and the implications for adaptation. *Global Environmental Change*. 15:151–163.

5. Stern, N. 2007. *The economics of climate change: the Stern Review*. Cambridge, UK: Cambridge University Press.

6. World Bank, 2009. *World development report: reshaping economic geography*. Washington, DC: World Bank.
    Lybbert, T., and Sumner, D. 2010. *Agricultural technologies for climate change mitigation and adaptation in developing countries: policy options for innovation and technology diffusion*. ICTSD-IPC Platform on Climate Change, Agriculture and Trade. Issue brief No. 6.

7. Erda, L., Conway D., Yue, L., and Calsamiglia-Mendlewicz, S., eds. 2008. *The impacts of climate change on Chinese agriculture—Phase II*. Climate change in Ningxia: scenarios and impacts, technical report. Final Report. UK: AEA Group. http://www .uea.ac.uk/polopoly_fs/1.147097!NingxiaRegionalReport_Issue_2.pdf.

8. NOAA. 2009. *State of the climate global analysis—annual 2008*. Washington, DC: National Oceanic and Atmospheric Administration, National Climate Data Center. http://www.ncdc.noaa.gov/sotc/index.php?report=global&year=2008&month=ann.

9. Keeling, R., Piper, S., Bollenbacher, A. and Walker, S. 2008. *Atmospheric carbon dioxide record from Mauna Loa*. La Jolla, CA: Carbon Dioxide Research Group, Scripps Institution of Oceanography (SIO), University of California. http://cdiac.ornl.gov/trends/co2/ sio-mlo.html; European Environment Agency. 2009. *CSI 013 – Atmospheric greenhouse gas concentrations*. http://www.eea.europa.eu/data-and-maps/indicators/atmospheric

-greenhouse-gas-concentrations/atmospheric-greenhouse-gas-concentrations-assess-ment-3.

10. Royal Society Climate Change Advisory Group. 2007. *Climate change controversies: a simple guide.* http://royalsociety.org/Report_WF.aspx?pageid=8030andterms=climate+change+controversies.

11. Stern. 2007. *The economics of climate change.*

12. O'Hare, G. Sweeney, J., and Wilby, R. 2005. *Weather, climate and climate change: human perspectives.* Harlow, England: Prentice Hall.
Lenton, T., Held, H., Kriegler, E., Hall, J., Lucht, W., Rahmstorf, S., and Schellnhuber, H. 2008. Tipping elements in the Earth's climate system. *Proceedings of the National Academy of Sciences of the United States of America* 105:1786–1893.
Verchot, L., Van Noordwijk, M., Kandji, S., Tomich, T., Ong, C., Albrecht, A., Mackensen, J., Bantilan, C., Anupama, K., and Palm, C. 2007. Climate change: linking adaptation and mitigation through agroforestry. *Mitigation and Adaptation Strategies for Global Change* 12:901–918.

13. Verchot et al. 2007. Climate change: linking adaptation and mitigation through agro-forestry.

14. O'Hare et al. 2005. *Weather, climate and climate change.*

15. NOAA. No date. *What is an El Niño?* http://www.pmel.noaa.gov/tao/elnino/el-nino-story.html.

16. Salinger, M., Allan. R., Bindoff, N., Hannah, J., Lavery, B., Lindesay, J. Nicholls, N., Plummer, N., and Torok, S. 1996. Observed variability and change in climate and sea level in Australia, New Zealand and the South Pacific. In: Bouma, W., Pearman, G., and Manning, M., eds. 1996. *Greenhouse: coping with climate change.* Melbourne: CSIRO; United Nations Framework Convention on Climate Change. No date. *National activities—PICCAP.* The Republic of the Marshall Islands Climate Change Web Site. http://unfccc.int/resource/ccsites/marshall/activity/piccap.htm.

17. Trenberth, K., Caron, J., Stepaniak, D., and Worley S. 2002. The evolution of ENSO and global atmospheric temperatures. *Journal of Geophysical Research* 107:4065.
Trenbath, K., Jones, P., Ambenje, P., Bojariu, R., Easterling, D., Klein Tank, A., Parker, D., Rahimzadeh, F., Renwick, J., Rusticucci, M., Soden, B., and Zhai, P. 2007. Chapter 3—observations: surface and atmospheric climate change. In: Solomon, S., Qin, D., Manning, M., Chen, Z., Marquis, M., Averyt, K., Tignor, M., and Miller, H., eds. *Climate change 2007: the physical science basis.* Contribution of Working Group I to the Fourth Assessment Report of the Intergovernmental Panel on Climate Change. Cambridge, UK and New York: Cambridge University Press.

18. May, W. 2004. Potential of future changes in the Indian summer monsoon due to greenhouse warming: analysis of mechanisms in a global time-slice experiment. *Climate Dynamics* 22:389–414.

19. Chase, T., Knaff, J., Pielke Sr., R., and Kalnay, E. 2003. Changes in global monsoon circulations since 1950. *Natural Hazards* 29:229–254.

Christensen, J., Hewitson, B., Busuioc, A., Chen, A., Gao, X., Held, I., Jones, R., Kolli, R., Kwon, W.-T., Laprise, R., Magaña Rueda, V., Mearns, L., Menéndez, C., Räisänen, J., Rinke, A., Sarr, A., and Whetton, P. 2007. Chapter 11—regional climate projections. In: Solomon et al., eds. *Climate change 2007.*

20. Chase et al. 2003. Changes in global monsoon circulations.

   Huijun, W. 2001. The weakening of the Asian monsoon circulation after the end of the 1970s. *Advances in Atmospheric Sciences* 18:3.

   Dash, S., Kulkarni, M., Mohanty, U., and Prasad, K. 2009. Changes in the characteristics of rain events in India, *Journal of Geophysical Research* 114.

21. NOAA Earth System Research Laboratory. 2011. *Multivariate ENSO Index.* U.S. National Oceanic and Atmospheric Administration (NOAA). http://www.cdc.noaa.gov/people/klaus.wolter/MEI/.

22. Solomon, S., Qin, D., Manning, M., Alley, R., Berntsen, T., Bindoff, N., Chen, Z., Chidthaisong, A., Gregory, J., Hegerl, G., Heimann, M., Hewitson, B., Hoskins, B., Joos, F., Jouzel, J., Kattsov, V., Lohmann, U., Matsuno, T., Molina, M., Nicholls, N., Overpeck, J., Raga, G., Ramaswamy, V., Ren, J., Rusticucci, M., Somerville, R., Stocker, T., Whetton, P., Wood, R., and Wratt, D. 2007. *Technical summary.* In: Solomon, et al., (eds). *Climate change 2007.*

23. Long, S., and Ort, D. 2010. More than taking the heat: crops and global change. *Current Opinion in Plant Biology* 13:241–248.

24. Challinor, A., Wheeler, T., Osborne, T., and Slingo, J. 2006. Assessing the vulnerability of crop productivity to climate change thresholds using an integrated crop-climate model. In: Schellnhuber, J., Cramer, W., Nakicenovic, N., Yohe, G., and Wigley, L., eds. *Avoiding dangerous climate change.* Cambridge, UK: Cambridge University Press.

25. Easterling, W., Aggarwal, P., Batima, P., Brander, K., Erda, L., Howden, S., Kirilenko, A., Morton, J., Soussana, J.-F., Schmidhuber, J., and Tubiello, F. 2007. Chapter 5—Food, fibre and forest products. In: Parry, M., Canziani, O., Palutikof, J., van der Linden, P., and Hanson, C., eds. *Climate change 2007: impacts, adaptation and vulnerability.* Working Group II Contribution to the Fourth Assessment Report of the Intergovernmental Panel on Climate Change, Cambridge, UK and New York: Cambridge University Press. Figure 5.2 (right panels).

26. Lobell, D., Bänziger, M., Magorokosho, C., and Vivek, B. 2011. Nonlinear heat effects on African maize as evidenced by historical yield trials. *Nature Climate Change.* Doi: 10.1038/nclimate1043.

27. Ericksen, P., Thornton, P., Notenbaert, A., Cramer, L., Jones, P., and Herrero, M. 2011. *Mapping hotspots of climate change and food insecurity in the global tropics.* CCAFS Report No. 5. Copenhagen: CGIAR Research Program on Climate Change, Agriculture and Food Security (CCAFS). http://www.ccafs.cgiar.org.

28. Easterling et al. 2007. Chapter 5—Food, fibre and forest products.

29. Labuschagné, I. 2004. Budbreak number as selection criterion for breeding apples adapted to mild winter climatic conditions: a review. *Acta Horticulture* (ISHS) 663:775–782.

30. Doering, D. 2005. *Public-private partnership to develop and deliver drought tolerant crops to food insecure farmers.* Draft document for discussion at the May 3–4, 2005, Strategy and Planning Meeting. Winrock International.

31. Ericksen et al. 2011. *Mapping hotspots of climate change and food security.*

32. Ericksen et al. 2011. *Mapping hotspots of climate change and food security.*

33. Long and Ort. 2010. More than taking the heat.

34. Long, S., Ainsworth, E., Leakey, A., Nösberger, J., and Ort, D. 2006. Food for thought: lower-than-expected crop yield simulation with rising CO_2 concentrations. *Science* 312:1918–1921.

    Long, S., Ainsworth, E., Leakey, A., Ort, D., Nösberger, J., and Schimel, D. 2007. Crop models, CO_2 and climate change—response. *Science* 315:460.

35. Warren, R., Arnell, N., Nicholls, R., Levy, P., and Price, J. 2006. *Understanding the regional impacts of climate change.* Research report prepared for the Stern Review on the Economics of Climate Change. Working Paper 90. Norwich, UK: Tyndall Centre for Climate Change Research, University of East Anglia.

36. Mader, T., and Davis, M. 2004. Effect of management strategies on reducing heat stress of feedlot cattle: feed and water intake. *Journal of Animal Science* 82:3077–3087.

37. Parson, D. Armstrong, A., Turnpenny, J., Matthews, A., Cooper, K., and Clark, J. 2001. Integrated models of livestock systems for climate change studies 1. Grazing systems. *Global Change Biology* 7:93–112.

    Landlearn NSW. No date. *The impacts of climate change on the livestock industry.* http://www.landlearnnsw.org.au/sustainability/climate-change/agriculture/livestock/impacts.

38. Easterling et al. 2007. Chapter 5—Food, fibre and forest products.

39. Toulmin, C. 1985. Chapter 2—The effects of drought. In: ILCA. *Livestock losses and post-drought rehabilitation in Sub-Saharan Africa.* LPU Working Paper No. 9. Addis Ababa: International Livestock Centre for Africa. http://www.fao.org/Wairdocs/ILRI/x5439E/x5439e00.htm.

40. Thornton, P., Herrero, M., Freeman, A., Mwai, O., Rege, E., Jones, P., and McDermott, J. 2007. Vulnerability, climate change and livestock—research opportunities and challenges for poverty alleviation. *ICRISAT SAT ejournal.* 14.

41. Seo, S., and Mendelsohn, R. 2006. *The impact of climate change on livestock management in Africa: a structural Ricardian analysis.* CEEPA Discussion Paper No. 23. Pretoria: Centre for Environmental Economics and Policy in Africa (CEEPA), University of Pretoria.

    World Bank. 2008. *Mali climate change: likely impacts on African crops, livestock and farm types.* Washington, DC: World Bank. http://web.worldbank.org/WBSITE/EXTERNAL/COUNTRIES/AFRICAEXT/MALIEXTN/0,,contentMDK:21793691~menuPK:50003484~pagePK:2865066~piPK:2865079~theSitePK:362183,00.html.

42. Easterling et al. 2007. Chapter 5—Food, fibre and forest products.

43. Thornton et al. 2007. Vulnerability, climate change and livestock.

44. Easterling et al. 2007. Chapter 5—Food, fibre and forest products.

45. Akita, S., and Moss, D. 1973. Photosynthetic response to CO_2 and light by maize and wheat leaves adjusted for constant stomatal apertures. *Crop Science* 13:234–237.

46. World Bank. 2008. *Focus F: adaptation to and mitigation of climate change in agriculture.* World development report 2008. Washington, DC: World Bank. http://siteresources .worldbank.org/INTWDR2008/Resources/2795087–1192112387976/WDR08 _15_Focus_F.pdf.

47. Boko, M., Niang, I., Nyong, A., Vogel, C., Githeko, A., Medany, M., Osman-Elasha, B., Tabo, R., and Yanda, P. 2007. Africa. In: Parry et al., eds. *Climate change 2007.* World Water Forum. 2000. *The Africa water vision for 2025: equitable and sustainable use of water for socioeconomic development.* UNWater/Africa.

48. Kundzewicz, Z., Mata, L., Arnell, N., Döll, P., Kabat, P., Jiménez, B., Miller, K., Oki, T., Sen, Z., and Shiklomanov, I. 2007. Freshwater resources and their management. In: Parry et al., eds. *Climate change 2007.*

49. Kundzewicz et al. 2007. Freshwater resources and their management.

50. Carter, M., Little, P., Mogues, T., and Nepatu, W. 2004. *Shock, sensitivity and resilience: tracking the economic impacts of environmental disaster on assets in Ethiopia and Honduras.* Wisconsin: BASIS.

51. FAO. 2004. *State of world food insecurity.* Rome: FAO.

52. Haile, M. 2005. Weather patterns, food security and humanitarian response in Sub-Saharan Africa. *Philosophical Transactions of the Royal Society, London.* 360:2169–2182.

53. Lybbert and Sumner. 2010. *Agricultural technologies for climate change mitigation and adaptation.*

54. IFAD. 2009. *Smallholder farming in transforming economies of Asia and the Pacific: challenges and opportunities.* Discussion Paper prepared for the side event organized during the Thirty third session of IFAD's Governing Council, 18 February. http://www.ifad .org/events/gc/33/roundtables/pl/pi_bg_e.pdf.

55. Lybbert and Sumner. 2010. *Agricultural technologies for climate change mitigation and adaptation.*

56. Smit, B., Pilifosova, O., Burton, I., Challenger, B., Huq, S., Klein, R., and Yohe, G. 2001. Adaptation to climate change in the context of sustainable development and equity. In: McCarthy, J., Canziani, O., Leary, N., Dokken, D., and White, K., eds. *Climate change 2001: impacts, adaptation and vulnerability.* Contribution of Working Group II to the third Assessment Report of the Intergovernmental Panel on Climate Change. Cambridge, UK and New York: Cambridge University Press.

57. Smit, B. et al. 2001. Adaptation to climate change in the context of sustainable development and equity.

58. Kurukulasuriya, P., and Mendelsohn, R. 2006. *Crop selection: adapting to climate change in Africa.* Pretoria: Centre for Environmental Economics and Policy in Africa (CEEPA). http://www.ceepa.co.za/docs/CDPNo26.pdf.

59. Osbahr, H., Twyman, C., Adger, W., and Thomas, D. 2008. Effective livelihood adaptation to climate change disturbance: scale dimensions of practice in Mozambique. *Geoforum* 39:1951–1964.

60. Conway, G. 2009. *The science of climate change in Africa: impacts and adaptation.* Grantham Institute for Climate Change Discussion Paper No. 1. London: Grantham Institute and Imperial College London. Page 15.

61. Conway. 2009. *The science of climate change in Africa.*

62. Lobell, D., and Burke, M. 2008. Why are agricultural impacts of climate change so uncertain? The importance of temperature relative to precipitation. *Environmental Research Letters.* 3. http://foodsecurity.stanford.edu/publications/why_are_agricultural_impacts_of_climate_change_so_uncertain_the_importance_of_temperature_relative_to_precipitation/.

63. Eriksen, S. 2005. The role of indigenous plants in household adaptation to climate change: the Kenyan experience. In: Pak Sum Low, ed. *Climate change and Africa.* Cambridge, UK: Cambridge University Press; Conway and Waage. 2010. *Science and innovation for development.*

64. Toulmin. 2007. *Africa's development prospects up in smoke?*

65. World Bank. 2008. *Focus F: adaptation to and mitigation of climate change in agriculture.*

66. Nelson, G., Rosegrant, M., Koo, J., Roberston, R., Sulser, T., Zhu, T., Ringler, C., Msangi, S., Palazzo, A., Batka, M., Magalhaes, M., Valmonte-Santos, R., Ewing, M., and Lee, D. 2009. *Climate change: impact on agriculture and costs of adaptation.* Washington, DC: IFPRI.

16. Reducing Greenhouse Gases

1. The Economist. 2010. *Flawed scientists.* http://www.economist.com/node/16539392.

2. Smith, P., Martino, D., Cai, Z., Gwary, D., Janzen, H., Kumar, P., McCarl, B., Ogle, S., O'Mara, F., Rice, C., Scholes, B., Sirotenko, O., Howden, M., McAllister, T., Pan, G., Romanenkov, V., Schneider, U., Towprayoon, S., Wattenbach, M., and Smith, J. 2008. Greenhouse gas mitigation in agriculture. *Philosophic Transactions of the Royal Society, B* 363:789–813.

3. Lybbert, T., and Sumner, D. 2010. *Agricultural technologies for climate change mitigation and adaptation in developing countries: policy options for innovation and technology diffusion.* ICTSD-IPC Platform on Climate Change, Agriculture and Trade. Issue brief No. May 6.

4. Le Quéré, C., Raupach, M., Canadell, J., Marland, G., Bopp, L., Ciais, P., Conway, T., Doney, S., Feely, R., Foster, P., Friedlingstein, F., Gurney, K., Houghton, R., House, J., Huntingford, C., Levy, P., Lomas, M., Majkut, J., Metzl, N., Ometto, J., Peters, G., Prentice, I. C., Randerson, J., Running, S., Sarmiento, J., Schuster, U., Sitch, S., Takahashi, T., Viovy, N., van der Werf, G. and Woodward, F. I. 2009. Trends in the sources and sinks of carbon dioxide. *Nature Geoscience* 2:831–836; Le

Quéré, C. 2010. Trends in the land and ocean carbon uptake. *Current Opinion in Environment Sustainability* 2:219–224.

5. Garnett, T. 2011. Where are the best opportunities for reducing greenhouse gas emissions in the food system (including the food chain). *Food Policy* 36:S23–S32.

    Smith, P., Martino, D., Zai, C., Gwary, D., Janzen, H., Kumar, P., McCarl, B., Ogle, S., O'Mara, F., Rice, C., Scholes, B., and Sirotenko, O. 2007. Agriculture. In: *Climate change 2007: mitigation of climate change*. Working Group III Contribution to the Fourth Assessment Report of the Intergovernmental Panel on Climate Change. Cambridge, UK and New York: Cambridge University Press.

6. Bellarby, J., Foereid, B., Hastings, A., and Smith, P. 2008. *Cool farming: climate impacts of agriculture and mitigation potential*. Amsterdam: Greenpeace International.

7. Bellarby et al. 2008. *Cool farming*.

8. Garnett. 2011. Where are the best opportunities for reducing greenhouse gas emissions.

9. Bellarby et al. 2008. *Cool farming*.

10. US-EPA. 2006. *Global anthropogenic non-CO2 greenhouse gas emissions: 1990–2020*. Washington, DC: United States Environmental protection Agency, EPA 430-R-06–003, June. http://www.epa.gov/climatechange/economics/international.html.

    FAO. 2003. *World agriculture: towards 2015/2030. An FAO perspective*. Rome: FAO.

11. IPCC. 2007. Summary for Policymakers. In: Metz et al. *Climate Change 2007*.

12. Bellarby et al. 2008. *Cool farming*.

13. Lal, R. 2004. Soil carbon sequestration impacts on global climate change and food security. *Science* 304:1623–1626.

14. Bellarby et al. 2008. *Cool farming*.

15. Based on *IPCC 2000: Special report of the Intergovernmental Panel on Climate Change on land use, land-use change, and forestry*, table 1. Cambridge, UK: Cambridge University Press.

16. FAO. 2010. *Global forest resources assessment 2010*. FAO Forestry Paper 163. Rome: FAO.

17. Lal. 2004. Soil carbon sequestration.

18. Lal. 2004. Soil carbon sequestration.

19. Lal, R. 2010. Beyond Copenhagen: mitigating climate change and achieving food security through soil carbon sequestration. *Food Security* 2:169–177.

20. Lal. 2004. Soil carbon sequestration.

21. Lal. 2004. Soil carbon sequestration. Reprinted with permission from the American Association for the Advancement of Science.

22. Lal. 2010. Beyond Copenhagen.

23. Lal. 2010. Beyond Copenhagen.

24. Lal. 2004. Soil carbon sequestration.

25. Smith et al. 2007. Agriculture.

26. Yan, H., Cao, M., Liu, J., and Tao, B. 2007. Potential and sustainability for carbon sequestration with improved soil management in agricultural soils of China. *Agriculture, Ecosystems and Environment* 121:325–335. Page 332.

27. Garnett. 2011. Where are the best opportunities for reducing greenhouse gas emissions.

28. Lal. 2004. Soil carbon sequestration.

29. Lal. 2004. Soil carbon sequestration.

30. FAO and Conservation Technology Information Center (CTIC). 2008. *Soil carbon sequestration in conservation agriculture: a framework for valuing soil carbon as a critical ecosystem service.* Conservation agriculture carbon offset consultation, West Lafayette, IN, October 2008. http://www.fao.org/ag/ca/doc/CA_SSC_Overview.pdf.

31. Smith et al. 2008. Greenhouse gas mitigation in agriculture.

32. Smith et al. 2008. Greenhouse gas mitigation in agriculture.

33. FAO. 2010. '*Climate-smart' agriculture: policies, practices and financing for food security, adaptation and mitigation.* Rome: FAO. http://www.fao.org/climatechange/climatesmart /66304/en/.
    World Agroforestry Centre. 2009. *Agroforestry: a global land use. Annual Report 2008–2009.* Nairobi: World Agroforestry Centre. http://www.worldagroforestry.org/ar2009 /Annual per cent20Report per cent202008–2009.pdf.
    Smith, G. 2010. *Faidherbia—Africa's fertiliser factory.* New Agriculturist. http://www .new-ag.info/en/developments/devItem.php?a=1036.

34. Makumba, W., Akinnifesi, F., Janssen, B., and Oenema, O. 2007. Long-term impact of a Gliricidia-maize intercropping system on carbon sequestration in southern Malawi. *Agriculture, Ecosystems and Environment* 118:237–243.
    Kaonga, M., and Bayliss-Smith, T. 2008. Carbon pools in tree biomass and the soil in improved fallows in eastern Zambia. *Agroforestry Systems* 76:37–51.

35. Verchot, L., Van Noordwijk, M., Kandji, S., Tomich, T., Ong, C., Albrecht, A., Mackensen, J., Bantilan, C., Anupama, K., and Palm, C. 2007. Climate change: linking adaptation and mitigation through agroforestry. *Mitigation and Adaptation Strategies for Global Change* 12:901–918.

36. Verchot et al. 2007. Climate change.

37. Verchot et al. 2007. Climate change. With kind permission from Springer Science + Business Media.

38. Smith et al. 2007. Agriculture.

39. Scherr, S., and Sthapit, S. 2009. *Mitigating climate change through food and land use.* Washington, DC: Worldwatch Report 179.

40. Scherr and Sthapit. 2009. *Mitigating climate change.*

41. Smith et al. 2008. Greenhouse gas mitigation in agriculture.

42. De Datta, S. 1986. Improving nitrogen fertiliser efficiency in lowland rice in tropical Asia. *Fertilizer Research* 9:171–186.

43. O'Brien, D., Sudjadi, M., Sri Adiningsih, J., and Irawan. 1987. Economic evaluation of deep placed urea for rice in farmers' fields: a pilot area approach, Ngawi, East Java, Indonesia. In: IRRI. *IRRI Efficiency of Nitrogen Fertilizers for Rice.* Los Baños, Philippines: International Rice Research Institute.

44. Powlson, D. 2011. *Greenhouse gas emissions associated with nitrogen fertiliser.* Presentation at Reducing Greenhouse Gases. February 28–March 1. London: Royal Society.
   Rankin, M. (no date). *Nitrification inhibitors and use.* Fond du Lac County, WI: University of Wisconsin Extensions. http://www.uwex.edu/ces/crops/ninhib.htm.

45. Nelson, D., and Huber, D. 1992. Nitrification inhibitors for corn production. *National Corn Handbook.* NCH-55.

46. Niggli, U., Fließbach, A., Hepperly, P., and Scialabba, N. 2009. *Low greenhouse gas agriculture: mitigation and adaptation potential of sustainable farming systems.* Rome: FAO, April, Rev. 2.

47. FAO. 2009. *The State of Food and Agriculture: Livestock in the Balance.* Rome: FAO.

48. Bellarby et al. 2008. *Cool farming.*

49. Garnett. 2011. Where are the best opportunities for reducing greenhouse gas.

50. Smith et al. 2007. Agriculture.

51. Bellarby et al. 2008. *Cool farming.*

52. IPCC. 1996. Agricultural options for mitigation of greenhouse gas emissions. In: IPCC. *Climate Change 1995: Impacts, adaptations and mitigations of climate change: scientific-technical analyses.* Contribution of Working Group II to the Second Assessment of the IPCC. Cambridge, UK: Cambridge University Press.

53. Leng, R. 1991. *Improving ruminant production and reducing methane emissions form ruminants by strategic supplementation.* USEPA Report 400/1–91/004. Washington, DC: Office of Air and Radiation, US Environmental Protection Agency.

54. Preston, T., and Leng, R. 1994. Agricultural technology transfer: Perspectives and case studies involving livestock. In: Anderson, J., ed. *Agricultural technology: policy issues for the international community.* Wallingford, UK: CAB International.

55. United Nations Framework Convention on Climate Change. 2008. *Challenges and opportunities for mitigation in the agricultural sector: Technical Paper.* FCCC/TP/2008/8. http://unfccc.int/resource/docs/2008/tp/08.pdf.

56. Bellarby et al. 2008. *Cool farming.*

57. Scherr and Sthapit. 2009. *Mitigating climate change.*

58. Garnett. 2011. Where are the best opportunities for reducing greenhouse gas emissions.

59. Sass, R., Fisher, Y., Wang, F., Turner, F., and Jud, M. 1992. Methane emission from rice fields: the effect of flood water management. *Global Biogeochemical Cycles* 6:249–262.

60. Wassman, R., Hosen, Y., and Sumfleth, K. 2009. *Agriculture and climate change: an agenda for negotiation in Copenhagen. Reducing methane emissions from irrigated rice.* Focus 16. Brief 3. Washington, DC: IFPRI.

61. Bellarby et al. 2008. *Cool farming.*

62. IRRI. 2009. *Aerobic rice.* Manila, Philippines: International Rice Research Institute. http://www.knowledgebank.irri.org/factsheetsPDFs/watermanagement_FSAero bicRice3.pdf; Hittalmani, S. 2010. *Aerobic rice.* Bangalore, India: University of Agricultural Sciences. http://aerobicrice.org/.

63. Wang, B. and Adachi, K. 2000. Differences among rice cultivars in root exudation, methane oxidation, and populations of methanogenic and methanotrophic bacteria in relation to methane emission. *Nutrient Cycling in Agroecosystems* 58:349–356.

64. Smith et al. 2007. Agriculture.

65. Brohé, A., Eyre, N., and Howarth, N. 2009. *Carbon markets: an international business guide.* London: Earthscan.

66. Brohé et al. 2009. *Carbon markets.*

67. Reyes, O., and Gilbertson, T. 2009. Carbon trading: how it works and why it fails. *Critical Currents.* Occasional Paper Series No. 7. Uppsala: Dag flammarskjöld Foundation.

68. Brohé et al. 2009. *Carbon markets.*

69. Reyes and Gilbertson. 2009. Carbon trading: how it works.

70. Smith et al. 2007. Agriculture.

71. World Bank. 2011. *Carbon finance at the World Bank: list of funds.* http://wbcarbonfinance .org/Router.cfm?Page=Funds&ItemID=24670.

72. Garnett. 2011. Where are the best opportunities for reducing greenhouse gas emissions.

73. Smith et al. 2007. Agriculture.

74. *Climate change 2007: mitigation of climate change.* Working Group III Contribution to the Fouth Assessment Report of the Intergovernmental Panel on Climate Change. Cambridge, UK: Cambridge University Press. Figure 8.4.

75. *Climate change 2007.* Figure 8.9.

76. UNFCCC. 2011. *Kyoto Protocol.* http://unfccc.int/kyoto_protocol/items/2830.php.

77. Smith et al. 2007. Agriculture.
    Smith, P., Martino, D., Cai, Z., Gwary, D., Janzen, H., Kumar, P., McCarl, B., Ogle, S., O'Mara, F., Rice, C., Scholes, B., Sirotenko, O., Howden, M., McAllister, T., Pan, G., Romanenkov, V., Schneider, U., and Towprayoon, S., 2007b. Policy and technological constraints to implementation of greenhouse gas mitigation options in agriculture. *Agriculture, Ecosystems and Environment* 118:6–28.

78. Mueller, A., Mann, W., and Lipper, L. 2009. *Climate change mitigation: tapping the potential of agriculture.* MEA Bulletin—Guest Article No. 65. Rome: FAO. http://www .iisd.ca/mea-l/guestarticle65.html.

79. Post, W., Amonette, J., Birdsey, R., Garten Jr, C., Cesar Izaurralde, I., Jardine, P., Jastrow, J., Lal, R., Marland, G., McCarl, B., Thomson, A., West, T., Wullschleger, S. and Metting, F. B. 2009. Terrestrial biological carbon sequestration: science for enhancement and implementation. In: McPherson, B. and Sundquist, E. (eds). Carbon Sequestration and Its Role in the Global Carbon Cycle. *Geophysical Monograph Series 183.* American Geophysical Union. Devon, UK: 73–88.

80. Smith et al. 2007. Agriculture.

81. Smith et al. 2007. Agriculture.

82. Tubiello, F., and Fischer, G. 2007. Reducing climate change impacts on agriculture: global and regional effects of mitigation, 2000–2080. *Technological Forecasting and Social Change* 74:1030–1056.

83. Smith, P., and Oleson, J. 2010. Synergies between the mitigation of, and adaptation to, climate change in agriculture. *Journal of Agricultural Science* 148:543–552.

84. Conway and Waage. 2010. *Science and innovation for development.*

85. EarthTrends. 2005. *Energy consumption by source 2005.* Paris: International Energy Agency. http://Earthtrends.wri.org/pdf_library/data_tables/ene2_2005.pdf.

86. Demirbas, A., and Demirbas, I. 2007. Importance of rural bioenergy for developing countries. *Energy Conversion and Management* 48:2386–2898.
    Kammen, D. 2006. *Bioenergy in developing countries: experiences and prospects.* 2020 Vision for Food, Agriculture and the Environment. IFPRI Brief 10 of 12. Washington, DC: IFPRI.

87. IEA. 2009. *Renewables information 2009.* IEA Statistics. Paris: International Energy Agency. © OECD / International Energy Agency 2009. Part 1, page 3, figure 2.

88. Conway and Waage. 2010. *Science and innovation for development.*

89. WHO. 2006. *Indoor air pollution—fuel for life: household energy and health.* Geneva: WHO. http://www.who.int/indoorair/publications/fuelforlife/en/index.html.

90. WHO. 2006. *Indoor air pollution.*
    Practical Action. 2009. *Fireless cooker.* Rugby, UK: Practical Action.Practicalaction.org /fireless-cooker.

91. Mandil, C., and Shihab-Eldin, A. 2010. *Assessment of biofuels potential and limitation.* A report commissioned by the International Energy Forum. http://www.ief.org/PDF %20Downloads/Bio-fuels%20Report.pdf.

92. Practical Action Consulting. 2009. *Small-scale bioenergy initiatives: brief description and preliminary lessons on livelihood impacts from case studies in Asia, Latin America and Africa.* Prepared for PISCES and FAO by Practical Action Consulting.
    Conway and Waage. 2010. *Science and innovation for development.*

93. Lars. 2007. *African farmers in Mali discover Jatropha weed as biofuels crop.* Practical Environmentalist. http://www.practicalenvironmentalist.com/gardening/african-farmers-in -mali-discover-jatropha-weed-as-biofuel-crop.htm.

94. Worldwatch Institute. 2011. *Eye on Mali: Jatropha oil lights up villages*. Washington, DC: Worldwatch Institute. http://www.worldwatch.org/node/5101.

95. OECD StatExtracts. 2011. *Biofuel OECD-FAO Agricultural outlook 2011–2020*. Paris: Organisation for Economic Co-operation and Development. http://stats.oecd.org /viewhtml.aspx?QueryId=30104&vh=0000&vf=0&l&il=blank&lang=en.
    Rosegrant, M., Zhu, T. Msangi, S., and Sulser, T. 2008. Global scenarios for biofuels: Impacts and implications. *Applied Economic Perspectives and Policy* 30:495–505.

96. Diaz-Chavez, R, Mutimba, S, Watson, H, Rodriguez-Sanchez, S., and Nguer, M. 2010. *Mapping food and bioenergy in Africa*. A report prepared on behalf of FARA. Ghana: Forum for Agricultural Research in Africa.

97. Goklany, I. 2011. Could biofuel policies increase death and disease in developing countries? *Journal of American Physicians and Surgeons* 16:9–13.

98. Mandil and Shihab-Eldin 2010. *Assessment of biofuels potential and limitation*.

99. Von Braun, J. 2009. Addressing the food crisis: governance, market functioning and investment in public goods. *Food Security* 1:9–15.

100. Economic Report on Africa. 2009. *Challenges to agricultural development in Africa*. Addis Ababa Ethiopia: African Union, United Nations Economic Commission for Africa.

101. Smith et al. 2007. Agriculture.

17. Can We Feed the World?

1. *New York Times*. 2009. *Transcript: Barack Obama's inaugural address*. http://www.nytimes .com/2009/01/20/us/politics/20text-obama.html.

2. Feed the Future. 2010. *The global commitment to food security*. http://www.feedthefuture .gov/commitment.html.

3. World Bank. 2010. *World development indicators 2010*. Washington DC: World Bank.

4. Coppard, D. 2010. *Agricultural development assistance: a summary review of trends and the challenges of monitoring progress*. London: Development Initiatives and One.

5. Ho, M., and Hanrahan. C. 2011. *U.S. global food security funding, FY2010–FY2012*. Washington, DC: Congressional Research Service Report for Congress.

6. One. 2011. *Agriculture accountability: holding donors to their L'Aquila promises*. One Reports. http://www.one.org/c/international/hottopic/3923/.

7. Dorward, A., Kydd, J., Morrison, J., and Urey, I. 2004. A policy agenda for pro-poor agricultural growth. *World Development* 32:73–89.

8. Kaminski, J. P. 2006. *The quotable Jefferson*. Princeton, NJ: Princeton University Press.

Index

Pages in italics refer to tables and figures.